W9-BJJ-577

Essentials of Algebra for College Students

Essentials of Algebra for College Students

Raymond A. Barnett

Department of Mathematics
Merritt College

McGraw-Hill Book Company

New York
St. Louis
San Francisco
Auckland
Bogotá
Düsseldorf
Johannesburg
London
Madrid
Mexico
Montreal
New Delhi
Panama
Paris
São Paulo
Singapore
Sydney
Tokyo
Toronto

ESSENTIALS OF ALGEBRA FOR COLLEGE STUDENTS

1234567890DODO7832109876

This book was set in Caledonia by Black Dot, Inc. The editors were A. Anthony Arthur and Shelly Levine Langman; the cover was designed by Joseph Gillians; the production supervisor was Dennis J. Conroy. The drawings were done by J & R Services, Inc.
R. R. Donnelley & Sons Company was printer and binder.

Library of Congress Cataloging in Publication Data

Barnett, Raymond A
Essentials of algebra for college students.

Includes index.
1. Algebra. I. Title.
QA152.2.B38 512.9 76-12491
ISBN 0-07-003756-6

Contents

Preface

This is an intermediate algebra book with substantial supportive material from elementary algebra—developments start at the elementary algebra level and are taken through intermediate algebra. The book is designed to prepare a student for courses in statistics, trigonometry, college algebra, finite mathematics, short calculus, and so on, and to do so in one quarter or in one semester.

Recommended uses:

1 *Intermediate algebra courses:* The book is well suited for use in a standard intermediate algebra course. Because of the supportive material from elementary algebra, forgotten basics can be reestablished quickly and intermediate-level material can then be covered with increased comprehension. As a consequence, many students who have had some algebra and have been away from it for a while (and have probably forgotten most of it) can enter an intermediate algebra course using this book without having to repeat an elementary algebra course first.

2 *Essential algebra courses:* This book is also well suited for colleges offering only one algebra course that is designed to remove algebraic deficiencies so that students can profitably enroll in courses such as statistics, college algebra, trigonometry, and finite mathematics. Students using this book will be able to remove these deficiencies in one semester or in one quarter, thus enabling them to return to their main course of study with minimal loss of time.

Other important features of the book are:

1 First, and foremost, the text is written for students. Each concept is illustrated with an example, and following the example is a parallel problem with an answer so that students can immediately check their understanding of the concept. These follow-up problems also serve to encourage an active rather than a passive reading of the text.
2 An informal style is used for exposition, statements of definitions, and proofs of theorems.
3 The text includes approximately 3,200 carefully selected and graded problems. Each set is divided into A, B, and C levels, and ranges from easy and routine (A section) to challenging and nonroutine (C section). By suitably choosing problems from these three categories (see chart, page xiii) courses may be designed to suit various class needs. For example, a student reasonably well prepared in elementary algebra would work a few A problems, a large number of B problems, and a few C problems. A weaker student would be encouraged to work many A problems followed by problems from the B group. A very strong student would place equal emphasis on the B and C groups. The C problems often fill in and extend theoretical developments in the text. Thus, the text is designed so that an average or below-average student will be able to experience success, and a very capable student will be challenged.
4 The subject matter is related to the real world through many carefully selected realistic applications from the social, natural, and physical sciences. Thus, the text is equally suited for students interested in each of these areas.
5 Set ideas and notation are kept on an informal basis and are used only where clarity results and not otherwise. Set-builder notation appears only in the C exercises and may be avoided altogether, if desired.
6 Brief historical remarks are included where appropriate to provide perspective.
7 If most students in a class are reasonably well prepared in elementary algebra, then many sections may be covered briefly or omitted. Those students who have forgotten or who have not had the omitted material will still be able to keep up by going through the omitted sections on their own.

8 A Keller-type plan for individualized instruction is also to be found in the Instructor's Manual. Included are pre- and posttests for each unit, and a final examination. Several forms of most tests are available, and all have easy-to-grade answer keys. An $8\frac{1}{2}$- by 11-inch format is used for ease of reproduction. Also included are brief discussions of Keller-type plans and a possible grading procedure.

Even though an author is solely responsible for the contents of the final form of a text, many people provide imput that influence his or her thinking. I am particularly grateful to the following prepublication reviewers for their many kind remarks and useful suggestions: Jerald Ball, Chabot College; Nancy Lieberman, Thompkins-Cortland Community College; Philip Montgomery, University of Kansas; and Ara B. Sullenberger, Tarrant County Junior College. A special thanks goes to Harold Engelsohn of Kingsborough Community College and Donald Ostberg of Northern Illinois University for their penetrating and detailed reviews of the total manuscript. In addition, I am indebted to Ikuko Workman and my daughters Margaret and Janet for making the manuscript legible through their expert typing. Answers to examples and exercises were checked by Margaret and by Vincent P. Houben, a very capable student. Both did outstanding work. Finally, I would like to thank the staff at McGraw-Hill for a job well done.

Raymond A. Barnett

POSSIBLE COURSES

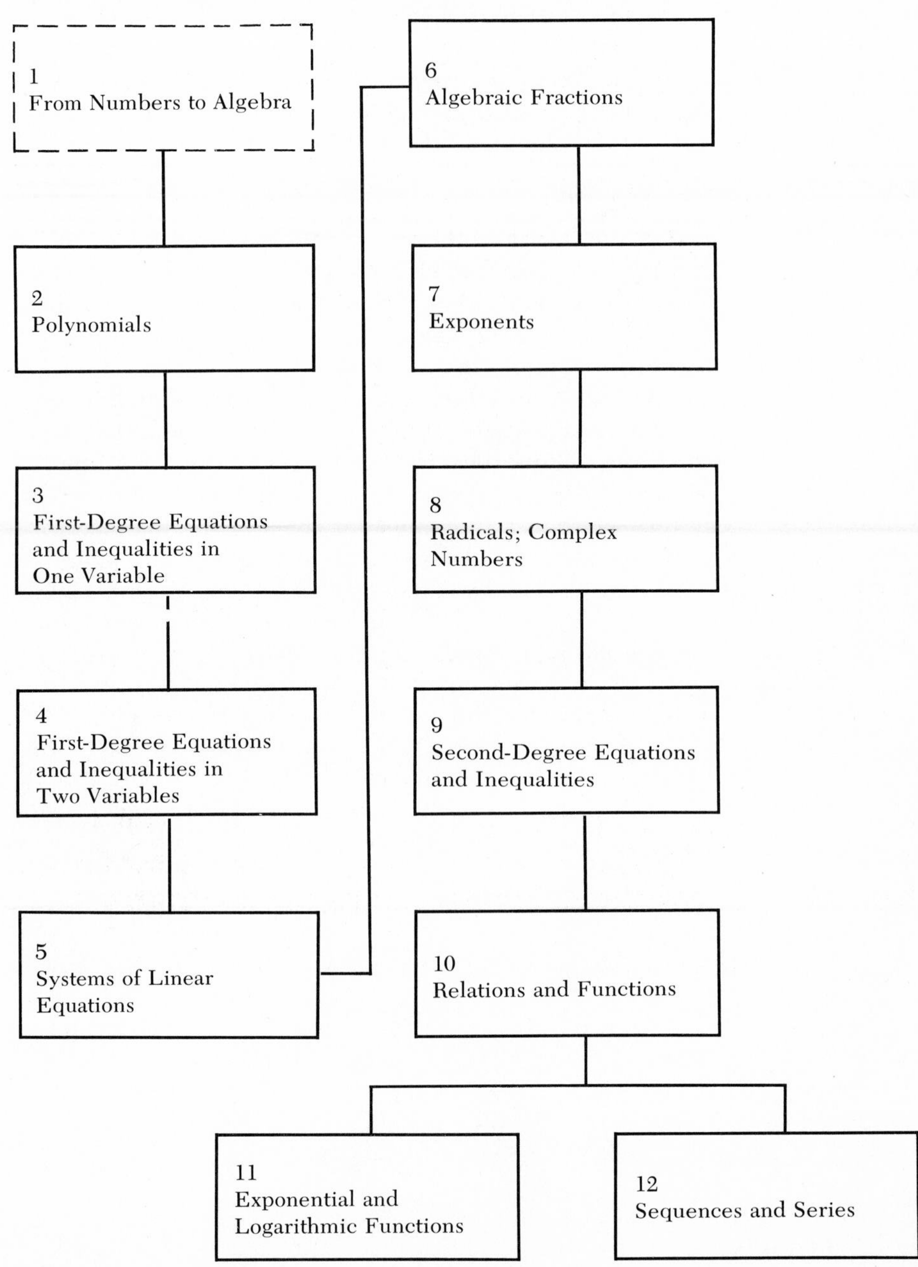

Sections that can be omitted without loss of continuity: 3.2, 3.4, 4.4, 5.2, 5.5, 6.3, 6.5, 6.6, 6.7, 9.4, 9.7, 10.4, 11.5, and 11.7.

USE OF A, B, and C EXERCISES

Type of Course	Exercise Emphasis		
	A	B	C
Light			
Average			
Strong			
	Simple mechanics	Strong mechanics Deeper understanding of basic principles	Theoretical Difficult mechanics

INDIVIDUALIZED ASSIGNMENTS (AVERAGE COURSE)

Student's preparation	Exercises		
	A	B	C
Below average			
Average			
Above average			

To the Student

The following suggestions are made to help you get the most out of the course and the most out of your efforts.

Using the text is essentially a five-step process.

For each section:

1. Read a mathematical development.
2. Read an illustrative example.
3. Work the matched problem.

(Repeat 1-2-3 cycle until section is finished.)

4. Review the main ideas in the section.
5. Work the assigned exercises at end of the section.

All of the above should be done with plenty of inexpensive paper, pencils, and a waste basket. No mathematics text should be read without pencil and paper in hand. This is not a spectator's sport!

If you have difficulty with the course, then, in addition to the regular assignments:

1. Spend more time on the examples and matched problems.
2. Work more A exercises, even if they are not assigned.

If you find the course too easy:

1. Work more problems from the C exercises, even if they are not assigned.
2. If the C exercises are consistently easy for you, then you should probably be in a more advanced course.

Raymond A. Barnett

Chapter 1

FROM NUMBERS TO ALGEBRA

In this chapter we will take a brief look at the set of real numbers and some of its important properties. We will see algebra develop naturally out of the real number system as a "generalized arithmetic."

1.1 Sets

Our use of the word "set" will not differ appreciably from the way it is used in everyday language. Words such as "set," "collection," "bunch," and "flock," all convey the same idea. Thus, we think of a *set* as any collection of objects with the important property that given any object, it is either a member of the set or it is not. If an object a is in a set A, we say that a is a *member* or *element* of A, and write

$$a \in A$$

If an object a is *not an element* of A, we write

$$a \notin A$$

Sets are often represented by listing their elements within braces { }. For example,

$\{2, 3, 5, 7\}$

represents the set with elements, 2, 3, 5, and 7.

Two sets A and B are said to be *equal*, and we write

$A = B$

if the two sets have exactly the same elements; the order of the elements does not matter.

From time to time we will be interested in sets within sets called subsets. We say that a set A is a *subset* of a set B if every element in A is in B. For example, the set of all girls in a mathematics class would form a subset of the set of all students in the class.

A set with no elements in it is called the *empty* or *null* set. It is denoted by

$\varnothing$

Additional ideas about sets will be introduced as needed throughout the book, however, our approach will be informal. We will use set concepts and notation only where clarity results, but not otherwise.

1.2 The Real Number System

The appropriate starting place for the study of algebra is with the set of real numbers (see Fig. 1). Note that the set of natural numbers is a subset of the set of integers; the set of integers is a subset of the set of rational numbers; and that the set of rational numbers is a subset of the set of real numbers. (Figure 1 in a slightly expanded form is printed again inside the front cover of the book for convenient reference.) We have a lot more to say about these numbers as the text progresses.

If you have trouble with some of these numbers, don't despair. It took mathematicians over 2,000 years to come up with a carefully formulated and workable definition of the concept of number, and this didn't happen until the last century. Fortunately, at this stage, we do not need to concern ourselves with this difficult problem. How symbols that name numbers are manipulated is at this point of greater interest to us than a precise knowledge of numbers as objects.

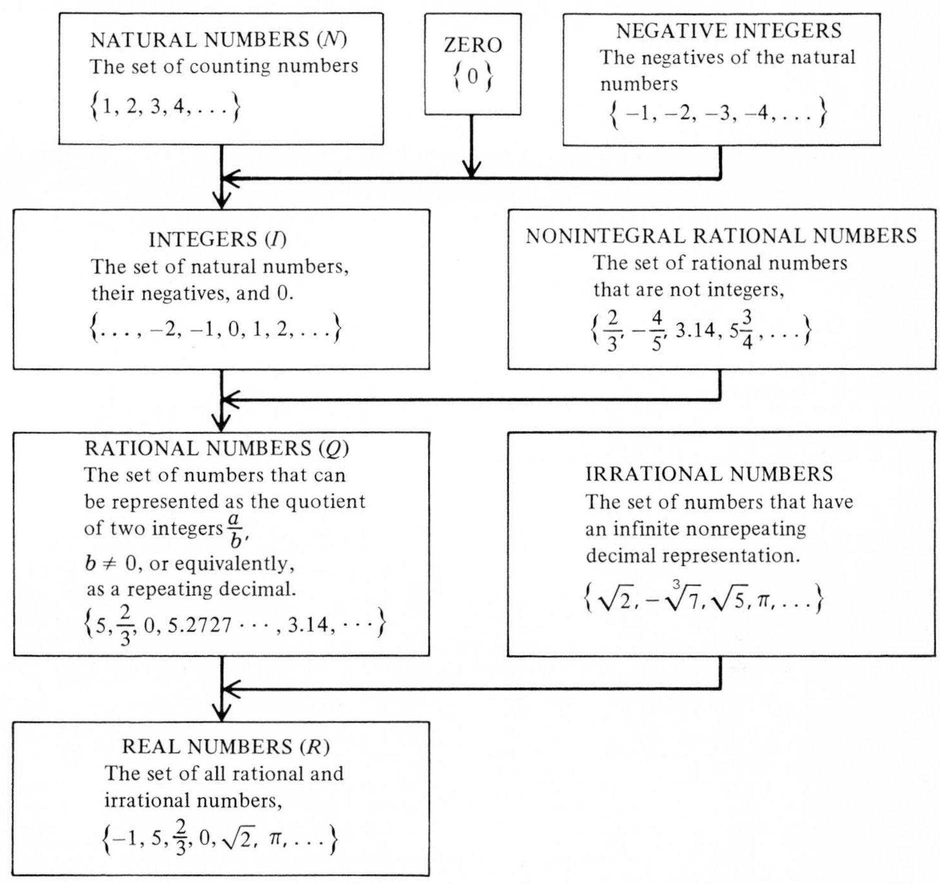

Figure 1 The real number system

1.3 Algebraic Expressions—Their Formulation and Evaluation

Consider the statement: "The perimeter of a rectangle is twice its length plus twice its width." If we let

$P =$ perimeter

$a =$ length

$b =$ width

b

a

then

$$P = 2a + 2b$$

has the same meaning as the original statement, but with increased clarity and a substantial decrease in the number of symbols used: 7 in the formula as compared to 57 in the written statement.

By the use of symbols that name numbers or are placeholders for numerals, we can form general statements relating many particular facts. The perimeter formula holds for *all* rectangles, not just one particular rectangle. This formula is an example of an algebraic expression.

VARIABLES AND CONSTANTS

In the perimeter formula above, the three letters P, a, and b can be replaced with many different numerals, depending on the size of the rectangle; hence, these letters are called variables. The symbol "2" names only one number and is consequently called a constant. In general, a *constant* is defined to be any symbol that names one particular thing; a *variable* is a symbol that holds a place for constants.

EXAMPLE 1 In the formula for the area of a circle, $A = \pi r^2$, A and r are variables, and π and 2 are constants.

PROBLEM 1 List the constants and variables in each formula:

(A) $P = 4s$ perimeter of a square

(B) $A = s^2$ area of a square

ANSWER Constants: (A) 4 (B) 2

Variables: (A) P and s (B) A and s

The introduction of variables into mathematics occurred about A.D. 1600. A French mathematician, Francois Vieta (1540–1603), is singled out as the one mainly responsible for this new idea. Many mark this point as the beginning of modern mathematics.

ALGEBRAIC EXPRESSIONS

An *algebraic expression* is a symbolic form involving constants; variables; mathematical operations such as addition, subtraction, multiplication, and division (other operations will be added later); and grouping symbols such as parentheses ( ), brackets [ ], and braces { }. For example,

$$8 + 7 \qquad 3 \cdot 5 - 6 \qquad 12 - 2(8 - 5)$$

$$3x - 5y \qquad 8(x - 3y) \qquad 3\{x - 2[x + 4(x + 3)]\}$$

are all algebraic expressions. [Note that $3 \cdot 5$, $3(5)$, $(3)(5)$, and 3×5 all represent the product of 3 and 5. The times sign "×" is not used very much in algebra because of its possible confusion with the letter x.]

It is also important to note that, unless otherwise indicated by symbols of grouping, *multiplication and division precede addition and subtraction.* This is a generally accepted procedure that results in the reduction of the use of grouping symbols, hence we obtain simpler looking algebraic expressions.

EXAMPLE 2 Evaluate each expression:

(*A*) $8 - 2 \cdot 3$

SOLUTION $8 - 2 \cdot 3 = 8 - 6 = 2$

(*B*) $9 - 2(5 - 3)$

SOLUTION $9 - 2(5 - 3) = 9 - 2 \cdot 2 = 9 - 4 = 5$

(*C*) $2[12 - 3(8 - 5)]$

SOLUTION $2[12 - 3(8 - 5)] = 2[12 - 3 \cdot 3] = 2(12 - 9) = 2 \cdot 3 = 6$

Note that we performed the operations within the grouping symbols, when present, first starting with the parentheses ( ), then the brackets [ ], and if braces { } were present, we would end with them. In short, we generally work from the inside out.

PROBLEM 2 Evaluate each expression:

(*A*) $2 \cdot 10 - 3 \cdot 5$ (*B*) $11 - 3(7 - 5)$ (*C*) $6[13 - 2(14 - 8)]$

ANSWER (*A*) 5 (*B*) 5 (*C*) 6

EXAMPLE 3 Evaluate each algebraic expression for $x = 10$ and $y = 3$:

(*A*) $2x - 3y$

SOLUTION $2(10) - 3(3) = 20 - 9 = 11$

(*B*) $x - 3(2y - 4)$

SOLUTION $10 - 3(2 \cdot 3 - 4) = 10 - 3(6 - 4) = 10 - 3 \cdot 2 = 10 - 6 = 4$

(*C*) $3[29 - x(y - 1)]$

SOLUTION $3[29 - 10(3 - 1)] = 3[29 - 10 \cdot 2] = 3(29 - 20) = 3 \cdot 9 = 27$

(*D*) $2\{2x - [3 + 2(3y - 8)]\}$

SOLUTION

$$
\begin{aligned}
2\{2 \cdot 10 - [3 + 2(3 \cdot 3 - 8)]\} &= 2\{20 - [3 + 2(9 - 8)]\} \\
&= 2\{20 - [3 + 2 \cdot 1]\} \\
&= 2\{20 - [3 + 2]\} = 2(20 - 5) \\
&= 2 \cdot 15 = 30
\end{aligned}
$$

PROBLEM 3 Evaluate each expression for $x = 11$ and $y = 2$:

(A) $x - 4y$ (B) $x - 2(3y - 3)$

(C) $2[35 - x(y + 1)]$ (D) $3\{3y + [1 + 2(x - 10)]\}$

ANSWER (A) 3 (B) 5 (C) 4 (D) 27

EXAMPLE 4 If x represents a real number, write an algebraic expression that represents each of the expressed numbers.

(A) A number 5 times as large as x

SOLUTION $5x$

(B) A number 5 more than x

SOLUTION $5 + x$ or $x + 5$

(C) A number 7 less than the product of 4 and x

SOLUTION $4x - 7$

(D) A number 3 times the number that is 2 less than x

SOLUTION $3(x - 2)$

PROBLEM 4 If y represents a real number, write an algebraic expression that represents each of the expressed numbers:

(A) A number 9 times as large as y (B) A number 9 more than y

(C) A number 8 less than the product of 7 and y (D) A number 5 times the number that is 3 less than twice y

ANSWER (A) $9y$ (B) $9 + y$ or $y + 9$ (C) $7y - 8$

(D) $5(2y - 3)$

EQUALITY

The equality sign "=" is very important in mathematics, and its correct use should be mastered early. We will always use *equality* in the sense of logical identity. That is, an *equality sign* will be used to join two expressions if the two expressions are names or descriptions of exactly the same thing. Thus,

$$a = b$$

means a and b are names for the same object. It is natural then, to define

$$a \neq b$$

to mean a and b do not name the same thing; that is, *a is not equal* to b.

If two algebraic expressions involving at least one variable are joined with an equal sign, the resulting form is called an *algebraic equation*. The following are algebraic equations in one or more variables:

$$2x - 3 = 3(x - 5)$$

$$a + b = b + a$$

$$3x + 5y = 7$$

Since a variable is a placeholder for constants, an equation is neither true nor false as it stands; it does not become so until the variable has been replaced by a constant. Formulating algebraic equations is an important first step in solving certain types of problems using algebraic methods.

EXAMPLE 5 Translate each statement into an algebraic equation using only one variable.

(*A*) 25 is 3 less than a certain number

SOLUTION Let x represent the certain number, then

$$25 = x - 3$$

(*B*) 5 times a number is 3 more than twice the number

SOLUTION Let n represent the certain number, then

$$5n = 2n + 3$$

PROBLEM 5 Translate each statement into an algebraic equation using only one variable:

(*A*) 7 is 3 more than a certain number

(*B*) 12 is 9 less than a certain number

(*C*) 3 times a certain number is 6 less than twice that number

(*D*) If 6 is subtracted from a certain number, the difference is twice the number that is 4 less than the original number.

ANSWER (*A*) $7 = 3 + x$ (*B*) $12 = x - 9$ (*C*) $3x = 2x - 6$

(*D*) $x - 6 = 2(x - 4)$

From the logical meaning of the equality sign, a number of rules or properties can easily be established for its use. We state these properties below for completion, and will return to them later when we discuss equations. They control a great deal of the activity related to the solving of equations.

PROPERTIES OF EQUALITY

If *a*, *b*, and *c* are names of objects, then:

1 $a = a$ reflexive property

2 If $a = b$, then $b = a$. symmetric property

3 If $a = b$ and $b = c$, then $a = c$. transitive property

4 If $a = b$, then either may replace the other in any expression without changing the truth or falsity of the statement. substitution principle

The importance of these four laws will not be fully appreciated until we start solving equations and simplifying algebraic expressions.

Exercise 1

A *Evaluate each expression.*

1. $9 - 4 \cdot 2$ **2.** $10 - 3 \cdot 2$

3. $3 \cdot 8 + 2 \cdot 4$ **4.** $2 \cdot 6 - 3 \cdot 3$

5. $(8 - 2) - (2 + 3)$ **6.** $(4 + 10) - (10 - 4)$

7. $12 - 2(7 - 5)$ **8.** $15 - 3(9 - 5)$

Evaluate each algebraic expression for $x = 9$ *and* $y = 2$.

9. $x - y$ **10.** $x + y$

11. $x - 3y$ **12.** $y + 2x$

13. $10y - 2x$ **14.** $2x - 8y$

15. $x - 3(4 - y)$ **16.** $x + 2(x - y)$

B *Evaluate each expression.*

17. $6(11 - 8) - 2 \cdot 5$ **18.** $3 \cdot 8 - 2(8 - 3)$

19. $2[12 - 4(8 - 6)]$

20. $3[18 - 6(12 - 10)]$

21. $2[(3 + 2) + 2(7 - 4) + 6 \cdot 2]$

22. $3[(6 - 4) + 4 \cdot 3 + 3(1 + 6)]$

Identify the constants and variables in the following algebraic expressions.

23. $A = \frac{1}{2}bh$ area of a triangle

24. $C = \pi D$ circumference of a circle

25. $P = 4s$ perimeter of a square

26. $A = s^2$ area of a square

27. $3u + 2(v - u)$

28. $6(x + 2y) - 3y$

Evaluate each for $u = 2$, $v = 3$, $w = 4$, and $x = 5$.

29. $4(x - u) - w$

30. $5w - 2(u + v)$

31. $2[x + 3(x - u)]$

32. $3[x - 2(w - v)]$

33. $3\{x - 3[9 - 4(x - v)]\}$

34. $2\{w + 2[7 - (u + v)]\}$

If x represents a real number, write an algebraic expression that represents each of the following numbers:

35. A number 7 more than x

36. A number 7 less than x

37. A number 7 times x

38. A number x less than 7

39. A number 8 less than twice x

40. A number 5 more than the product of 3 and x

Translate each statement into an algebraic equation using only the variable x.

41. 18 is 3 times a certain number

42. 26 is 12 less than a certain number

43. 80 is 3 more than twice a certain number

44. 43 is 7 less than the product of 4 and a certain number

45. 4 times a certain number is 3 less than the product of 12 and the number

46. If 4 is subtracted from the product of 3 and a certain number, the difference is 6 times the number.

C *Evaluate each expression.*

47. $2\{12 - 2[7 - 2(12 - 10)]\} - 3(9 - 6)$

48. $2(3 + 2 \cdot 3) + 2\{4 + 2[2 \cdot 5 - (7 - 5)]\}$

Evaluate each expression for $u = 12$, $v = 3$, $w = 8$, and $x = 5$:

49. $5\{18 - 2[u - v(x - v)]\} - vw$

50. $ux - 2\{2[(w - v) + (u - x)] + 1\}$

Translate each statement in Problems 51 and 52 into an algebraic equation using only the variable x.

51. 5 less than twice a given number is 3 times the number that is 2 less than the given number.

52. 8 more than 6 times a given number is twice the number that is 3 less than 4 times the given number

53. Sound travels through the air at approximately 1,120 feet/second. (*A*) Write a formula for the distance *d* that sound travels in *t* seconds. (*B*) Identify the constants and the variables. (*C*) How many feet more or less than a mile (5,280 feet) will sound from lightning travel in 5 seconds?

54. An earthquake produces several types of waves that travel through the earth. One is called a shear wave, which travels at about 2 miles/second and causes the earth to move at right angles to the direction of motion of the wave. (*A*) Write a formula that indicates the distance *d* that the wave travels in *t* seconds. (*B*) Identify the constants and variables. (*C*) How far will the wave travel in 10 seconds?

1.4 Real Numbers and the Rules of Algebra

In the last section you learned how to recognize and form algebraic expressions and to evaluate them for numerical replacements of the variables. For algebra to be of greater use to you, you will need to know how to change algebraic expressions into "equivalent forms" and know how to "solve" algebraic equations.

With this end in mind, we start with a basic list of properties that we assume all real numbers share. These unproved statements are referred to as *axioms*, and they constitute the basic rules for the "game of algebra" you are about to play.

BASIC (FIELD) PROPERTIES OF THE SET OF REAL NUMBERS

Let R be the set of real numbers, and let a, b, and c be any three real numbers.

Addition Properties

CLOSURE: $a + b$ is a unique element of R

COMMUTATIVE: $a + b = b + a$

ASSOCIATIVE: $(a + b) + c = a + (b + c)$

IDENTITY: $0 + a = a + 0 = a$ for each real number a
0 is called the *identity element for addition*

INVERSE: For each real number a, there exists a unique inverse, denoted by $-a$, such that $a + (-a) = (-a) + a = 0$.

Multiplication Properties

CLOSURE: ab is a unique element of R

COMMUTATIVE: $ab = ba$

ASSOCIATIVE: $(ab)c = a(bc)$

IDENTITY: $1a = a1 = a$ for each real number a
1 is called the *identity element for multiplication*

INVERSE: For each real number a, $a \neq 0$, there exists a unique inverse, denoted by $1/a$, such that $a(1/a) = (1/a)a = 1$.

Combined Property

DISTRIBUTIVE: $a(b + c) = ab + ac$

Any mathematical system with addition and multiplication defined that satisfies the above 11 axioms is called a *field*. This is why these axioms are referred to as the field properties of the real numbers.

Don't let the names of these properties frighten you. Actually, most of the ideas represented here are fairly simple. In fact, you have been using many of these properties in arithmetic for a long time. We will discuss a few of these properties now, and others as needed. The whole list is reproduced inside the back cover of the text for easy reference.

Certainly, you are familiar with the commutative properties for addition and multiplication. They simply indicate that the order in which addition or multiplication are performed doesn't matter: $2+3=3+2$ and $2 \cdot 3 = 3 \cdot 2$.

EXAMPLE 6 If x and y are real numbers, then

(A) $x + 9 = 9 + x$

(B) $7y = y7$

(C) $2x + 3y = 3y + 2x$

(D) $xy = yx$

PROBLEM 6 If a and b are real numbers, use the commutative axioms to write each of the following in an equivalent form.

(A) $a + 3$ (B) $x3$

(C) ab (D) $3a + 9$

ANSWER (A) $3 + a$ (B) $3x$ (C) ba

(D) $9 + 3a$ or $a3 + 9$ or $9 + a3$

Does the commutative property hold relative to subtraction and division? That is, does $a - b = b - a$ and $a \div b = b \div a$ for all real numbers a and b, division by 0 excluded? The answer is no, since, for example, $5 - 3 \neq 3 - 5$ and $8 \div 4 \neq 4 \div 8$.

When computing

$$4 + 3 + 5$$

or

$$4 \cdot 3 \cdot 5$$

why don't we need parentheses to show us which two numbers are to be added or multiplied first? The answer is to be found in the associative axioms. These axioms allow us to write

$$(4 + 3) + 5 = 4 + (3 + 5)$$

$$(4 \cdot 3) \cdot 5 = 4 \cdot (3 \cdot 5)$$

and we see that it doesn't matter how we group relative to either operation.

EXAMPLE 7 If x, y, and z are real numbers, then

(A) $(x + 7) + 2 = x + (7 + 2)$

(B) $3(5y) = (3 \cdot 5)y$

(C) $x + (x + 3) = (x + x) + 3$

(D) $(2y)y = 2(yy)$

PROBLEM 7 If a, b, and c are real numbers, replace each question mark with an appropriate expression.

(A) $(a + 3) + 9 = a + (\ ?\ )$

(B) $6(3c) = (\ ?\ )c$

(C) $(5 + b) + b = 5 + (\ ?\ )$

(D) $(8a)a = 8(\ ?\ )$

(E) $(a + b) + c = a + (\ ?\ )$

(F) $(ab)c = a(\ ?\ )$

ANSWER (*A*) $3+9$ (*B*) $6 \cdot 3$ (*C*) $b+b$ (*D*) aa (*E*) $b+c$ (*F*) bc

Does the associative axiom hold for subtraction and division? The answer is no, since, for example, $(8-4)-2 \neq 8-(4-2)$ and $(8 \div 4) \div 2 \neq 8 \div (4 \div 2)$. Carry out the arithmetic in both examples to see why.

CONCLUSION

1 Relative to addition (but not subtraction) the commutative and associative axioms permit us to rearrange and regroup symbols that represent real numbers as we please.

2 Relative to multiplication (but not division) the commutative and associative axioms permit us to rearrange and regroup symbols that represent real numbers as we please.

The commutative and associative axioms for addition and multiplication provide us with our first tools to transform algebraic expressions into other equivalent forms. (Two algebraic expressions are *equivalent* if they are equal for each replacement of the variables by numbers for which each expression is defined.) Notice the use of these axioms in the following example.

EXAMPLE 8 (*A*)

$$\begin{aligned}(x+7)+(y+3) &= x+7+y+3\\ &= x+y+7+3\\ &= x+y+10\end{aligned}$$

(*B*)

$$\begin{aligned}(3x)(5y) &= 3x5y\\ &= 3 \cdot 5xy\\ &= 15xy\end{aligned}$$

PROBLEM 8 Simplify, using the commutative and associative axioms:

(*A*) $(a+9)+(b+7)$ (*B*) $(8a)(2b)$

ANSWER (*A*) $a+b+16$ (*B*) $16ab$

The identity axiom for addition states that 0 is the real number having the property that when it is added to any real number we get that number back again. Similarly, the number 1 plays the same role relative to multiplication. That is, when any number is multiplied by 1, we get that number back again. Thus we can write

$1x = x$

$xy = 1xy$

$0 + 3x = 3x$

$(x + y) + 0 = x + y$

The other field axioms will be discussed later as needed.

Exercise 2

All variables represent real numbers.

A *State the justifying field axiom for each statement.*

1. $12 + w = w + 12$

2. $2x + 3 = 3 + 2x$

3. $m + (n + 3) = (m + n) + 3$

4. $(3x + y) + 5 = 3x + (y + 5)$

5. $20x = x20$

6. $MN = NM$

7. $4(8y) = (4 \cdot 8)y$

8. $(12u)v = 12(uv)$

9. $3x + 0 = 3x$

10. $0 + (2x + 3) = 2x + 3$

11. $1m = m$

12. $uv = 1uv$

Remove parentheses and simplify:

13. $(x + 7) + 2$

14. $3 + (5 + m)$

15. $4(5y)$

16. $6(8n)$

17. $12 + (u + 3)$

18. $(4 + x) + 13$

19. $(3x)7$

20. $4(y3)$

21. $0 + 1x$

22. $(1y + 3) + 0$

B *State the justifying field axiom for each statement.*

23. $2 + (y + 3) = 2 + (3 + y)$

24. $(3m)n = n(3m)$

25. $7(y4) = 7(4y)$

26. $5 + (y + 2) = (y + 2) + 5$

27. $3x + 2y = 2y + 3x$

28. $3x + (2x + 5y) = (3x + 2x) + 5y$

29. $(2x)(x + 3) = 2[x(x + 3)]$

30. $(x + 3) + (2 + y) = x + [3 + (2 + y)]$

Remove parentheses and simplify:

31. $(x + 7) + (y + 4) + (z + 1)$

32. $(7 + m) + (8 + n) + (3 + p)$

33. $(3x + 5) + (4y + 6)$

34. $(3a + 7) + (5b + 2)$

35. $0 + (1x + 3) + (y + 2)$

36. $1(x + 3) + 0 + (y + 2)$

37. $(12m)(3n)(1p)$

38. $(8x)(4y)(2z)$

C **39.** Indicate whether true (T) or false (F), and for each false statement find real number replacements for a and b that will illustrate its falseness: For all real numbers a and b

(A) $a + b = b + a$ (B) $a - b = b - a$

(C) $ab = ba$ (D) $a \div b = b \div a$

40. Indicate whether true (T) or false (F), and for each false statement find real number replacements for a, b, and c that will illustrate its falseness: For all real numbers a, b, and c

(A) $(a + b) + c = a + (b + c)$ (B) $(a - b) - c = a - (b - c)$

(C) $a(bc) = (ab)c$ (D) $(a \div b) \div c = a \div (b \div c)$

41. Supply a reason for each step.

	STATEMENT		REASON
1	$(x + 3) + (y + 4) = (x + 3) + (4 + y)$	1	
2	$= x + [3 + (4 + y)]$	2	
3	$= x + [(3 + 4) + y]$	3	
4	$= x + (7 + y)$	4	
5	$= x + (y + 7)$	5	
6	$= (x + y) + 7$	6	

42. Supply a reason for each step.

	STATEMENT	REASON
1	$(5x)(2y) = (x5)(2y)$	1
2	$= x[5(2y)]$	2
3	$= x[(5 \cdot 2)y]$	3
4	$= x(10y)$	4
5	$= (x10)y$	5
6	$= (10x)y$	6
7	$= 10(xy)$	7

1.5 Addition and Subtraction of Signed Numbers

POSITIVE AND NEGATIVE REAL NUMBERS, AND ZERO

The set of real numbers can be divided into the following three important subsets that do not overlap:

The set of positive real numbers

Zero

The set of negative real numbers

These numbers are often pictured graphically on a real number line (Fig. 2).

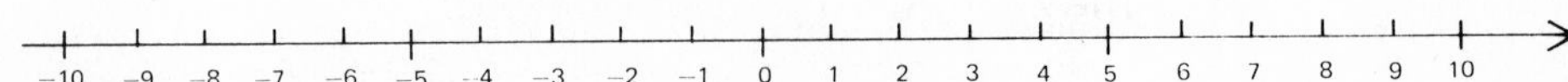

Figure 2 The real number line

The positive real numbers are associated with points to the right of zero and the negative real numbers are associated with the points to the left of zero. A positive number such as five may be denoted by either 5 or +5, depending on the emphasis desired.

Preliminary to discussing addition and subtraction for signed numbers, we first define two important operations on single real numbers called "the negative of" and "the absolute value of."

THE NEGATIVE OF A NUMBER

The *negative of a number* x (which is not to be confused with "a negative number") is an operation on x, symbolized by

$-x$

that produces another number: It changes the sign of x if x is not 0, and if x is 0 it leaves it alone.

EXAMPLE 9 (A) $-(+5) = -5$

(B) $-(-7) = +7$ or 7

(C) $-(0) = 0$

(D) $-[-(-4)] = -(+4) = -4$

PROBLEM 9 Find:

(A) $-(+10)$ (B) $-(-8)$

(C) $-(0)$ (D) $-[-(+3)]$

ANSWER (A) -10 (B) $+8$ or 8 (C) 0 (D) $+3$ or 3

As a consequence of the definition of "the negative of a number," we see that the negative of a number is not necessarily negative: $-x$ names a positive number if x is negative and a negative number if x is positive.

It is important to notice that the minus sign "−" is now being used in three distinct ways:

1 As the operation "subtract": $9 \overset{\downarrow}{-} 3 = 6$
2 As the operation "the negative of": $\overset{\downarrow}{-}(-8) = +8$
3 As part of a number symbol: $\overset{\downarrow}{-}4$

THE ABSOLUTE VALUE OF A NUMBER

The *absolute value of a number* x is an operation on x, denoted symbolically by

$$|x|$$

(not square brackets) that produces another number. What number? If x is positive or 0 it leaves it alone; if x is negative it makes it positive. Symbolically, and more formally,

$$|x| = \begin{cases} x & \text{if } x \text{ is positive or } 0 \\ -x & \text{if } x \text{ is negative} \end{cases}$$

EXAMPLE 10 (A) $|+3| = +3$ or 3

(B) $|-5| = +5$ or 5

(C) $|0| = 0$

PROBLEM 10 Find:

(A) $|+7|$ (B) $|-6|$ (C) $|0|$

ANSWER (A) $+7$ or 7 (B) $+6$ or 6 (C) 0

We note that

The absolute value of a number is never negative.

"The absolute value" and "the negative of" are often used in combination and it is important to perform the operations in the right order—from the inside out.

EXAMPLE 11

(A) $|-(-5)|=|+5|=+5$ or 5

(B) $-|-5|=-(+5)=-5$

(C) $-(|-8|-|-3|)=-(8-3)=-(5)=-5$

PROBLEM 11 Evaluate:

(A) $|-(+2)|$ (B) $-|+6|$ (C) $-(|-6|-|+2|)$

ANSWER (A) $+2$ or 2 (B) -6 (C) -4

ADDITION OF SIGNED NUMBERS

How should addition be defined so that we can assign numbers to each of the following sums?

$(+3)+(+5)=?$

$(+3)+(-5)=?$

$(-3)+(+5)=?$

$(-3)+(-5)=?$

$(+6)+0=?$

$0+(-4)=?$

To start, we assume that the arithmetic of positive real numbers and its properties are known. Based on this assumption we state the following theorem on addition of signed numbers. This theorem follows logically from the field axioms for the real numbers, and the definition of absolute value and negative of a number.

THEOREM 1 **ADDITION**

(A) NUMBERS WITH LIKE SIGNS If a and b are positive integers, add as in arithmetic. If a and b are both negative, their sum is the negative of the sum of their absolute values.

(*B*) NUMBERS WITH UNLIKE SIGNS To add two numbers with unlike signs subtract the smaller absolute value from the larger absolute value, then attach the sign of the number with the larger absolute value to the difference. If the numbers have the same absolute value, their sum is 0.

(*C*) ZERO The sum of any number and 0 is that number; the sum of 0 and any number is that number.

It may relieve you to know that no one (not even the professional mathematician) in his or her everyday routine calculations goes through the steps precisely as described in the theorem; mechanical shortcuts of the type described below soon take over. You should not forget, however, that these mechanical rules are justified on the basis of the above theorem and not vice versa.

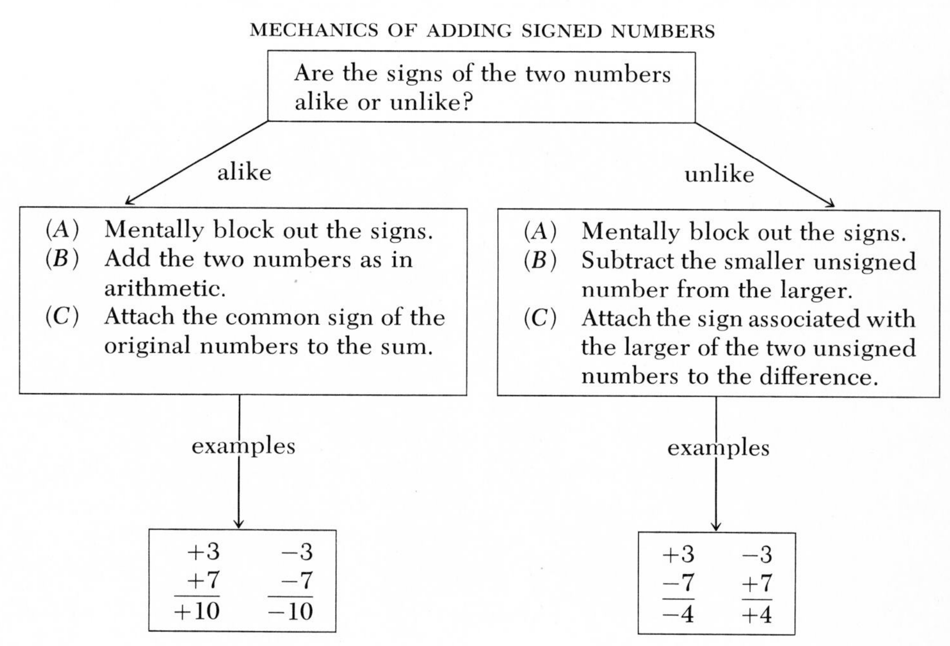

EXAMPLE 12

(*A*) $3 + 5 = 8$

(*B*) $3 + (-5) = -2$

(*C*) $(-3) + 5 = 2$

(*D*) $(-3) + (-5) = -8$

(*E*) $3 + (-3) = 0$

(*F*) $7 + 0 = 7$

(*G*) $0 + (-3) = -3$

PROBLEM 12 Add:

(A) $7+2$ (B) $7+(-2)$ (C) $(-7)+2$

(D) $(-7)+(-2)$ (E) $(-9)+9$ (F) $0+(-6)$

ANSWER (A) 9 (B) 5 (C) −5 (D) −9 (E) 0

(F) −6

To add three or more real numbers, add all of the positive numbers together, add all of the negative numbers together (the commutative and associative axioms justify this procedure), and then add the two resulting sums as above.

EXAMPLE 13

$$\begin{aligned} 3+(-6)+8+(-4)+(-5) &= (3+8)+[(-6)+(-4)+(-5)] \\ &= 11+(-15) = -4 \end{aligned}$$

PROBLEM 13 Add: $6+(-8)+(-4)+10+(-3)+1$

ANSWER 2

We state for future reference the following important theorem:

THEOREM 2 For all real numbers a and b

(A) $a+(-a)=0$

(B) $a+b=0$ if and only if $b=-a$ and $a=-b$

SUBTRACTION OF SIGNED NUMBERS

From subtraction of positive real numbers we know that

$$(+8)-(+5)=+3$$

but what can we write for

$$(+5)-(+8)=?$$

$$(+8)-(-5)=?$$

$$(-8)-(+5)=?$$

$$(-8)-(-5)=?$$

$$(-5)-(-8)=?$$

$$0-(-5)=?$$

To answer this question, we generalize on our ideas about subtraction from the arithmetic of positive real numbers. In subtracting 5 from 8, instead of thinking of taking 5 away from 8 to get 3, we can, instead, ask what must be added to 5 to produce 8. Note that the answer is the same for each approach. Since we now know how to add signed numbers, we use the second approach to formally define subtraction.

DEFINITION **SUBTRACTION**

For all real numbers a and b, we write

$$a - b = c \qquad \text{if and only if} \qquad a = b + c$$

This formal definition leads to an easy mechanical rule for subtraction through the following theorem.

THEOREM 3 To subtract b from a, add the negative of b to a; symbolically,

$$a - b = a + (-b)$$

To prove this theorem, we must show that the sum of b and $a + (-b)$ is a (see Problem 63, Exercise 3). If you look at the theorem very carefully and recall that when we take the negative of a number we change its sign, then you should arrive at the following simple *mechanical rule for subtraction:*

> To subtract one number from another, change the sign of the number being subtracted and add.

EXAMPLE 14

(*A*) $(+5) - (+8) = (+5) + (-8) = -3$

(*B*) $(+8) - (-5) = (+8) + (+5) = 13$

(*C*) $(-8) - (+5) = (-8) + (-5) = -13$

(*D*) $(-8) - (-5) = (-8) + (+5) = -3$

(*E*) $(-5) - (-8) = (-5) + (+8) = 3$

(*F*) $0 - (-5) = 0 + (+5) = 5$

You should try to do problems of this type mentally without writing down the middle step.

PROBLEM 14 Subtract:

(*A*) $4 - 7$ (*B*) $7 - (-4)$ (*C*) $(-7) - 4$

(*D*) $(-7) - (-4)$ (*E*) $(-4) - (-7)$ (*F*) $0 - (-4)$

ANSWER (*A*) -3 (*B*) 11 (*C*) -11 (*D*) -3 (*E*) 3 (*F*) 4

When three or more terms are combined by addition or subtraction and symbols of grouping are omitted, we convert (mentally) any subtraction to addition (Theorem 3) and add. Thus,

$$8 - 5 + 3 \;\; \underbrace{= 8 + (-5) + 3}_{think} \;\; = 6$$

EXAMPLE 15 (*A*) $2 - 3 - 7 + 4 \;\; \underbrace{= 2 + (-3) + (-7) + 4}_{think} \;\; = -4$

(*B*) $-4 - 8 + 2 + 9 \;\; \underbrace{= (-4) + (-8) + 2 + 9}_{think} \;\; = -1$

PROBLEM 15 Evaluate:

(*A*) $5 - 8 + 2 - 6$ (*B*) $-6 + 12 - 2 - 1$

ANSWER (*A*) -7 (*B*) 3

Exercise 3

A *Evaluate:*

1. $-(+7)$ **2.** $-(+12)$ **3.** $-(-6)$

4. $-(-8)$ **5.** $|+2|$ **6.** $|+9|$

7. $|-27|$ **8.** $|-32|$ **9.** $|0|$

10. $-(0)$ **11.** $(-7) + (-3)$ **12.** $(-7) + (+3)$

13. $(+7) + (-3)$ **14.** $(-12) + (+8)$ **15.** $(+3) - (+9)$

16. $(+3) + (+9)$ **17.** $(+9) + (+3)$ **18.** $(-9) + (+3)$

19. The negative of a number is (*always*, *sometimes*, *never*) a negative number.

20. The absolute value of a number is (*always*, *sometimes*, *never*) a positive number.

B *Evaluate:*

21. $-[-(-3)]$ **22.** $-[-(+6)]$

23. $-|-(+2)|$ **24.** $-|-(-3)|$

25. $-(|+9| - |+3|)$ **26.** $-(|+14| - |+8|)$

27. $(-2) + (-6) + 3$

28. $(-2) + (-8) + 5$

29. $5 + 7 - 3$

30. $3 - 2 + 4$

31. $-7 + 6 - 4$

32. $-4 + 7 - 6$

33. $-2 + 3 + 6 + 2$

34. $-4 + 7 - 3 - 2$

35. $6 - [3 + (-9)]$

36. $(-10) - [(-6) + 3]$

37. $[6 + (-8)] - [(-8) + 6]$

38. $[3 - 5] + [(-5) - (-2)]$

Replace each question mark with an appropriate real number.

39. $-(\ ?\ ) = 5$

40. $-(\ ?\ ) = -8$

41. $|\ ?\ | = 7$

42. $|\ ?\ | = -4$

43. $(-3) + ? = -8$

44. $? + 5 = -6$

45. $(-3) - ? = -8$

46. $? - (-2) = -4$

Evaluate for $x = 3$, $y = -8$, and $z = -2$.

47. $x + y$

48. $y + z$

49. $(x - z) + y$

50. $y - (z - x)$

51. $|(-z) - |y||$

52. $||-y| - 12|$

53. You own a stock that is traded on the New York Stock Exchange. On Monday it closed at $23 per share; it fell $3 on Tuesday and another $6 on Wednesday; it rose $2 on Thursday; and it finished strongly on Friday by rising $7. Use addition of signed numbers to determine the closing price of the stock on Friday.

54. Find, using subtraction of signed numbers, the difference in the height between the highest point in the United States, Mount McKinley (20,270 feet) and the lowest point in the United States, Death Valley (−280 feet).

C *Which of the following hold for all integers a, b, and c? Illustrate each false statement with an example that shows that it is false.*

55. $a + b = b + a$

56. $a + (-a) = 0$

57. $a - b = b - a$

58. $a - b = a + (-b)$

59. $(a + b) + c = a + (b + c)$

60. $(a - b) - c = a - (b - c)$

61. $|a + b| = |a| + |b|$

62. $|a - b| = |a| - |b|$

63. Supply the reasons for each of the following steps in the proof of Theorem 3.

$$\begin{aligned} b + [a + (-b)] &= b + [(-b) + a] \\ &= [b + (-b)] + a \\ &= 0 + a \\ &= a \end{aligned}$$

64. Supply the reasons for each step.

$$\begin{aligned}(a+b)+[(-a)+(-b)] &= (b+a)+[(-a)+(-b)]\\ &= [(b+a)+(-a)]+(-b)\\ &= \{b+[a+(-a)]\}+(-b)\\ &= (b+0)+(-b)\\ &= b+(-b)\\ &= 0\end{aligned}$$

Therefore, $-(a+b) = (-a)+(-b)$

1.6 Multiplication and Division of Signed Numbers

Having discussed addition and subtraction of signed numbers, we now turn to multiplication and division.

MULTIPLICATION OF SIGNED NUMBERS

We assume that multiplication of positive real numbers and its properties are known. For example, we know that

$$(+2)(+7) = +14$$

but what numbers should be assigned to the products

$$(+2)(-7) = ?$$

$$(-2)(+7) = ?$$

$$(-2)(-7) = ?$$

$$0(-7) = ?$$

To see what numbers must be assigned to these products, let us return to the field axioms for the real numbers. Consider the following argument, assuming $a \cdot 0 = 0$ for all real numbers a (see Problem 63, Exercise 4).

$(+7)+(-7)=0$	Inverse axiom for addition
$(+2)[(+7)+(-7)]=0$	Property of equality (If $a=b$, then $ca=cb$) that follows from laws of equality
$(+2)(+7)+(+2)(-7)=0$	Distributive axiom
$(+14)+(+2)(-7)=0$	Substitution principle for equality
$(+2)(-7)=-14$	Theorem 2

Thus, we see that if we assume the field axioms for all real numbers, then the product of (+2) and (−7) must be −14. There is no other choice! Similar arguments are used in the general proof of the following theorem.

THEOREM 4 **MULTIPLICATION**

(*A*) NUMBERS WITH LIKE SIGNS. The product of two numbers with like signs is a positive number and is found by multiplying the absolute value of the two numbers.

(*B*) NUMBERS WITH UNLIKE SIGNS. The product of two numbers with unlike signs is a negative number and is found by taking the negative of the product of the absolute values of the two numbers.

(*C*) ZERO. The product of any real number and 0 is 0.

EXAMPLE 16

(*A*) $(+2)(+7) = +14$

(*B*) $(+2)(-7) = -14$

(*C*) $(-2)(+7) = -14$

(*D*) $(-2)(-7) = +14$

(*E*) $0(-7) = 0$

PROBLEM 16 Evaluate:

(*A*) $(+4)(-3)$ (*B*) $(-4)(+3)$

(*C*) $(-4)(-3)$ (*D*) $0(-3)$

ANSWER (*A*) -12 (*B*) -12 (*C*) $+12$ (*D*) 0

Several more important sign properties for multiplication are indicated in the following theorem.

THEOREM 5

(*A*) $(-1)a = -a$

(*B*) $(-a)b = -(ab)$

(*C*) $(-a)(-b) = ab$

(See Problem 64, Exercise 4 for a proof of part *A*.)

PROOF OF PART B

$(-a)b = [(-1)a]b$	Theorem 5A
$= (-1)(ab)$	Associative axiom for multiplication
$= -(ab)$	Theorem 5A

EXAMPLE 17 Evaluate $(-a)b$ and $-(ab)$ for $a=-5$ and $b=+4$.

SOLUTION $(-a)b = [-(-5)](+4) = (+5)(+4) = +20$

$-(ab) = -[(-5)(+4)] = -(-20) = +20$

PROBLEM 17 Evaluate $(-a)(-b)$ and ab for $a=-5$ and $b=+4$.

ANSWER Both are -20.

Expressions of the form

$$-ab$$

occur frequently and at first glance are confusing to students. If you were asked to evaluate $-ab$ for $a=-3$ and $b=+2$, how would you proceed? Would you take the negative of a and then multiply it by b, or multiply a and b first and then take the negative of the product? Actually it does not matter! Because of Theorem 5 we get the same result either way since $(-a)b = -(ab)$. If, in addition, we consider other material in this section, we find that

$$-ab = \begin{cases} (-a)b \\ a(-b) \\ -(ab) \\ (-1)ab \end{cases}$$

and we are at liberty to replace any one of these five forms with another from the same group.

DIVISION OF SIGNED NUMBERS

Just as in multiplication, we assume division of positive real numbers and its properties are known. We know, for example, that

$$(+12) \div (+3) = +4$$

but what numbers do we assign to

$$(+12) \div (-3) = ?$$

$$(-12) \div (+3) = ?$$

$$(-12) \div (-3) = ?$$

$$0 \div (-3) = ?$$

$$(-12) \div 0 = ?$$

In dividing 12 by 3 you "automatically" know that the answer is 4. That is, that there are four 3s in 12, or 4 times 3 is equal to 12. Notice that we can convert division into multiplication by asking "what times 3 is 12?"

It is this latter approach that we are going to take to generalize division so that we can write answers for the various problems stated above.

DEFINITION

DIVISION

$$a \div b = Q \qquad \text{if and only if} \qquad a = bQ \text{ and } Q \text{ is unique}$$

Let us use this definition to find

$$(+12) \div (-3) = ? \qquad \text{or} \qquad -3\overline{)+12}^{\;?}$$

We ask, "what must (-3) be multiplied by to produce $(+12)$?" From our discussion on multiplication, we know the answer must be -4. Thus, we write

$$(+12) \div (-3) = -4 \qquad \text{or} \qquad -3\overline{)+12}^{\;-4}$$

since $(-3)(-4) = +12$.

What about

$$0 \div (-3), \qquad (+12) \div 0, \qquad \text{and} \qquad 0 \div 0$$

The first is assigned a value of 0, since $0(-3) = 0$. The second is not defined, since no number times 0 is $+12$. In the third case, any number could be the quotient, since 0 times any number is 0, and so the quotient is not unique. We thus conclude that

Zero cannot be used as a divisor—ever!

The two division symbols "$\div$" and "$\overline{)}$" from arithmetic are not used a great deal in algebra and higher mathematics. The horizontal bar "—" and slash mark "/" are the symbols most frequently used. Thus

$$a/b, \qquad \frac{a}{b}, \qquad a \div b, \qquad \text{and} \qquad b\overline{)a}$$

all name the same number (assuming the quotient is defined), and we can write

$$a/b = \frac{a}{b} = a \div b = b\overline{)a}$$

The following theorem leads to a simple mechanical rule for carrying out division of signed numbers.

THEOREM 6 (A) If a and b are real numbers with like signs, then

$$\frac{a}{b} = \frac{|a|}{|b|} \qquad b \neq 0$$

(*B*) If a and b are real numbers with unlike signs, then

$$\frac{a}{b} = -\frac{|a|}{|b|} \qquad b \neq 0$$

(*C*) Is a is any nonzero real number, then

$$\frac{0}{a} = 0 \qquad \frac{a}{0} \text{ is not defined} \qquad \frac{0}{0} \text{ is not defined}$$

A careful look at this theorem leads directly to the *simple mechanical rule for division.*

> If neither number is 0, use the same rule of signs as in multiplication (that is, quotients of numbers with like signs are positive and quotients of numbers with unlike signs are negative) and divide their absolute volume as in arithmetic, attaching the appropriate sign to the result.
>
> Zero divided by a nonzero number is always zero.
>
> Division by 0 is not defined.

EXAMPLE 18 Divide:

(*A*) $\dfrac{+36}{-9} = -4$

(*B*) $\dfrac{-36}{+9} = -4$

(*C*) $\dfrac{-36}{-9} = 4$

(*D*) $\dfrac{0}{-9} = 0$

(*E*) $\dfrac{-36}{0}$ is not defined

(*F*) $\dfrac{0}{0}$ is not defined

PROBLEM 18 Divide:

(A) $\dfrac{+24}{-8}$ (B) $\dfrac{-24}{+8}$ (C) $\dfrac{-24}{-8}$

(*D*) $\dfrac{0}{-8}$ (*E*) $\dfrac{-24}{0}$ (*F*) $\dfrac{0}{0}$

ANSWER (*A*) -3 (*B*) -3 (*C*) 3 (*D*) 0 (*E*) not defined

(*F*) not defined

EXAMPLE 19 Perform the indicated operations and simplify.

(*A*) $$\frac{-12}{-3} - \frac{24}{-4} = 4 - (-6) = 4 + 6 = 10$$

(*B*) $$\frac{3-9}{-2-(-3)} - \frac{(-6)(4)}{-8} = \frac{-6}{1} - \frac{-24}{-8} = -6 - 3 = -9$$

PROBLEM 19 Perform the indicated operations and simplify.

(*A*) $\dfrac{-16}{8} - \dfrac{-8}{-2}$ (*B*) $\dfrac{(-3)(-8)}{-12} - \dfrac{-5+(+7)}{3+5}$

ANSWER (*A*) -6 (*B*) -1

Exercise 4

A *Perform the indicated operations.*

1. $(-3)(-5)$ **2.** $(-7)(-4)$ **3.** $(-18) \div (-6)$

4. $(-20) \div (-4)$ **5.** $(-2)(+9)$ **6.** $(+6)(-3)$

7. $\dfrac{-9}{+3}$ **8.** $\dfrac{+12}{-4}$ **9.** $0(-7)$

10. $(-6)0$ **11.** $0/5$ **12.** $0/(-2)$

13. $3/0$ **14.** $-2/0$ **15.** $0 \div 0$

16. $\dfrac{0}{0}$ **17.** $\dfrac{-21}{3}$ **18.** $\dfrac{-36}{-4}$

19. $(-4)(-2) + (-9)$ **20.** $(-7) + (-3)(+2)$

21. $(+5) - (-2)(+3)$ **22.** $(-7) - (-3)(-4)$

23. $5 - \dfrac{-8}{2}$ **24.** $7 - \dfrac{-16}{-2}$

25. $(-1)(-8)$ and $-(-8)$ **26.** $(-1)(+3)$ and $-(+3)$

B 27. $-12 + \frac{-14}{-7}$

28. $\frac{-10}{5} + (-7)$

29. $\frac{6(-4)}{-8}$

30. $\frac{5(-3)}{3}$

31. $\frac{22}{-11} - (-4)(-3)$

32. $3(-2) - \frac{-10}{-5}$

33. $\frac{-16}{2} - \frac{3}{-1}$

34. $\frac{27}{-9} - \frac{-21}{-7}$

35. $(+5)(-7)(+2)$

36. $(-6)(-3)(+4)$

37. $(-22)(+36)(0)$

38. $(+19)(0)(-35)$

39. $[(+2) + (-7)][(+8) - (+10)]$

40. $[(-3) - (+8)][(+4) + (-2)]$

Evaluate for $w = +2$, $x = -3$, $y = 0$, and $z = -24$.

41. z/w

42. z/x

43. w/y

44. y/x

45. $\frac{z}{x} - wz$

46. $wx - \frac{z}{w}$

47. $\frac{xy}{w} - xyz$

48. $wxy - \frac{y}{z}$

49. $-|w||x|$

50. $(|x||z|)$

51. $\frac{|z|}{|x|}$

52. $-\frac{|z|}{|w|}$

53. Any integer divided by 0 is (*always, sometimes, never*) 0

54. 0 divided by *any* integer is (*always, sometimes, never*) 0

55. A product made up of an odd number of negative factors is (*sometimes, always, never*) negative.

56. A product made up of an even number of negative factors is (*sometimes, always, never*) negative.

C *Evaluate for $w = +2$, $x = -3$, $y = 0$, and $z = -24$.*

57. $wx + \frac{z}{wx} + wz$

58. $xyz + \frac{y}{z} + x$

59. $\frac{8x}{z} - \frac{z - 6x}{wx}$

60. $\frac{w - x}{w + x} - \frac{z}{2x}$

61. If the quotient $\frac{x}{y}$ exists, when is it equal to $\frac{-|x|}{|y|}$?

62. If the quotient $\frac{x}{y}$ exists, when is it equal to $\frac{|x|}{|y|}$?

63. Provide the reason(s) for each step in the proof that $a0 = 0$ for all real numbers a (Theorem 4C).

	STATEMENT	REASON
1	$a0 = a(0+0)$	1
2	$a0 = a0 + a0$	2
3	$a0 + [-(a0)] = (a0 + a0) + [-(a0)]$	3
4	$0 = a0 + \{a0 + [-(a0)]\}$	4
5	$0 = a0 + 0$	5
6	$0 = a0$	6
7	$a0 = 0$	7

64. Provide the reason(s) for each step in the proof that $(-1)a = -a$ (Theorem 5A).

	STATEMENT	REASON
1	$a + (-1)a = 1a + (-1)a$	1
2	$= a[1 + (-1)]$	2
3	$= a \cdot 0$	3
4	$= 0$	4

Therefore,

5	$(-1)a = -a$	5

1.7 Inequalities and Line Graphs

Earlier in this chapter we discussed the equality relation and some of its properties. In this section we will introduce another important relation called *the order relation*. This relation has to do with "less than" and "greater than."

Just as we use "=" to replace the words "is equal to," we will use the *inequality symbols* "<" and ">" to represent "is less than" and "is greater than," respectively. Thus, we can write the following equivalent forms:

$a < b$	a is less than b
$a > b$	a is greater than b
$a \leq b$	a is less than or equal to b
$a \geq b$	a is greater than or equal to b

Note that the small end (the point) of the inequality symbol is directed toward the smaller of the two numbers.

It no doubt seems obvious to you that

$$2 < 3$$

is a true statement. But does it seem equally obvious that

$$-5 < -1$$

$$-12 < 3$$

$$0 > -7$$

$$-1 > -1{,}000$$

are also true statements? To make the order relation precise so that we can interpret it relative to *all* real numbers, we need a more careful definition of the concept.

DEFINITION

INEQUALITY RELATION

If a and b are real numbers, then we write

$$a < b$$

if there exists a positive real number p such that $a + p = b$. We write

$$a > b$$

if $b < a$.

Certainly, one would expect that if a positive number were added to *any* real number, the sum would be larger than the original. That is essentially what the definition states.

EXAMPLE 20

(A) $2 < 3$ since $2 + 1 = 3$

(B) $-5 < -1$ since $-5 + 4 = -1$

(C) $-12 < 3$ since $-12 + 15 = 3$

(D) $0 > -7$ since $-7 < 0$

(E) $-1 > -1{,}000$ since $-1{,}000 < -1$

PROBLEM 20

Replace each question mark with either $<$ or $>$:

(A) 4 ? 6 (B) 6 ? 4 (C) −6 ? −4

(D) −8 ? 8 (E) 3 ? −9 (F) 0 ? −4

ANSWER (A) $<$ (B) $>$ (C) $<$ (D) $<$ (E) $>$ (F) $>$

The inequality symbols have a very clear geometric interpretation on the real number line. If $a < b$, then a is to the left of b; if $c > d$, then c is to the right of d (Fig. 3).

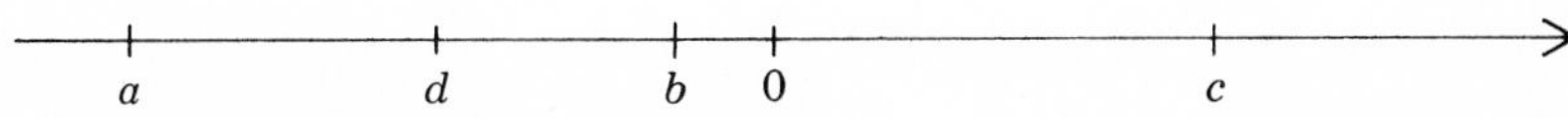

Figure 3 $a < b, c > d$

EXAMPLE 21 Refer to Fig. 3.

(A) $a < d$ since a is to the left of d

(B) $c > 0$ since c is to the right of 0

(C) $d < c$ since d is to the left of c

(D) $a < 0$ since a is to the left of 0

PROBLEM 21 Referring to Fig. 3, replace each question mark with either $<$ or $>$.

(A) $b\ ?\ d$ (B) $0\ ?\ b$

(C) $a\ ?\ c$ (D) $d\ ?\ 0$

ANSWER (A) $>$ (B) $>$ (C) $<$ (D) $<$

Now let us turn to simple inequalities of the form

$x > 2$ $\quad -2 < x \le 3$

$x \le -3$ $\quad 0 \le x < 5$

To solve such inequalities is to find the set of all replacements of the variable x (from a specified set of numbers) that makes the inequality true. This set is called the *solution set* for the inequality. For example, the solution set for

$x \le -3$ $\quad x$ an integer

is the set of all integers less than or equal to -3. Graphically, the general solution can be represented as follows:

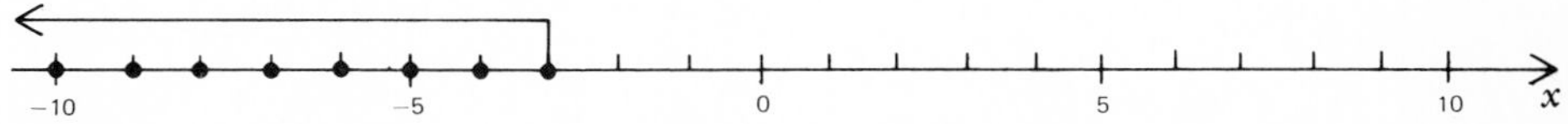

The solution set for

$x \le -3$ $\quad x$ a real number

is the set of all real numbers less than or equal to -3. Graphically, this includes all the points to the left of and at -3 (a solid line):

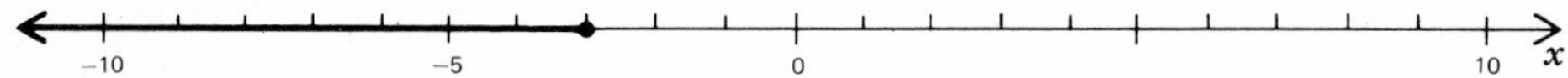

The double inequality

$2 < x \leq 3$

is a short way of writing

$-2 < x$ and $x \leq 3$

which means that x is greater than -2 and at the same time x is less than or equal to 3. In other words, x can be any number between -2 and 3, including 3, but not including -2. The graphs of the solution set in the integers and the solution set in the real numbers are as follows:

x an integer

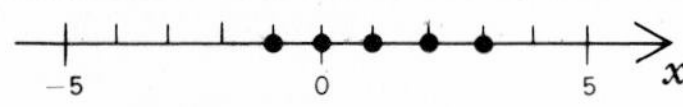

x a real number

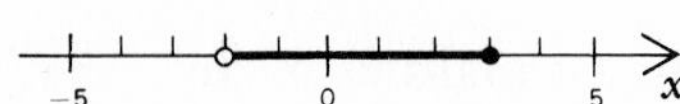

Note that a hollow circle indicates that an endpoint is not included and a solid circle indicates that an endpoint is included.

EXAMPLE 22 Graph each inequality statement on a real number line.

(*A*) $x \geq -3$ x an integer

SOLUTION

(*B*) $-4 \leq x < 2$ x a real number

SOLUTION

PROBLEM 22 Graph each inequality statement on a real number line.

(*A*) $x > -3$ x a real number (*B*) $-4 \leq x < 2$ x an integer

ANSWER (*A*) (*B*)

We will continue the discussion of inequalities and their properties in Chapter 3, at which time you will learn how to solve inequality statements of a more complicated nature.

Exercise 5

A *Write in symbolic form.*

1. 18 is greater than 4

2. 7 is greater than 3

3. -5 is less than 9

4. 11 is less than 12

5. x is greater than or equal to -8

6. x is greater than or equal to -5

7. x is less than or equal to 3

8. x is less than or equal to 12

Write in verbal form.

9. $12 > 11$ **10.** $5 > -2$ **11.** $11 < 12$

12. $-2 < 5$ **13.** $x \leq 4$ **14.** $x \geq 3$

Replace each question mark with $<$ *or* $>$.

15. 6 ? 3 **16.** 5 ? 7 **17.** -3 ? -6

18. -7 ? -5 **19.** -6 ? -3 **20.** -5 ? -7

21. 5 ? 0 **22.** 0 ? 8 **23.** -5 ? 0

24. 0 ? -8 **25.** -6 ? -3 **26.** -7 ? -5

Referring to

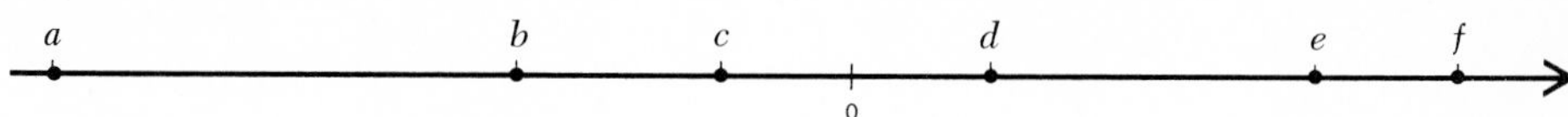

replace each question mark in Problems 27 to 32 with either $<$ *or* $>$.

27. e ? a **28.** a ? d **29.** c ? b

30. e ? f **31.** 0 ? d **32.** 0 ? a

B **33.** If we add a positive real number to any real number, will the sum be greater than or less than the original number?

34. If we add a positive real number to any real number, will the sum be to the right or left of the original number on a number line?

Graph on a real number line for x an integer.

35. $-5 < x \leq -1$

36. $-4 \leq x < 1$

37. $-1 < x < 3$

38. $-2 \leq x \leq 2$

Graph on a real number line for x a real number.

39. $-5 < x \leq -1$

40. $-4 \leq x < 1$

41. $-1 < x < 3$

42. $-2 \leq x \leq 2$

Replace each question mark with < or >.

43. $5 + \dfrac{-18}{3}$? 0

44. $\dfrac{-6}{2} - 4$? -2

45. $\dfrac{24}{-3} - (4)(-2)$? -6

46. $(-7)(2) - \dfrac{-12}{3}$? 9

Find the solution set for each inequality for x restricted to $\{-6, -4, -2, 0, 2, 4, 6\}$.

47. $2x < -2$

48. $3x > 0$

49. $4 - x > 6$

50. $2 - x \geq 0$

51. $-2x - \dfrac{12}{x} \leq 10$

52. $\dfrac{x}{-2} + x \leq 0$

53. A rectangular lot is to be fenced with 400 feet of wire. (*A*) If the lot is x feet wide, write a formula for the area of the lot in terms of x. (*B*) What real number values can x assume? Write the answer in terms of a double inequality statement.

54. A flat sheet of cardboard in the shape of an 18- by 12-inch rectangle is to be used to make an open-topped box by cutting an x- by x-inch square out of each corner and folding the remaining part appropriately. (*A*) Write an equation for the volume of the box in terms of x. (*B*) What real number values can x assume? Formulate the answer in terms of a double inequality statement.

C The statement "A is the set of all natural numbers n such that $5 < n < 100$" is symbolized $A = \{n \in N \mid 5 < n < 100\}$ where N is the set of natural numbers and the symbol "$\in$" means "is an element of." This is not as forbidding as it first looks; each part of the symbolic expression has a meaning as indicated below.

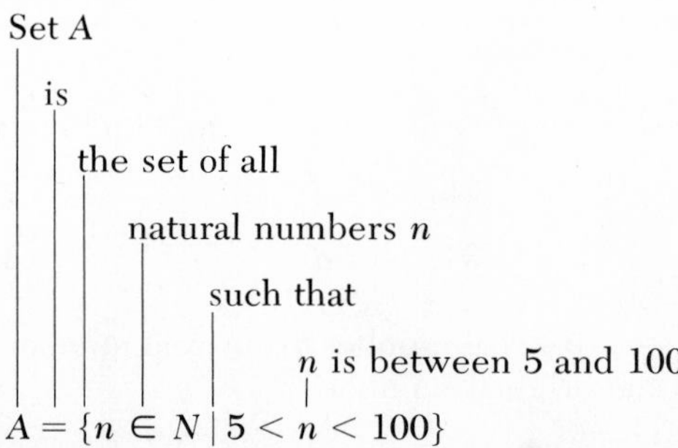

EXAMPLE

(*A*) $\{n \in N \mid n < 8\} = \{1, 2, 3, 4, 5, 6, 7\}$

(*B*) $\{x \in I \mid -2 \leq x < 3\} = \{-2, -1, 0, 1, 2\}$

Graph each set:

55. $\{x \in I \mid -3 < x \leq 2\}$

56. $\{n \in N \mid n < 5\}$

57. $\{x \in R \mid -3 < x \leq 2\}$

58. $\{y \in R \mid y < 5\}$

59. Show that if $a < b$, then $b - a$ is a positive number.

60. Show that if $b - a$ is a positive number, then $a < b$.

Exercise 6 Chapter Review

A

1. True (T) or false (F): (*A*) $3 \cdot 7 - 4 = 9$; (*B*) $7 + 2 \cdot 3 = 13$

2. If y is a certain number, write an algebraic expression for a number (*A*) 8 times as large as y; (*B*) 8 more than y.

3. State the justifying field axiom for (*A*) $3x = x3$; (*B*) $(x+3)+2 = x+(3+2)$.

4. Remove parentheses and simplify: (*A*) $7 + (x + 3)$; (*B*) $5(x2)$; (*C*) $0 + (1x + 3) + 2$

5. Evaluate: (*A*) $-(+4)$; (*B*) $|-8|$; (*C*) $-(-2)$; (*D*) $|+9|$

Evaluate:

6. $(-8) + (+3)$

7. $(-9) + (-4)$

8. $(-3) - (-9)$

9. $4 - 7$

10. $0 - (-3)$

11. $(-12) - 0$

12. $(-7)(-4)$

13. $3(-6)$

14. $\frac{-16}{4}$

15. $\frac{-12}{-2}$

16. $\frac{-6}{0}$

17. $\frac{0}{2}$

18. $(-6) + (-3) + (+7) + 0 + (-1)$

19. $10 - 3(6 - 4)$

20. $(-8) - (-2)(-3)$

21. $(-9) - (-12)/3$

22. $(3 - 2) - (4 - 7)$

23. $(4 - 8) + (-2)(+4)$

Replace each question mark with either $<$ *or* $>$.

24. 7 ? 2

25. -7 ? -2

26. 0 ? -5

27. -525 ? -2

B *Evaluate:*

28. $(-54) + 44$

29. $(-62) + (-18) + 0 + 20$

30. $2[9 - 3(3 - 1)]$

31. $[(-3) - (-3)] - (-4)$

32. $[-(-4)] + (-|-3|)$

33. $2\{9 - 2[(3 + 1) - (1 + 1)]\}$

34. $3[(5 - 7) - (6 - 9)] + (-2)(3)$

35. $\frac{-16}{2} - (-3)(4)$

36. $(-2)(-4)(-3) - \frac{-36}{(-2)(+9)}$

37. $(5 - 7)[(-2) - (-2)]$

38. Evaluate $3[14 - x(x + 1)]$ for $x = 3$.

39. Evaluate for $x = -8$ and $y = 3$: (A) $-x$; (B) $-(x + y)$

40. Evaluate $(x + y) - z$ for $x = 6$, $y = -8$, and $z = 4$.

41. Evaluate $(wx - z/x) - w/x$ for $w = -10$, $x = -2$, and $z = 0$.

42. Evaluate $7\{x + 2[(x + y) - (x - y)]\}$ for $x = 3$, and $y = 1$.

43. Evaluate $\frac{(xyz + xz) - z}{z}$ for $x = -6$, $y = 0$, and $z = -3$.

State the field axiom that justifies each statement.

44. $5 + (x + 3) = 5 + (3 + x)$

45. $5 + (3 + x) = (5 + 3) + x$

46. $5(x3) = 5(3x)$

47. $5(3x) = (5 \cdot 3)x$

48. Remove parentheses and simplify: (A) $(m + 5) + (n + 7) + (p + 2)$; (B) $(4x)(7y)$

49. Translate into an algebraic equation using only x as a variable: 3 times a certain number is 8 more than that number.

50. If the length of a rectangle is 5 inches more than twice the width x, write an algebraic expression for its perimeter.

51. For a an integer, (A) $-a + ? = 0$; (B) $a + (-a) = ?$.

52. True (*T*) or false (*F*): (A) $2 \cdot 3 + 5 > 15$; (B) $12 - 5 \cdot 2 \leq 2$

53. Find the solution set for each inequality statement for x restricted to $\{-2, -1, 0, 1, 2\}$: (A) $-3 \leq x + 2 < 3$, (B) $-4 < 2x < 2$

54. Graph $-4 \leq x < -1$ (A) for x an integer, (B) for x a real number.

55. If $a + p = b$ for some positive number b, then a is (*greater than, less than*) b.

56. Find all solutions: (A) $|x| = 5$, (B) $-x = +7$

C **57.** Evaluate: $(-3) - 2\{5 - 3[2 - 2(3 - 6)]\}$

58. Evaluate $uv - 3\{x - 2[(x + y) - (x - y)] + u\}$ for $u = -2$, $v = 3$, $x = 2$, and $y = -3$.

59. Evaluate $\frac{5w}{x-7} - \frac{wx-4}{x-w}$ for $w=-4$, $x=2$, and $y=4$.

60. Describe the elements in the set $\{x \in I \mid |x| = x\}$.

61. Translate into an algebraic equation using only one variable: Subtracting 3 from a given number and multiplying the difference by 2 yields a number that is 5 more than twice the given number.

62. Replace the question marks in $C = \{x \in N \mid ? \leq x \; ? \; 7\} = \{3, 4, 5, 6\}$ with appropriate symbols.

63. Supply a reason for each step in the proof that $a + (-a) = 0$ for all real numbers a (Theorem 2A).

	STATEMENT		REASON
1	$a + (-a) = a + (-1)a$	1	
2	$= 1a + (-1)a$	2	
3	$= a[1 + (-1)\}$	3	
4	$= a \cdot 0$	4	
5	$= 0$	5	

Chapter 2

POLYNOMIALS

2.1 Polynomials in One or More Variables

In this section we will introduce a very important type of algebraic expression, called a polynomial, which is encountered with great frequency throughout mathematics. First, however, we must generalize the idea of exponent to include all natural numbers. Recall that

$$b^2 = bb$$

and that

$$b^3 = bbb$$

In general we define b^n where n is any natural number and b is any real number as follows:

$$b^n = \underbrace{bb \cdots b}_{n \text{ factors of } b}$$

b is called the *base* and n the *power* or *exponent*. In addition, we define

$$b^1 = b$$

and usually use b in place of b^1.

EXAMPLE 1 (A) $x^3 = xxx$ $\qquad m^1 = m$

$3^4 = 3 \cdot 3 \cdot 3 \cdot 3$ $\qquad 2x^4y^2 = 2xxxxyy$

(B) $mmmm = m^4$ $\qquad 3uvv = 3uv^2$

$2 \cdot 2 \cdot 2 = 2^3$ $\qquad 5xxxyy = 5x^3y^2$

PROBLEM 1 (A) Write in nonexponent form as in Example 1A: x^5, 2^4, $4x^2y^3$

(B) Write in exponent form: mm, $7 \cdot 7 \cdot 7 \cdot 7 \cdot 7$, $6xxxyyyy$

ANSWER (A) $xxxxx$, $2 \cdot 2 \cdot 2 \cdot 2$, $4xxyyy$ (B) m^2, 7^5, $6x^3y^4$

If we take exponent forms of the type illustrated in Example 1 and combine them using addition and subtraction, we obtain a mathematical form called a polynomial. We may have polynomials in one variable, two variables, and so on. In general, a *polynomial* is any algebraic expression that can be formed by using only the operations of addition, subtraction, and multiplication on variables and real number constants. For example,

$$x \qquad 3x^2 \qquad 2x - 1 \qquad 3x^2 + 2x - 5$$

$$x^2 - 3xy + 2y^2 \qquad 5x^3 - 2x^2y + xy^2 - 2y^2$$

are polynomials in one and two variables. The algebraic expressions

$$\frac{2x - 7}{3x + y} \qquad x^3 - 2\sqrt{x} + \frac{1}{x^3}$$

are not polynomials, since they cannot be formed by using only the operations referred to above.

It is convenient to identify certain types of polynomials for more efficient study. The concept of degree is used for this purpose. If a term in a polynomial has only one variable as a factor, then the *degree of that term* is the power of the variable. If two or more variables are present in a term as factors, then the *degree of the term* is the sum of the powers of the variables. The *degree of a polynomial* is the degree of the nonzero term with the highest degree in the polynomial. (Recall that we add or subtract *terms* and multiply *factors*.)

EXAMPLE 2

(A) $4x^3$ is of degree 3

(B) $3x^3y^2$ is of degree 5

(C) In $3x^5 - 2x^4 + x^2 - 3$, the highest-degree term is the first, with degree 5; thus, the degree of the polynomial is 5.

(D) The degree of each of the first three terms in $x^2 - 2xy + y^2 + 2x - 3y + 2$ is 2; the fourth and fifth terms each is of degree 1; thus, this is a second-degree polynomial.

PROBLEM 2

(A) What is the degree of $7x^5$? Of $3x^3y^4$?

(B) What is the degree of the polynomial $5x^3 - 2x^2 + x - 9$? Of $4x^2 - 3xy + y^2 + 2x - y + 6$?

ANSWER (A) 5, 7 (B) 3, 2

NOTE: A nonzero constant is defined to be a polynomial of degree 0 for reasons that will become clear later on, and the number 0 is not assigned a degree.

2.2 Addition and Subtraction of Polynomials

DISTRIBUTIVE AXIOM

Basic to adding and subtracting polynomials, as well as many other operations on algebraic expressions, is the distributive axiom

$$\boxed{a(b + c) = ab + ac}$$

where a, b, and c are any real numbers. In words, this axiom states that multiplication distributes over addition.

EXAMPLE 3

Multiply, using the distributive axiom.

(A) $3(x + y) = 3x + 3y$

(B) $5(2x^2 + 3) = 10x^2 + 15$

(C) $m(u + v) = mu + mv$

PROBLEM 3

Multiply, using the distributive axiom.

(A) $5(m + n)$ (B) $4(3y^3 + 5)$ (C) $a(u + v)$

ANSWER (A) $5m + 5n$ (B) $12y^3 + 20$ (C) $au + av$

It should be clear from the properties of equality in Chapter 1 that we can also write the distributive axiom in the form

$$ab + ac = a(b + c)$$

that is, a factor common to each term can be taken out.

EXAMPLE 4 Take out factors common to both terms.

(A) $6x + 6y = 6(x + y)$

(B) $7x^2 + 14 = 7(x^2 + 2)$

(C) $ax + ay = a(x + y)$

PROBLEM 4 Take out factors common to both terms.

(A) $9m + 9n$ (B) $12y^3 + 6$ (C) $du + dy$

ANSWER (A) $9(m + n)$ (B) $6(2y^3 + 1)$ (C) $d(u + y)$

Several other useful distributive forms follow from the basic one above and other properties discussed in the first chapter:

$$(ab + ac) = (b + c)a$$

$$(ab - ac) = a(b - c)$$

$$= (b - c)a$$

$$a(b + c + d + \cdots + f) = ab + ac + ad + \cdots + af$$

$$ab + ac + ad + \cdots + af = a(b + c + d + \cdots + f)$$

$$= (b + c + d + \cdots + f)a$$

COMBINING LIKE TERMS

A constant present as a factor in a term is called the *numerical coefficient* (or simply the *coefficient*) of the term. If no constant appears in the term, then the coefficient is understood to be 1. The coefficient of a term in a polynomial includes the sign that precedes it.

EXAMPLE 5 Since $3x^4 - 2x^3 + x^2 - x + 3$ $= 3x^4 + (-2x^3) + 1x^2 + (-1)x + 3$

think

the coefficient of

(A) the first term is 3

(B) the second term is -2

(*C*) the third term is 1

(*D*) the fourth term is -1

PROBLEM 5 In $5x^4 - x^3 - 3x^2 + x - 7$, what is the coefficient of the

(*A*) first term? (*B*) second term?

(*C*) third term? (*D*) fourth term?

ANSWER (*A*) 5 (*B*) -1 (*C*) -3 (*D*) 1

Two terms are called *like terms* if they are exactly alike, except the numerical coefficients may or may not be the same.

EXAMPLE 6 (*A*) In $7x + 3y - 6x$, the first and third terms are like terms.

(*B*) In $9x^2y - 3xy - 2x^2y + x^2y$, the first, third, and fourth terms are like terms.

PROBLEM 6 List the like terms in

(*A*) $7m - 3m + 5n$ (*B*) $3xy - 7x^2y + xy + 2xy^2$

ANSWER (*A*) first and second (*B*) first and third

If an algebraic expression contains two or more like terms, these terms can always be combined into a single term. The distributive axiom is the principal tool behind this process.

EXAMPLE 7 (*A*) $3x + 7x \boxed{= (3 + 7)x} = 10x$

(*B*) $6m - 9m \boxed{= (6 - 9)m} = -3m$

(*C*)
$$\begin{aligned} 5x^3y - 2xy - x^3y - 2x^3y &= 5x^3y - x^3y - 2x^3y - 2xy \\ &= (5 - 1 - 2)x^3y - 2xy \\ &= 2x^3y - 2xy \end{aligned}$$

PROBLEM 7 Combine like terms proceeding as in Example 7:

(*A*) $5y + 4y$ (*B*) $2u - 6u$ (*C*) $6mn^2 - m^2n - 3mn^2 - mn^2$

ANSWER (*A*) $9y$ (*B*) $-4u$ (*C*) $2mn^2 - m^2n$

It should be clear that free use was made of the axioms discussed in the first chapter. Most of the steps illustrated in the "dotted boxes" are done mentally. The process is quickly mechanized as follows:

> Like terms are combined by adding their numerical coefficients.

EXAMPLE 8 Combine like terms mentally.

(A) $3x - 5y + 6x + 2y = 9x - 3y$

(B) $x^3y^2 - 2x^2y^3 + 5x^2y^2 - 4x^2y^3 - x^3y^2 - 5x^2y^2 = -6x^2y^3$

PROBLEM 8 Combine like terms mentally.

(A) $7m + 8n - 5m - 10n$ (B) $2u^4v^2 - 3uv^3 - u^4v^2 + 6u^4v^2 + 2uv^3 - 6u^4v^2$

ANSWER (A) $2m - 2n$ (B) $u^4v^2 - uv^3$

REMOVING SYMBOLS OF GROUPING

How can we simplify expressions such as

$$2(3x - 5y) - 2(x + 3y)$$

You no doubt would guess that we could rewrite this expression, using the various forms of the distributive property, as

$$6x - 10y - 2x - 6y$$

and combine terms to obtain

$$4x - 16y$$

and your guess would be correct.

EXAMPLE 9 (A) $2(3x^2 - 2x + 5) + (x^2 + 3x - 7)$ $= 2(3x^2 - 2x + 5) + 1(x^2 + 3x - 7)$ *think*

$$= 6x^2 - 4x + 10 + x^2 + 3x - 7$$

$$= 7x^2 - x + 3$$

(B) $(x^3 - 2x - 6) - (2x^3 - x^2 + 2x - 3)$

$= 1(x^3 - 2x - 6) - 1(2x^3 - x^2 + 2x - 3)$ *think*

$$= x^3 - 2x - 6 - 2x^3 + x^2 - 2x + 3$$

$$= -x^3 + x^2 - 4x - 3$$

PROBLEM 9 Remove parentheses and simplify.

(A) $3(u^2 - 2v^2) + (u^2 + 5v^2)$

(B) $(m^3 - 3m^2 + m - 1) - (2m^3 - m + 3)$

ANSWER (A) $4u^2 - v^2$ (B) $-m^3 - 3m^2 + 2m - 4$

ADDITION AND SUBTRACTION OF POLYNOMIALS

Addition and subtraction of polynomials can be thought of in terms of removing parentheses and combining like terms, as illustrated in Example 9 above. Sometimes a vertical arrangement is useful. Both processes are illustrated in the next two examples. You should be able to work either way, letting the situation dictate the choice.

EXAMPLE 10 Add:

$$x^4 - 3x^3 + x^2, \qquad -x^3 - 2x^2 + 3x, \qquad \text{and} \qquad 3x^2 - 4x - 5$$

SOLUTION Add horizontally:

$$\begin{aligned}(x^4 - 3x^3 + x^2) + (-x^3 - 2x^2 + 3x) + (3x^2 - 4x - 5)&\\ &= x^4 - 3x^3 + x^2 - x^3 - 2x^2 + 3x + 3x^2 - 4x - 5\\ &= x^4 - 4x^3 + 2x^2 - x - 5\end{aligned}$$

or vertically by lining up like terms and adding their coefficients:

$$\begin{array}{rrrrr} x^4 & -\ 3x^3 & +\ \ x^2 & & \\ & -\ \ x^3 & -\ 2x^2 & +\ 3x & \\ & & 3x^2 & -\ 4x & -\ 5 \\ \hline x^4 & -\ 4x^3 & +\ 2x^2 & -\ \ x & -\ 5 \end{array}$$

PROBLEM 10 Add horizontally and vertically.

$$3x^4 - 2x^3 - 4x^2, \qquad x^3 - 2x^2 - 5x, \qquad \text{and} \qquad x^2 + 7x - 2$$

ANSWER $3x^4 - x^3 - 5x^2 + 2x - 2$

EXAMPLE 11 Subtract:

$$4x^2 - 3x + 5 \qquad \text{from} \qquad x^2 - 8$$

SOLUTION

$$\begin{aligned}&(x^2 - 8) - (4x^2 - 3x + 5)\\ &= x^2 - 8 - 4x^2 + 3x - 5\\ &= -3x^2 + 3x - 13\end{aligned} \qquad \text{or} \qquad \begin{array}{rrr} x^2 & & -8 \\ 4x^2 & -\ 3x & +\ 5 \\ \hline -3x^2 & +\ 3x & -\ 13 \end{array}$$

PROBLEM 11 Subtract:

$2x^2 - 5x + 4$ from $5x^2 - 6$

ANSWER $3x^2 + 5x - 10$

Exercise 7

A *Given the polynomial $7x^4 - 3x^3 - x^2 + x - 3$, indicate:*

1. The coefficient of the second term

2. The coefficient of the third term

3. The exponent of the variable in the second term

4. The exponent of the variable in the fourth term

5. The coefficient of the fourth term

6. The coefficient of the first term

Simplify by removing parentheses, if any, and combining like terms.

7. $9x + 8x$

8. $7x + 3x$

9. $9x - 8x$

10. $7x - 3x$

11. $5x + x + 2x$

12. $2x + 5x + x$

13. $4t + 8t + 9t$

14. $2x + 5x + x$

15. $4y + 3x + y$

16. $2x + 3y + 5x$

17. $5m + 3n - m - 9n$

18. $2x + 8y - 7x - 5y$

19. $3(u - 2v) + 2(3u + v)$

20. $2(m + 3n) + 4(m - 2n)$

21. $4(m - 3n) - 3(2m + 4n)$

22. $2(x - y) - 3(3x - 2y)$

23. $(2u - v) + (3u - 5v)$

24. $(x + 3y) + (2x - 5y)$

25. $(2u - v) - (3u - 5v)$

26. $(x + 3y) - (2x - 5y)$

Add:

27. $6x + 5$ and $3x - 8$

28. $3x - 5$ and $2x + 3$

29. $7x - 5$, $-x + 3$, and $-8x - 2$

30. $2x + 3$, $-4x - 2$, and $7x - 4$

31. $5x^2 + 2x - 7$, $2x^2 + 3$, and $-3x - 8$

32. $2x^2 - 3x + 1$, $2x - 3$, and $4x^2 + 5$

Subtract:

33. $3x - 8$ from $2x - 7$

34. $4x - 9$ from $2x + 3$

35. $2y^2 - 6y + 1$ from $y^2 - 6y - 1$

36. $x^2 - 3x - 5$ from $2x^2 - 6x - 5$

B *Simplify by removing symbols of grouping, if any, and combining like terms.*

37. $-x^2y + 3x^2y - 5x^2y$

38. $-4r^3t^3 - 7r^3t^3 - 7r^3t^3 + 9r^3t^3$

39. $y^3 + 4y^2 - 10 + 2y^3 - y + 7$

40. $3x^2 - 2x + 5 - x^2 + 4x - 8$

41. $a^2 - 3ab + b^2 + 2a^2 + 3ab - 2b^2$

42. $2x^2y + 2xy^2 - 5xy + 2xy^2 - xy - 4x^2y$

43. $x - 3y - 4(2x - 3y)$

44. $a + b - 2(a - b)$

45. $y - 2(x - y) - 3x$

46. $x - 3(x + 2y) + 5y$

47. $-2(-3x + 1) - (2x + 4)$

48. $-3(-t + 7) - (t - 1)$

49. $2(x - 1) - 3(2x - 3) - (4x - 5)$

50. $-2(y - 7) - 3(2y + 1) - (-5y + 7)$

51. $4t - 3[4 - 2(t - 1)]$

52. $3x - 2[2x - (x - 7)]$

Replace each question mark with an appropriate algebraic expression.

53. $5 + m - 2n = 5 + (?)$

54. $2 + 3x - y = 2 + (?)$

55. $5 + m - 2n = 5 - (?)$

56. $2 + 3x - y = 2 - (?)$

57. $w^2 - x + y - z = w^2 + (?)$

58. $w^2 - x + y - z = w^2 - (?)$

Add:

59. $2x^4 - x^2 - 7$, $3x^3 + 7x^2 + 2x$, and $x^2 - 3x - 1$

60. $3x^3 - 2x^2 + 5$, $3x^2 - x - 3$, and $2x + 4$

Subtract:

61. $5x^3 - 3x + 1$ from $2x^3 + x^2 - 1$

62. $3x^3 - 2x^2 - 5$ from $2x^3 - 3x + 2$

63. Subtract the sum of the first two polynomials from the sum of the last two: $3m^3 - 2m + 5$, $4m^2 - m$, $3m^2 - 3m - 2$, and $m^3 + m^2 + 2$

64. Subtract the sum of the last two polynomials from the sum of the first two: $2x^2 - 4xy + y^2$, $3xy - y^2$, $x^2 - 2xy - y^2$, and $-x^2 + 3xy - 2y^2$

65. The width of a rectangle is 5 inches less than its length. If x is the length of the rectangle, write an algebraic expression that represents the perimeter of the rectangle and simplify the expression.

66. Repeat Problem 65 if the length of the rectangle is 3 inches more than twice its width.

C *Remove symbols of grouping and combine like terms.*

67. $2t - 3\{t + 2[t - (t + 5)] + 1\}$

68. $x - \{x - [x - (x - 1)]\}$

69. $w - \{x - [z - (w - x) - z] - (x - w)\} + x$

70. $3x^2 - 2\{x - x[x + 4(x - 3)] - 5\}$

71. A pile of coins consists of nickels, dimes, and quarters. There are 5 less dimes than nickels and 2 more quarters than dimes. If x equals the number of nickels, write an algebraic expression that represents the value of the pile in cents. Simplify the expression. HINT: If x represents the number of nickels, then what do $x - 5$ and $(x - 5) + 2$ represent?

72. A coin purse contains dimes and quarters only. There are 4 more dimes than quarters. If x equals the number of dimes, write an algebraic expression that represents the value of the money in the purse. Simplify the expression. HINT: If x represents the number of dimes, then what does $x - 4$ represent?

2.3 Multiplication of Polynomials

Before taking up the general problem of multiplying polynomials, it is useful to introduce an important property of exponents.

FIRST PROPERTY OF EXPONENTS

Recall that for n a natural number and b any real number

$$b^n = \underbrace{bb \cdots b}_{n \text{ factors of } b}$$

that is, a natural number exponent indicates how many times a base is to be taken as a factor. Thus,

$$x^3y^5 = xxxyyyyy$$

If we multiply two exponent forms with the same base, something interesting happens:

$$\begin{aligned} x^3x^5 &= (xxx)(xxxxx) \\ &= xxxxxxxx \\ &= x^8 \end{aligned}$$

which we could get by simply adding the exponents in x^3x^5.

In general, for any natural numbers m and n, and any real number b

$$\boxed{b^m b^n = b^{m+n}}$$

This is the *first property of exponents*, one of five exponent properties you will get to know well before the end of the book.

EXAMPLE 12 (A) $x^3x^5 = x^{3+5} = x^8$

(B) $(3m^{12})(5m^{23}) = 3 \cdot 5m^{12+23} = 15m^{35}$

(C) $(-3x^3y^4)(2x^2y^3) = (-3)(2)x^{3+2}y^{4+3} = -6x^5y^7$

PROBLEM 12 Simplify as in Example 12.

(A) y^4y^7 (B) $(9x^4)(3x^2)$ (C) $(4u^3v^2)(-3uv^3)$

ANSWER (A) y^{11} (B) $27x^6$ (C) $-12u^4v^5$

MULTIPLICATION OF POLYNOMIALS

In Example 12 we were actually multiplying single-term polynomials called *monomials*. How do we multiply polynomials with more than one term? Again, the distributive axiom plays the central role in the process, and leads directly to the mechanical rule:

> To multiply two polynomials, multiply each term of one by each term of the other, and add like terms.

EXAMPLE 13 (A) $3x^2(2x^2 - 3x + 4) = 6x^4 - 9x^3 + 12x^2$

(B) $(2x - 3)(3x^2 - 2x + 3)$

$= 2x(3x^2 - 2x + 3) - 3(3x^2 - 2x + 3)$

$= 6x^3 - 4x^2 + 6x - 9x^2 + 6x - 9$

$= 6x^3 - 13x^2 + 12x - 9$

or

$$\begin{array}{rrrr} & 3x^2 & -\ 2x & +\ 3 \\ & & 2x & -\ 3 \\ \hline 6x^3 & -\ 4x^2 & +\ 6x & \\ & -\ 9x^2 & +\ 6x & -\ 9 \\ \hline 6x^3 & -\ 13x^2 & +\ 12x & -\ 9 \end{array}$$

Note that either way, each term in $3x^2 - 2x + 3$ is multiplied by each term in $2x - 3$. In the vertical arrangement we start with $2x$ first, then like terms line up more conveniently.

PROBLEM 13 Multiply:

(A) $2m^3(3m^2 - 4m - 3)$ (B) $(3x - 4)(2x^2 + 3x - 1)$

ANSWER (A) $6m^5 - 8m^4 - 6m^3$ (B) $6x^3 + x^2 - 15x + 4$

SPECIAL PRODUCTS OF THE FORM $(ax + b)(cx + d)$ AND $(ax + by)(cx + dy)$

For reasons that will become clear shortly, it is essential that you learn to multiply first-degree polynomials of the type $(2x - 3)(3x + 1)$ and $(3x - y)(x + 2y)$ mentally. To discover relationships that will make this possible, let us first multiply $(2x - 3)$ and $(3x + 1)$ using a vertical arrangement.

$$\begin{array}{lll} 2x & - 3 & \\ 3x & + 1 & \\ \hline 6x^2 & - 9x & \\ & 2x & - 3 \\ \hline 6x^2 & - 7x & - 3 \end{array}$$

Notice that the two first-degree terms $-9x$ and $2x$ combine into the single term $-7x$. Also, notice that the product of two first-degree polynomials is a second-degree polynomial. From these observations we state a simple three-step process for mentally multiplying first-degree polynomials of the same type:

EXAMPLE 14 (*A*)

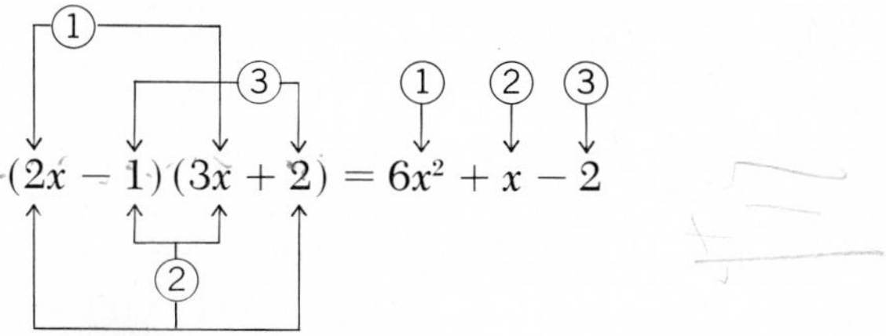

The like terms are picked up in step 2 by multiplying the inner and outer products, and are combined in your head.

(*B*)

1 3 1 2 3

$(2a - b)(a + 3b) = 2a^2 + 5ab - 3b^2$

2

(*C*)

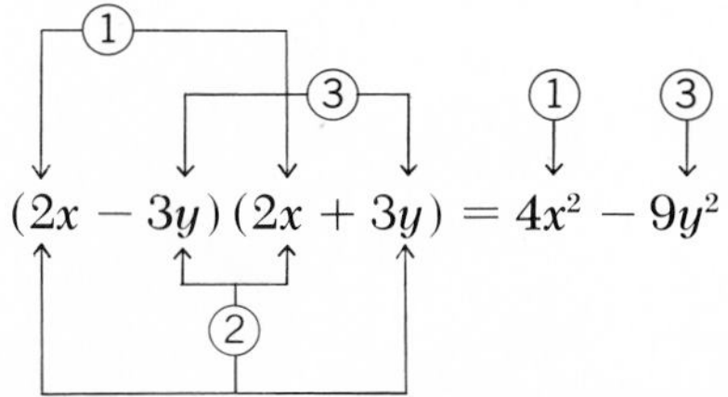

Notice that the middle term dropped out since its coefficient is 0.

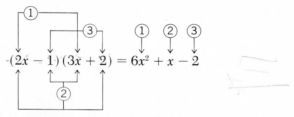

PROBLEM 14 Mentally multiply:

(A) $(3x - 2)(2x + 1)$ (B) $(a - 3b)(2a + b)$ (C) $(5x - y)(5x + y)$

ANSWER (A) $6x^2 - x - 2$ (B) $2a^2 - 5ab - 3b^2$ (C) $25x^2 - y^2$

In the next section we will consider the reverse problem: Given a second-degree polynomial, such as $2x^2 - 5x - 3$, find first-degree factors with integer coefficients that will produce this second-degree polynomial as a product. To be able to factor second-degree polynomial forms with any degree of efficiency, it is important that you know how to mentally multiply first-degree factors of the types illustrated in this section quickly and accurately.

EXERCISE 8

A *Use the first property of exponents to simplify.*

1. y^2y^3

2. x^3x^2

3. $(5y^4)(2y)$

4. $(2x)(3x^4)$

5. $(8x^{11})(-3x^9)$

6. $(-7u^9)(5u^7)$

7. $(-3u^4)(2u^5)(-u^7)$

8. $(2x^3)(-3x)(-4x^5)$

9. $(cd^2)(c^2d^2)$

10. $(a^2b)(ab^2)$

11. $(-3xy^2z^3)(-5xyz^2)$

12. $(-2xy^3z)(3x^3yz)$

Multiply:

13. $y(y + 7)$

14. $x(1 + x)$

15. $5y(2y - 7)$

16. $3x(2x - 5)$

17. $3a^2(a^3 + 2a^2)$

18. $2m^2(m^2 + 3m)$

19. $2y(y^2 + 2y - 3)$

20. $2x(2x^2 - 3x + 1)$

21. $7m^3(m^3 - 2m^2 - m + 4)$

22. $3x^2(2x^3 + 3x^2 - x - 2)$

23. $5uv^2(2u^3v - 3uv^2)$

24. $4m^2n^3(2m^3n - mn^2)$

25. $2cd^3(c^2d - 2cd + 4c^3d^2)$

26. $3x^2y(2xy^3 + 4x - y^2)$

B **27.** $(3y + 2)(2y^2 + 5y - 3)$

28. $(2x - 1)(x^2 - 3x + 5)$

29. $(m + 2n)(m^2 - 4mn - n^2)$

30. $(x - 3y)(x^2 - 3xy + y^2)$

31. $(2m^2 + 2m - 1)(3m^2 - 2m + 1)$

32. $(x^2 - 3x + 5)(2x^2 + x - 2)$

33. $(a+b)(a^2-ab+b^2)$

34. $(a-b)(a^2+ab+b^2)$

35. $(2x^2-3xy+y^2)(x^2+2xy-y^2)$

36. $(a^2-2ab+b^2)(a^2+2ab+b^2)$

Multiply mentally:

37. $(x+3)(x+2)$

38. $(m-2)(m-3)$

39. $(a+8)(a-4)$

40. $(m-12)(m+5)$

41. $(t+4)(t-4)$

42. $(u-3)(u+3)$

43. $(m-n)(m+n)$

44. $(a+b)(a-b)$

45. $(4t-3)(t-2)$

46. $(3x-5)(2x+1)$

47. $(3x+2y)(x-3y)$

48. $(2x-3y)(x+2y)$

49. $(2m-7)(2m+7)$

50. $(3y+2)(3y-2)$

51. $(6x-4y)(5x+3y)$

52. $(3m+7n)(2m-5n)$

53. $(2s-3t)(3s-t)$

54. $(2x-3y)(3x-2y)$

Since $(a+b)^2=a^2+2ab+b^2$ and $(a-b)^2=a^2-2ab+b^2$, we can formulate a simple mechanical rule for squaring any binomial: The first and last terms in the product are the squares of first and second terms in the binomial, respectively; the middle term in the product is twice the product of the two terms in the binomial. Use this rule to find the following squares.

55. $(3x+2)^2$

56. $(2x-3)^2$

57. $(3x+4y)^2$

58. $(2x-5y)^2$

C *Simplify:*

59. $(x+2y)^3$

60. $(2m-n)^3$

61. $(3x-1)(x+2)-(2x-3)^2$

62. $(2x+3)(x-5)-(3x-1)^2$

63. $2(x-2)^3-(x-2)^2-3(x-2)-4$

64. $(2x-1)^3-2(2x-1)^2+3(2x-1)+7$

65. $-3x\{x[x-x(2-x)]-(x+2)(x^2-3)\}$

66. $2\{(x-3)(x^2-2x+1)-x[3-x(x-2)]\}$

67. $(2x-1)(2x+1)(3x^3-4x+3)$

68. $(x-1)(x-2)(2x^3-3x^2-2x-1)$

69. The length of a rectangle is 8 feet more than its width. If y is the length of the rectangle, write an algebraic expression that represents its area. Change the expression to a form without parentheses.

70. If you are given two polynomials, one of degree m and another of degree n, $m>n$, then (*A*) what is the degree of their sum? (*B*) their product?

2.4 Factoring

COMMON FACTORS

Just as the distributive property allows us to distribute multiplication over an algebraic sum

$$a(b + c + d + \cdots + f) = ab + ac + ad + \cdots + af$$

it also allows us to take out a factor common to all terms and write the original expression as a product

$$ab + ac + ad + \cdots + af = a(b + c + d + \cdots + f)$$

We refer to the first process as multiplication and the second process as factoring.

EXAMPLE 15 Factor out all common factors.

(A) $5x - 5 = 5(x - 1)$

(B) $6x^2y + 12xy^2 = 6xy(x + y)$

(C) $2m^3n - 8m^2n^2 - 6mn^3 = 2mn(m^2 - 4mn - 3n^2)$

PROBLEM 15 Factor out all common factors.

(A) $3u^2 + u$ (B) $8m^3n^2 - 12m^2n^3$

(C) $6x^4y - 4x^3y + 8x^2y$

ANSWER (A) $u(3u + 1)$ (B) $4m^2n^2(2m - 3n)$ (C) $2x^2y(3x^2 - 2x + 4)$

NOTE: You can check your work by multiplying back again to see if you get the original expression.

FACTORING SECOND-DEGREE POLYNOMIALS

You should now be able to perform the following multiplications without much effort:

$$(x - 2)(x + 3) = x^2 + x - 6$$

$$(x - 2y)(x + 3y) = x^2 + xy - 6y^2$$

but can you reverse the process? Can you, for example, find integers a, b, c, and d so that

$$3x^2 - 6x - 7 = (ax + b)(cx + d)$$

Factoring a second-degree polynomial with integer coefficients as the product of two first-degree polynomials with integer coefficients is not as easy as multiplying first-degree polynomials. There is, however, a systematic approach to the problem that will enable you to find first-degree factors of special second-degree polynomials if they exist.

To start, we will consider a very simple second-degree polynomial whose first-degree factors you will likely guess right away:

$$x^2 - 7x + 6$$

Our problem is to find two first-degree factors, if they exist, with integers as coefficients. Let us begin by writing

$$x^2 - 7x + 6 = (\ x \qquad)(\ x \qquad)$$

leaving space for the coefficients of x and the constant terms. First, we insert what we know for certain:

If we choose the coefficients of both x's to be 1, then the factors of 6 must both be negative. (Why?)

Thus,

$$x^2 - 7x + 6 = (x - \quad)(x - \quad)$$

Since the constant terms both must be negative integers and factors of 6, the only possibilities are

$$(-2)(-3)$$

$$(-1)(-6)$$

We test each pair—now you know why you must be able to multiply first-degree factors mentally—and we find that the second pair produces the middle term. Hence,

$$x^2 - 7x + 6 = (x - 1)(x - 6)$$

Because of the commutative property of real numbers, we can reverse the factors, if we wish.

Before you get the impression that all second-degree polynomials with integers as coefficients have first-degree factors with integers as coefficients, consider the following simple polynomial:

$$x^2 + x + 2$$

Proceeding as above, we write

$$x^2 + x + 2 = (x + \quad)(x + \quad)$$

but find that no pair of integer factors of 2 produce the middle term (try them). Hence, we conclude that $x^2 + x + 2$ does not have first-degree factors with integer coefficients.

EXAMPLE 16 Factor each polynomial in the integers.

(A) $x^2 + 7x + 12$

SOLUTION $x^2 + 7x + 12 = (x + \quad)(x + \quad)$

Positive factors of 12:

$(1)(12)$

$(2)(6)$

$(3)(4)$

The last pair produces the middle term; thus

$$x^2 + 7x + 12 = (x + 3)(x + 4)$$

(B) $2x^2 + 3xy - 2y^2$

SOLUTION $2x^2 + 3xy - 2y^2 = (2x \quad y)(x \quad y)$

The signs of the coefficients of y must be opposite. (Why?) Possible factors of -2 are:

$(+1)(-2)$

$(-1)(+2)$

The second pair produces the middle term; thus

$$2x^2 + 3xy - 2y^2 = (2x - y)(x + 2y)$$

(C) $x^2 - 3x + 4$

SOLUTION $x^2 - 3x + 4 = (x - \quad)(x - \quad)$

Possible factors of 4:

$(-1)(-4)$

$(-2)(-2)$

Neither pair works; hence, $x^2 - 3x + 4$ is not factorable in the integers.

PROBLEM 16 Factor each polynomial in the integers. If it is not factorable, say so.

(A) $x^2 - 8x + 12$ (B) $x^2 + 2x + 5$ (C) $2x^2 + 7xy - 4y^2$

ANSWER (A) $(x - 2)(x - 6)$ (B) Not factorable in the integers

(C) $(2x - y)(x + 4y)$

Now let us look at an example with more involved coefficients.

EXAMPLE 17 Factor $6x^2 + 5xy - 4y^2$, if possible, in the integers.

SOLUTION We start as before, and write

$$6x^2 + 5xy - 4y^2 = (\ \ x \qquad y)(\ \ x \qquad y)$$

The coefficients of x must be integer factors of 6 and the coefficients of y must be integer factors of -4. Thus, we write

6	-4
(2)(3)	(−2)(2)
(3)(2)	(2)(−2)
(1)(6)	(−1)(4)
(6)(1)	(1)(−4)
	(−4)(1)
	(4)(−1)

We need not consider negative integer factors of 6. Any pair out of the first column combined with any pair out of the second column will produce the first and last terms in $6x^2 + 5xy - 4y^2$, but which combination produces the middle term? Try the top pair on the left with each pair on the right. If none works, continue by combining the second pair on the left with each pair on the right, and so on (24 possibilities) until either one combination produces the middle term or all combinations fail. If the latter happens, then the polynomial is not factorable in the integers. In the case above, we find that the combination (2)(3) and (−1)(4) produces the middle term. Thus,

$$6x^2 + 5xy - 4y^2 = (2x - y)(3x + 4y)$$

PROBLEM 17 Factor $4x^2 - 4xy - 3y^2$, if possible, in the integers.

ANSWER $(2x - 3y)(2x + y)$

ac TEST†

In the second-degree polynomials

$$ax^2 + bx + c$$

$$ax^2 + bxy + cy^2$$

as a and c get larger with the possibility of more factors, the trial-and-error procedure outlined above gets more tedious. It would be helpful to know at the start if factors exist. It can be proved that these polynomials are factorable in the integers if and only if ac can be written as the product of two integers whose sum is b; that is, if and only if there exist integers m and n such that

† Optional.

$$ac = mn$$

and

$$m + n = b$$

We will call this *the ac test*. Use it if it helps you. Avoid it and come back to it later if it tends to confuse you.

It is useful to point out that if a, b, and c are selected at random out of the integers, the probability that

$$ax^2 + bx + c$$

is not factorable in the integers is much greater than the probability that it is. But even being able to factor some second-degree polynomials leads to marked simplification of some algebraic expressions and an easy way to solve some second-degree equations, as will be seen later.

COMBINED FORMS

We conclude this section by considering an example that involves common factors as well as second-degree forms that can be factored. In fact, *one should always take out common factors first*, if they exist, before using other methods.

EXAMPLE 18 (A)

$$\begin{aligned} 4x^3 - 14x^2 + 6x &= 2x(2x^2 - 7x + 3) \\ &= 2x(x - 3)(2x - 1) \end{aligned}$$

take out all common factors first, then factor further, if possible

(B)

$$\begin{aligned} 8x^3y + 20x^2y^2 - 12xy^3 &= 4xy(2x^2 + 5xy - 3y^2) \\ &= 4xy(2x - y)(x + 3y) \end{aligned}$$

PROBLEM 18 Factor as far as possible in the integers:

(A) $3x^3 - 15x^2y + 18xy^2$ (B) $3x^3y + 3x^2y - 36xy$

ANSWER (A) $3x(x - 2y)(x - 3y)$ (B) $3xy(x - 3)(x + 4)$

Exercise 9

A *Factor out all common factors.*

1. $7m + 7$

2. $9x - 9$

3. $14x^3y - 21x^2y$

4. $12mn^3 + 9m^2n^2$

5. $3y - 6y^2 - 12y^3$

6. $6x^3 - 4x^2 + 2x$

7. $10x^3y - 5x^2y^2 - 20xy^3$

8. $12m^2n + 4mn^2 - 8n^3$

Factor as far as possible in the integers. If not factorable, say so.

9. $x^2 + 7x + 10$

10. $x^2 + 5x + 6$

11. $x^2 - 7x + 10$

12. $m^2 - 7m + 12$

13. $x^2 - 2y + 3$

14. $z^2 + 3z + 5$

15. $a^2 + 8ab + 15b^2$

16. $u^2 + 9uv + 20v^2$

17. $x^2 - 10xy + 16y^2$

18. $x^2 - 10xy + 21y^2$

19. $u^2 + 5uv + 3v^2$

20. $u^2 + 4uv + v^2$

21. $2x^2 - 7x + 6$

22. $3x^2 - 7x + 4$

23. $2x^2 - 7xy + 6y^2$

24. $3x^2 - 11xy + 6y^2$

25. $n^2 + 2n - 8$

26. $n^2 - 2n - 8$

B

27. $m^2 - 4mn - 12n^2$

28. $m^2 - 4mn - 12n^2$

29. $12x^2 + 17xy + 6y^2$

30. $6x^2 + 11xy + 3y^2$

31. $3u^2 + 2uv - 3v^2$

32. $2a^2 - 3ab - 4b^2$

33. $x^2 - 4xy - 12y^2$

34. $x^2 + 4xy - 12y^2$

35. $6x^2 + 7xy - 3y^2$

36. $6x^2 - xy - 12y^2$

37. $6m^2 + 48m + 72$

38. $4y^2 - 28y + 48$

39. $x^3 - 11x^2 + 24x$

40. $2t^4 - 24t^3 + 40t^2$

41. $3m^3 - 6m^2 + 15m$

42. $2y^3 - 2y^2 + 8y$

43. $8x^2 + 6x - 9$

44. $12x^2 + 16x - 3$

45. $2x^2 - 3xy - 4y^2$

46. $3x^2 + 2xy - 3y^2$

47. $4u^3v + 14u^2v^2 + 6uv^3$

48. $3m^3n - 15m^2n^2 + 18mn^3$

49. $15x^2 + 17xy - 4y^2$

50. $12x^2 - 40xy - 7y^2$

C

51. $9x^2y + 3xy^2 - 30y^3$

52. $12x^3 + 16x^2y - 16xy^2$

53. $24x^2 - 31xy - 15y^2$

54. $12x^2 + 19xy - 10y^2$

55. $60x^4 + 68x^3y - 16x^2y^2$

56. $60x^2y^2 - 200xy^3 - 35y^4$

57. Find all positive integers p under 15 so that $x^2 - 7x + p$ can be factored.

58. Find all integers p such that $x^2 + px + 12$ can be factored.

2.5 More Factoring

SUM AND DIFFERENCE OF TWO SQUARES

If we multiply $(A - B)$ and $(A + B)$ we obtain

$$(A - B)(A + B) = A^2 - B^2$$

a difference of two squares. Writing this result from right to left, we obtain the very useful factoring formula

$$\boxed{A^2 - B^2 = (A - B)(A + B)}$$

which finds far wider use than you might expect. The sum of two squares

$$A^2 + B^2$$

does not factor in the integers. Try it to see why. The difference-of-two-squares formula should be used any time you encounter a difference-of-two-squares form.

EXAMPLE 19

(A) $x^2 - y^2 = (x - y)(x + y)$

(B) $4x^2 - 9 = (2x)^2 - 3^2 = (2x - 3)(2x + 3)$

(C) $18x^3 - 8x = 2x(9x^2 - 4) = 2x(3x - 2)(3x + 2)$

(D) $x^2 + y^2$ is not factorable in the integers

PROBLEM 19

Factor as far as possible in the integers:

(A) $x^2 - 9$ (B) $4x^2 - 25y^2$

(C) $3x^3 - 48x$ (D) $4x^2 + 9$

ANSWER (A) $(x - 3)(x + 3)$ (B) $(2x - 5y)(2x + 5y)$

(C) $3x(x - 4)(x + 4)$ (D) not factorable

SUM AND DIFFERENCE OF TWO CUBES

It is easy to verify by direct multiplication the following factoring formulas for the *sum and difference of two cubes:*

$$A^3 + B^3 = (A + B)(A^2 - AB + B^2) \quad (1)$$

$$A^3 - B^3 = (A - B)(A^2 + AB + B^2) \quad (2)$$

These formulas are used in the same way as the factoring formula for the difference of two squares. (Notice that $A^2 - AB + B^2$ and $A^2 + AB + B^2$ do not factor further in the integers.)

To factor

$$y^3 - 27$$

we first note that it can be written in the form

$$y^3 - 3^3$$

which identifies the expression as the difference of two cubes. If in the factoring formula (2) we let $A = y$ and $B = 3$, we obtain

$$y^3 - 27 = y^3 - 3^3 = (y-3)(y^2+3y+9)$$

EXAMPLE 20 (A) $8x^3 + 27 = (2x)^3 + 3^3 = (2x+3)(4x^2 - 6x + 9)$

(B) $2t^4 - 16t = 2t(t^3 - 8) = 2t(t^3 - 2^3) = 2t(t-2)(t^2+2t+4)$

PROBLEM 20 Factor as far as possible in the integers.

(A) $u^3 + 8$ (B) $3x^4 - 24xy^3$

ANSWER (A) $(u+2)(u^2-2u+4)$ (B) $3x(x-2y)(x^2+2xy+4y^2)$

FACTORING BY GROUPING

Occasionally some polynomial forms can be factored by appropriate grouping of terms. For example, to factor

$$x^2 + xy + 2x + 2y$$

we group the first two terms and the last two terms

$$(x^2 + xy) + (2x + 2y)$$

and factor out common factors from each group to obtain

$$x(x+y) + 2(x+y)$$

This last expression is of the form

$$xA + 2A$$

which factors as

$$A(x+2)$$

Thus, in the same way (letting $A = x + y$),

$$x(x+y) + 2(x+y) = (x+y)(x+2)$$

EXAMPLE 21 (A)

$$\begin{aligned} x^2 + 4x - xy - 4y &= (x^2 + 4x) - (xy + 4y) \\ &= x(x+4) - y(x+4) \\ &= (x+4)(x-y) \end{aligned}$$

(*B*)
$$\begin{aligned} 2ax - 3bx + 10ay - 15by &= (2ax - 3bx) + (10ay - 15by) \\ &= x(2a - 3b) + 5y(2a - 3b) \\ &= (2a - 3b)(x + 5y) \end{aligned}$$

PROBLEM 21 Factor by grouping:

(*A*) $x^2 - 2x - xy + 2y$ (*B*) $x^3 + 2x^2 + 2x + 4$

ANSWER (*A*) $(x - 2)(x - y)$ (*B*) $(x + 2)(x^2 + 2)$

Exercise 10

Factor as far as possible in the integers.

A

1. $y^2 - 25$
2. $t^2 - 36$
3. $9x^2 - 4$
4. $4x^2 - 1$
5. $x^2 + 25$
6. $x^2 + 36$
7. $25x^2 - 1$
8. $9x^2 - 16y^2$
9. $3x^2 - 3$
10. $2x^2 - 8$
11. $x^3 + 1$
12. $y^3 - 1$
13. $x^3 + 27$
14. $x^3 - 8$
15. $xy + 2x + y^2 + 2y$
16. $x^2 + 3x + xy + 3y$
17. $x^2 - 5x + xy - 5y$
18. $x^2 - 3x - xy + 3y$

B

19. $a^2b^2 - c^2$
20. $25x^2 - 16y^2$
21. $4x^2 + 9y^2$
22. $9u^2 + 4v^2$
23. $4x^3y - xy^3$
24. $u^3v - 9uv^3$
25. $2x^3 + 8x$
26. $3x^4 + 27x^2$
27. $8y^3 - 1$
28. $27m^3 + 1$
29. $a^3b^3 + 8$
30. $27 - x^3y^3$
31. $4u^3 + 32v^3$
32. $54x^3 - 2y^3$
33. $ax - 2xb - ay + 2by$
34. $mx + my - 2nx - 2ny$
35. $15ac - 20ad + 3bc - 4bd$
36. $2am - 3an + 2bm - 3bn$
37. $x^3 - 2x^2 + x - 2$
38. $x^3 - 2x^2 - x + 2$
39. $(y - x)^2 - y + x$
40. $x^2(x - 1) - x + 1$

C

41. $r^4 - s^4$

42. $16a^4 - b^4$

43. $x^4 - 3x^2 - 4$

44. $x^4 - 7x^2 - 18$

45. $(a - b)^2 - 4(c - d)^2$

46. $(x + 2)^2 - 9$

47. $25(4x^2 - 12xy + 9y^2) - 9a^2b^2$

48. $18a^3 - 8a(x^2 + 8x + 16)$

49. $x^6 - 1$

50. $a^6 - 64b^6$

51. $2x^3 - x^2 - 8x + 4$

52. $4y^3 - 12y^2 - 9y + 27$

53. $25 - a^2 - 2ab - b^2$

54. $x^2 - 2xy + y^2 - 9$

55. $16x^4 - x^2 + 6xy - 9y^2$

56. $x^4 - x^2 + 4x - 4$

2.6 Division of Polynomials

There are times when it is useful to find quotients of polynomials by a long-division process similar to that used in arithmetic. Several examples will illustrate the process.

EXAMPLE 22 (A) Divide: $2x^2 + 5x - 12$ by $x + 4$

SOLUTION	COMMENTS
$x + 4\,\overline{)\,2x^2 + 5x - 12}$	Both polynomials are arranged in descending powers of the variable if this is not already done.
$\begin{array}{r} 2x \\ x + 4\,\overline{)\,2x^2 + 5x - 12} \end{array}$	Divide the first term of the divisor into the first term of the dividend, i.e., what must x be multiplied by so that the product is exactly $2x^2$?
$\begin{array}{r} 2x \qquad\quad \\ x + 4\,\overline{)\,2x^2 + 5x - 12} \\ \underline{2x^2 + 8x} \qquad \\ -3x - 12 \end{array}$	Multiply the divisor by $2x$, line up like terms, subtract as in arithmetic, and bring down -12.
$\begin{array}{r} 2x - 3 \quad\; \\ x + 4\,\overline{)\,2x^2 + 5x - 12} \\ \underline{2x^2 + 8x} \qquad \\ -3x - 12 \\ \underline{-3x - 12} \\ 0 \end{array}$	Repeat the process above until the degree of the remainder is less than that of the divisor.

CHECK: $(x + 4)(2x - 3) = 2x^2 + 5x - 12$

(*B*) Divide: $x^3 + 8$ by $x + 2$

SOLUTION

$$\begin{array}{r} x^2 - 2x + 4 \\ x + 2\overline{)x^3 + 0x^2 + 0x + 8} \\ \underline{x^3 + 2x^2} \qquad\qquad \\ -2x^2 + 0x \qquad \\ \underline{-2x^2 - 4x} \qquad \\ 4x + 8 \\ \underline{4x + 8} \\ 0 \end{array}$$

COMMENT

Insert, with 0 coefficients, any missing terms of lower degree than 3, and proceed as in part *A*.

Can you check this problem?

(*C*) Divide: $3 - 7x + 6x^2$ by $3x + 1$

SOLUTION

$$\begin{array}{r} 2x - 3 \\ 3x + 1\overline{)6x^2 - 7x + 3} \\ \underline{6x^2 + 2x} \qquad \\ -9x + 3 \\ \underline{-9x + 3} \\ 6 = \text{R} \quad \text{(remainder)} \end{array}$$

COMMENT

Arrange $3 - 7x + 6x^2$ in descending powers of x, then proceed as above until the degree of the remainder is less than the degree of the divisor.

CHECK: Just as in arithmetic, when there is a remainder we check by adding the remainder to the product of the divisor and quotient. Thus

$$(3x + 1)(2x - 3) + 6 \stackrel{?}{=} 6x^2 - 7x + 3$$

$$6x^2 - 7x - 3 + 6 \stackrel{?}{=} 6x^2 - 7x + 3$$

$$6x^2 - 7x + 3 \stackrel{\checkmark}{=} 6x^2 - 7x + 3$$

PROBLEM 22 Divide, using the long-division process, and check:

(*A*) $(2x^2 + 7x + 3)/(x + 3)$ (*B*) $(x^3 - 8)/(x - 2)$

(*C*) $(2 - x + 6x^2)/(3x - 2)$

ANSWER (*A*) $2x + 1$ (*B*) $x^2 + 2x + 4$ (*C*) $2x + 1$, $R = 4$

Exercise 11

Divide, using the long-division process. Check the answers.

A **1.** $(3x^2 - 5x - 2)/(x - 2)$ **2.** $(2x^2 + x - 6)/(x + 2)$

3. $(2y^3 + 5y^2 - y - 6)/(y + 2)$ **4.** $(x^3 - 5x^2 + x + 10)/(x - 2)$

5. $(3x^2 - 11x - 1)/(x - 4)$
6. $(2x^2 - 3x - 4)/(x - 3)$
7. $(8x^2 - 14x + 3)/(2x - 3)$
8. $(6x^2 + 5x - 6)/(3x - 2)$
9. $(6x^2 + x - 13)/(2x + 3)$
10. $(6x^2 + 11x - 12)/(3x - 2)$
11. $(x^2 - 4)/(x - 2)$
12. $(y^2 - 9)/(y + 3)$

B 13. $(12x^2 + 11x - 2)/(3x + 2)$
14. $(8x^2 - 6x + 6)/(2x - 1)$
15. $(8x^2 + 7)/(2x - 3)$
16. $(9x^2 - 8)/(3x - 2)$
17. $(-7x + 2x^2 - 1)/(2x + 1)$
18. $(13x - 12 + 3x^2)/(3x - 2)$
19. $(x^3 - 1)/(x - 1)$
20. $(a^3 + 27)/(a + 3)$
21. $(x^4 - 81)/(x - 3)$
22. $(x^4 - 16)/(x + 2)$
23. $(4a^2 - 22 - 7a)/(a - 3)$
24. $(8c + 4 + 5c^2)/(c + 2)$
25. $(x + 5x^2 - 10 + x^3)/(x + 2)$
26. $(5y^2 - y + 2y^3 - 6)/(y + 2)$
27. $(3 + x^3 - x)/(x - 3)$
28. $(3y - y^2 + 2y^3 - 1)/(y + 2)$

C 29. $(9x^4 - 2 - 6x - x^2)/(3x - 1)$
30. $(4x^4 - 10x - 9x^2 - 10)/(2x + 3)$
31. $(8x^2 - 7 - 13x + 24x^4)/(3x + 5 + 6x^2)$
32. $(16x - 5x^3 - 8 + 6x^4 - 8x^2)/(2x - 4 + 3x^2)$
33. $(9x^3 - x + 2x^5 + 9x^3 - 2 - x)/(2 + x^2 - 3x)$
34. $(12x^2 - 19x^3 - 4x - 3 + 12x^5)/(4x^2 - 1)$
35. Given $P(x) = x^3 - 6x^2 + 12x - 4$ and $D(x) = x^2 - 3x + 2$, find $Q(x)$ and $R(x)$ such that $P(x) = D(x)Q(x) + R(x)$ and the degree of $R(x)$ is less than the degree of $D(x)$ or $R(x) = 0$.
36. Repeat the preceding problem for $P(x) = x^4 + x^3 - 4x^2 + 7x + 2$ and $D(x) = x^2 - x + 1$.

Exercise 12 Chapter Review

A 1. Add: $3x^2 - 2x + 1$, $3x - 2$, and $2x^2 - 3$
2. Subtract: $5x^2 - 2x + 5$ from $3x^2 - x - 2$
3. Multiply: $3x^2y(2x^3 - 3x^2y + y^2)$
4. Multiply: $(3x - 2)(2x + 5)$
5. Multiply: $(2x - 1)(x^2 - 3x + 5)$
6. Divide, using long division: $(6x^2 + 5x - 2)/(2x - 1)$

Factor as far as possible in the integers.

7. $4x^2y - 6xy^2$

8. $x^2 - 9x + 14$

9. $9x^2 - 12x + 4$

10. $t^2 + 4t - 6$

11. $u^2 - 64$

12. $3x^2 - 10x + 8$

13. $x^3 - 5x^2 + 6x$

14. $x(x + y) + y(x + y)$

B **15.** Multiply: $(9x^2 - 4)(3x^2 + 7x - 6)$

16. Subtract $2x^2 - 5x - 6$ from the product $(2x - 1)(2x + 1)$.

17. Divide, using long division: $(2 - 10x + 9x^3)/(3x - 2)$

18. Given the polynomial $3x^5 - 2x^3 + 7x^2 - x + 2$, (*A*) what is its degree? (*B*) what is the degree of the second term?

Factor as far as possible in the integers.

19. $m^2 - 3mn - 4n^2$

20. $2m^2 - 8n^2$

21. $12x^3y + 27xy^3$

22. $2x^2 - xy - 3y^2$

23. $6n^3 - 9n^2 - 15n$

24. $3x^2 + 2xy - 7y^2$

25. $xp + xq + yp + yq$

26. $x^2 - xy - 4x + 4y$

27. $(y - b)^2 - y + b$

28. $3x^3 - 24y^3$

C **29.** Multiply: $(4x^2 - 1)(3x^3 - 4x + 3)$

30. Divide, using long division: $(5x - 5 + 8x^4)/(x + 2x^2 - 1)$

31. Simplify: $[(3x^2 - x + 1) - (x^2 - 4)] - [(2x - 5)(x + 3)]$

32. Simplify: $-2x\{(x^2 + 2)(x - 3) - x[x - x(3 - x)]\}$

Factor as far as possible in the integers.

33. $36x^3y + 24x^2y^2 - 45xy^3$

34. $12u^4 - 12u^3v - 20u^2v^2$

35. $(x - y)^2 - x^2$

36. $a^4 - 2a^2b^2 + b^4$

37. $m^6 - n^6$

38. $4x^2 - 9m^2 + 6m -$

39. Find all integers p such that $x^2 + px + 8$ can be factored in the integers.

40. Describe the elements in the set.

$A = \{x \mid -3x^3 - 6x = -3x(x^2 + 2),\ x \text{ a real number}\}$

41. Given $P(x) = 4x^3 - 8x^2 + 8x - 2$ and $D(x) = 2x^2 - x + 2$, find $Q(x)$ and $R(x)$ such that $P(x) = D(x)Q(x) + R(x)$ where the degree of $R(x)$ is less than the degree of $D(x)$ or $R(x) = 0$.

Chapter 3

FIRST-DEGREE EQUATIONS AND INEQUALITIES IN ONE VARIABLE

3.1 First-Degree Equations in One Variable

We have now arrived at the place where we will learn how to solve algebraic equations and a variety of real-world problems. Here you will see the power of knowing even a little algebra.

A *solution* or *root* of an equation involving a single variable is a replacement of the variable by a constant that makes the left side of the equation equal to the right. The set of all solutions is called the *solution set*. To *solve an equation* is to find its solution set.

Knowing what we mean by the solution set of an equation is one thing, finding it is another. Our objective now is to develop a systematic method of solving equations that is free from guess work. We start by introducing the idea of equivalent equations. We say that two equations are *equivalent* if they both have the same solution set.

The basic idea in solving equations is to perform operations on equations that produce simpler equivalent equations and to continue the process until we reach an equation whose solution is obvious—generally, an equation such as

$$x = -3$$

With a little practice you will find the methods that we are going to develop very easy to use and very powerful. Before proceeding further, it is recommended that you briefly review the properties of equality discussed in Sec. 1.3. The following important theorem is a direct consequence of these properties.

THEOREM 1 (*Further properties of equality.*) For a, b, and c real numbers,

(A) If $a = b$, then $a + c = b + c$ — addition property

(B) If $a = b$, then $a - c = b - c$ — subtraction property

(C) If $a = b$, then $ca = cb$ — multiplication property

(D) If $a = b$ and $c \neq 0$, then $\dfrac{a}{c} = \dfrac{b}{c}$ — division property

The proofs of these properties are very easy. We will prove part A and leave the others as exercises:

$a + c = a + c$	reflexive property of equality
$a = b$	given
$a + c = b + c$	substitution principle

The next theorem, which we will freely use but not prove, provides us with our final instructions for solving simple equations.

THEOREM 2 An equivalent equation will result if

(A) An equation is changed in any way by use of Theorem 1, except for multiplication or division by 0.

(B) Any algebraic expression in an equation is replaced by its equal (substitution principle).

We are now ready to solve equations! Several examples will illustrate the process.

EXAMPLE 1 Solve $x - 5 = -2$ and check.

SOLUTION	COMMENTS
$x - 5 = -2$	How can we eliminate the -5 from the left side?

SOLUTION

$$x - 5 + 5 = -2 + 5$$

Add 5 to each side (addition property).

$$x = 3$$

Solution is obvious.

CHECK

$$3 - 5 \stackrel{?}{=} -2$$

To check, replace x with 3 in the original equation to see if the left side equals the right side.

$$-2 \stackrel{\checkmark}{=} -2$$

PROBLEM 1 Solve $x + 8 = -6$ and check.

ANSWER $x = -14$

EXAMPLE 2 Solve $-3x = 15$ and check.

SOLUTION

$$-3x = 15$$

How can we make the coefficient of x a positive 1?

$$\frac{-3x}{-3} = \frac{15}{-3}$$

Divide each side by -3 (division property).

$$x = -5$$

Solution is obvious.

CHECK

$$(-3)(-5) \stackrel{?}{=} 15$$

$$15 \stackrel{\checkmark}{=} 15$$

PROBLEM 2 Solve $5x = -20$ and check.

ANSWER $x = -4$

The following examples are a little more difficult, but each can be converted into one of the above types simply by following a two-step process:

1 Simplify the left- and right-hand sides of the equation by removing parentheses and combining like terms.

2 Perform operations on the resulting equations that will get all of the variable terms on one side (usually the left) and all of the constant terms on the other side (usually the right), then solve as in Example 2.

EXAMPLE 3 Solve $2x - 8 = 5x + 4$ and check.

SOLUTION

$$2x - 8 = 5x + 4$$

$$\boxed{2x - 8 + 8 = 5x + 4 + 8}$$

$$2x = 5x + 12$$

$$\boxed{2x - 5x = 5x + 12 - 5x}$$

$$-3x = 12$$

$$\boxed{\frac{-3x}{-3} = \frac{12}{-3}}$$

$$x = -4$$

CHECK

$$2(-4) - 8 \stackrel{?}{=} 5(-4) + 4$$

$$-8 - 8 \stackrel{?}{=} -20 + 4$$

$$-16 \stackrel{\checkmark}{=} -16$$

PROBLEM 3 Solve $3x - 9 = 7x + 3$ and check.

ANSWER $x = -3$

EXAMPLE 4 Solve $3x - 2(2x - 5) = 2(x + 3) - 8$ and check.

SOLUTION This equation is not as difficult as it might at first appear; simplify the expressions on each side of the equal sign first, and then proceed as in the preceding example. (Note that some steps in the following solution are done mentally.)

$$3x - 2(2x - 5) = 2(x + 3) - 8$$

$$3x - 4x + 10 = 2x + 6 - 8$$

$$-x + 10 = 2x - 2$$

$$-x = 2x - 12$$

$$-3x = -12$$

$$x = 4$$

CHECK

$$3(4) - 2[2(4) - 5] \stackrel{?}{=} 2[(4) + 3] - 8$$

$$12 - 2(8 - 5) \stackrel{?}{=} 2(7) - 8$$

$$12 - 2(3) \stackrel{?}{=} 14 - 8$$

$$12 - 6 \stackrel{?}{=} 6$$

$$6 \stackrel{\checkmark}{=} 6$$

PROBLEM 4 Solve $8x - 3(x - 4) = 3(x - 4) + 6$ and check.

ANSWER $x = -9$

If all terms in any one of the above equations had been transferred to the left of the equal sign (leaving 0 on the right) and like terms combined, the equation would have been of the form

$$ax + b = 0$$

a *first-degree equation in one variable*. Any equation of this form with

$a \neq 0$ always has a unique solution, namely $-b/a$. Showing that $-b/a$ is a solution is left to the reader. To show that this solution is unique, we proceed as follows: Assume p and q are both solutions of $ax + b = 0$, then

$$ap + b = aq + b \qquad \text{Why?}$$

$$ap = aq \qquad \text{Why?}$$

$$p = q \qquad \text{Why?}$$

In general, it is important to know under what conditions an equation has a solution and how many solutions are possible. We have now answered both of these questions for equations of the type $ax + b = 0$ with $a \neq 0$. Other types of equations will be studied later which have more than one solution; for example,

$$x^2 - 4 = 0$$

has two solutions, -2 and $+2$.

Exercise 13

A *Solve and check:*

1. $x + 7 = 2$

2. $x + 4 = -6$

3. $x - 7 = -9$

4. $x - 4 = 3$

5. $x - 5 = 0$

6. $y + 13 = 0$

7. $7x = -21$

8. $9x = 36$

9. $-9x = -27$

10. $-2x = 18$

11. $-5m = 0$

12. $3y = 0$

13. $-4m + 3 = -9$

14. $2w + 18 = -2$

15. $3y = 7y + 8$

16. $2n = 5n + 12$

17. $4y + 8 = 2y - 6$

18. $3x - 8 = x + 6$

19. $3x - 4 = 6x - 19$

20. $2t + 9 = 5t - 6$

21. $2y + 8 = 2y - 6$

22. $x - 3 = x + 7$

B **23.** $3(x + 2) = 5(x - 6)$

24. $5x + 10(x - 2) = 40$

25. $5 + 4(t - 2) = 2(t + 7) + 1$

26. $5x - (7x - 4) - 2 = 5 - (3x + 2)$

27. $10x + 25(x - 3) = 275$

28. $x + (x + 2) + (x + 4) = 54$

29. $5x - (7x - 4) - 2 = 5 - (3x + 2)$

30. $-3(4 - t) = 5 - (t + 1)$

31. $x(x - 1) + 5 = x^2 + x - 3$

32. $x(x + 2) = x(x + 4) - 12$

33. $x(x - 4) - 2 = x^2 - 4(x + 3)$

34. $t(t - 6) + 8 = t^2 - 6t - 3$

35. $-2\{3 + [2x - (x - 4)]\} = 2[(x + 2) - 3]$

36. $-2\{2 - [1 - 2(x + 1)]\} = 2(x + 5) - 4$

37. $0.4(x + 5) - 0.3x = 17$

38. $0.1(x - 7) + 0.05x = 0.8$

39. Which of the following are equivalent to $2x + 5 = x - 3$:

$2x = x - 8$, $2x = x + 2$, $3x = -8$, $x = -8$?

40. Which of the following are equivalent to $3x - 6 = 6$:

$3x = 12$, $3x = 0$, $x = 4$, $x = 0$?

C *Which of the following are true? (Recall that the symbol $\emptyset$ represents the empty or null set, a set with no elements.)*

41. $\{t \in I \mid 2t - 1 = 19\} = \{10\}$

42. $\{x \in I \mid 3x = 5\} = \emptyset$

43. $\{x \in I \mid x^2 = 4\} = \{-2, 2\}$

44. $\{x \in I \mid 3x + 11 = 5, x > 0\} = \emptyset$

45. Prove the multiplication property in Theorem 1.

46. Prove the subtraction property in Theorem 1.

3.2 Applications

To start, we will solve a fairly simple problem dealing with numbers. Through this problem you will learn a method of attack that can be applied to many other problems.

EXAMPLE 5 Find 3 consecutive integers whose sum is 66.

SOLUTION

Let

$x =$ the first integer

then

$x + 1 =$ the next integer

and

$x + 2 =$ the third integer

COMMENTS: Identify one of the unknowns with a letter, and then write other unknowns in terms of this letter.

$$x + (x + 1) + (x + 2) = 66$$

Write an equation that relates the unknown quantities with other facts in the problem.

$$x + x + 1 + x + 2 = 66$$
$$3x + 3 = 66$$
$$3x = 63$$
$$x = 21$$

Solve the equation.

$$x + 1 = 22$$
$$x + 2 = 23$$

Write all answers requested.

CHECK

21
22
23

66

Thus we have found three consecutive integers whose sum is 66.

Checking back in the equation is not enough since you might have made a mistake in setting up the equation; a final check is provided only if the conditions in the original problem are satisfied.

PROBLEM 5 Find 3 consecutive integers whose sum is 54.

ANSWER 17, 18, 19

EXAMPLE 6 Find 3 consecutive even numbers such that twice the second plus 3 times the third is 7 times the first.

SOLUTION Let

$x =$ the first even number

then

$x + 2 =$ the second even number

and

$x + 4 =$ the third even number

Twice the second even number	+	three times the third even number	=	seven times the first even number

$$2(x + 2) + 3(x + 4) = 7x$$
$$2x + 4 + 3x + 12 = 7x$$
$$5x + 16 = 7x$$
$$-2x = -16$$
$$x = 8$$
$$x + 2 = 10$$
$$x + 4 = 12$$

CHECK

8, 10, and 12 are three consecutive even numbers

$$2 \cdot 10 + 3 \cdot 12 \stackrel{?}{=} 7 \cdot 8$$
$$20 + 36 \stackrel{?}{=} 56$$
$$56 \stackrel{\checkmark}{=} 56$$

PROBLEM 6 Find 3 consecutive even numbers such that the second plus twice the third is 4 times the first.

ANSWER 10, 12, 14

EXAMPLE 7 Find the dimensions of a rectangle with a perimeter of 52 inches if its length is 5 inches more than twice its width.

SOLUTION

2(length) + 2(width) = perimeter

$$2(2x + 5) + 2x = 52$$
$$4x + 10 + 2x = 52$$
$$6x = 42$$
$$x = 7 \qquad \text{width}$$
$$2x + 5 = 19 \qquad \text{length}$$

CHECK

19 is 5 more than twice 7

$$2 \cdot 19 + 2 \cdot 7 \stackrel{?}{=} 52$$
$$38 + 14 \stackrel{?}{=} 52$$
$$52 \stackrel{\checkmark}{=} 52$$

PROBLEM 7 Find the dimensions of a rectangle with a perimeter of 30 feet if its length is 7 feet more than its width.

ANSWER 4 by 11 feet

EXAMPLE 8 In a pile of coins containing only dimes and nickels, there are 7 more dimes than nickels. If the total value of all of the coins in the pile is \$1, how many of each type of coin is in the pile?

SOLUTION Let

x = the number of nickels in the pile

then

$x + 7$ = the number of dimes in the pile

$$\text{Value of nickels in cents} + \text{value of dimes in cents} = \text{value of pile in cents}$$

$$5x + 10(x + 7) = 100$$
$$5x + 10x + 70 = 100$$
$$15x = 30$$
$$x = 2 \qquad \text{nickels}$$
$$x + 7 = 9 \qquad \text{dimes}$$

CHECK

9 dimes is seven more than 2 nickels

Value of nickels in cents =	10
Value of dimes in cents =	90
Total value =	100

PROBLEM 8 A person has dimes and quarters worth $1.80 in his pocket. If there are twice as many dimes as quarters, how many of each does he have?

ANSWER 4 quarters and 8 dimes

EXAMPLE 9 An airplane flew out to an island from the mainland and back in 5 hours. How far is the island from the mainland if the pilot averaged 600 miles/hour going to the island and 400 miles/hour returning?

SOLUTION In this problem we will find it convenient to find the time out to the island first. The formula $d = rt$, which relates distance, rate, and time, will be of great use to us here. Let

$x =$ the time it took to get to the island

then, since the total round-trip time is 5 hours, we subtract the time it took to get to the island, x, from the total round trip time, 5, to obtain the return-trip time. Thus,

$5 - x =$ the time to return

Distance out = distance back

$$600x = 400(5 - x)$$

$$600x = 2{,}000 - 400x$$

$$1{,}000x = 2{,}000$$

$$x = 2 \text{ hours} \qquad \text{time going}$$

$$5 - x = 3 \text{ hours} \qquad \text{time returning}$$

Since distance = (rate)(time), the distance to the island from the mainland is $600 \cdot 2 = 1{,}200$ miles.

CHECK

Time going + time returning = 5

$$\frac{1{,}200}{600} + \frac{1{,}200}{400} \stackrel{?}{=} 5$$

$$2 + 3 \stackrel{\checkmark}{=} 5$$

PROBLEM 9 An airplane flew from San Francisco to a distressed ship out at sea and back in 7 hours. How far was the ship from San Francisco if the pilot averaged 400 miles/hour going and 300 miles/hour returning?

ANSWER 1,200 miles

You are now beginning to see the strengths of an algebraic approach. It was an historic occasion when it was realized that a solution to a problem that was difficult to obtain by arithmetic computation could be obtained instead by a deductive process involving conditions that the solution was required to satisfy.

There are many different types of algebraic applications, so many, in fact, that no single approach will apply to all. The following suggestions, however, may be of help to you:

1 Read the problem very carefully—a second and third time if necessary.
2 Write down important facts and relationships on a piece of scratch paper.
3 Identify unknown quantities in terms of a single letter if possible.
4 Write an equation that relates these unknown quantities and the facts in the problem.
5 Solve the equation.
6 Write down all of the solutions asked for in the original problem.
7 Check the solution(s) in the original problem.

Remember, mathematics is not a spectator sport! Just reading examples is not enough; you must set up and solve problems yourself.

Exercise 14

A **1.** Find three consecutive numbers whose sum is 96.

2. Find three consecutive numbers whose sum is 78.

3. Find three consecutive even numbers whose sum is 42.

4. Find three consecutive even numbers whose sum is 54.

5. If you drove from Berkeley to Lake Tahoe, a distance of 200 miles, in 4 hours, what is your average speed? ($d = rt$)

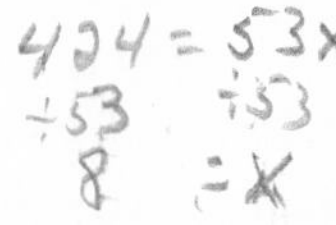

6. How long would it take you to drive from San Francisco to Los Angeles, a distance of about 424 miles, if you could average 53 miles/hour? ($d = rt$)

7. A chord called an octave can be produced by dividing a stretched string into two parts so that one part is twice as long as the other part. How long will each part of the string be if the total length of the string is 57 inches?

8. About 8 times as much of an iceberg is under water as is above the water.

If the total height of an iceberg from bottom to top is 117 feet, approximately how much is above and how much is below the surface?

9. You are asked to construct a triangle with two equal angles so that the third angle is twice the size of either of the two equal ones. How large should each angle be? NOTE: The sum of the three angles in any triangle is 180°.

10. The sun is about 390 times as far from the earth as the moon. If the sun is approximately 93,210,000 miles from the earth, how far is the moon from the earth?

B **11.** Find three consecutive odd numbers such that the sum of the second and third is 1 more than 3 times the first.

12. Find three consecutive odd numbers such that the sum of the first and third is twice the second.

13. Find the dimension of a rectangle with perimeter 128 inches if its length is 6 inches less than four times the width.

14. Find the dimensions of a rectangle with perimeter 66 feet if its length is 3 feet more than twice the width.

15. If you have 20 dimes and nickels in your pocket worth $1.40, how many of each do you have?

16. In a pile of coins containing only quarters and dimes, there are 3 less quarters than dimes. If the total value of the pile is $2.75, how many of each type of coin is in the pile?

17. A toy rocket shot vertically upward with an initial velocity of 160 feet/second has at time t a velocity given by the equation $v = 160 - 32t$, where air resistance is neglected. In how many seconds will the rocket reach its highest point? HINT: Find t when $v = 0$.

18. In the preceding problem, when will the rocket's velocity be 32 feet/second?

19. A mechanic charges $6/hour for labor and $4/hour for his assistant's labor. On a repair job the bill was $190 with $92 for labor and $98 for parts. If the assistant worked 2 hours less than the mechanic, how many hours did each work?

20. Air temperature drops approximately 5°F per 1,000 feet in altitude above the surface of the earth up to 30,000 feet. If T represents temperature and A represents altitude in thousands of feet, and if the temperature on the ground is 60°F, then we can write

$$T = 60 - 5A \qquad 0 \leq A \leq 30$$

If you were in a balloon, how high would you be if the thermometer registered −50°F?

21. If an adult with pure-brown eyes marries an adult with blue eyes, their children, because of the dominance of brown, will all have brown eyes but will be carriers of the gene for blue. If the children marry others with the same type of parents, then according to Mendel's laws of heredity, we would expect the third generation (the children's children) to include 3 times as many with brown eyes as with blue. Out of a sample of 1,748 third-generation children with second-generation parents as described, how many brown-eyed children and blue-eyed children would you expect?

22. In a recent election involving five candidates, the winner beat the opponents by 805, 413, 135, and 52, respectively. If the total number of votes cast was 10,250, how many votes did each receive?

C **23.** You are at a river resort and rent a motor boat for 5 hours at 7 A.M. You are told that the boat will travel at 8 miles/hour upstream and 12 miles/hour returning. You decide that you would like to go as far up the river as you can and still be back at noon. At what time should you turn back, and how far from the resort will you be at that time?

24. A man in a canoe went up a river and back in 6 hours. If his rate up the river was 2 miles/hour and back 4 miles/hour, how far did he go up the river?

25. At 8 A.M. your mother left by car on a long business trip. An hour later you find that she has left her purse behind. You decide to take another car to try to catch up with her. From past experience you know that she averages about 48 miles/hour. If you can average 54 miles/hour, how long will it take you to catch up to her?

26. One ship leaves England and another leaves the United States at the same time. If the distance between the two ports is 3,150 miles, the ship from the United States averages 25 miles/hour and the one from England 20 miles/hour, and they both travel the same route, how long will it take the ships to reach a rendezvous point, and how far from the United States will they be at that time?

27. In a computer center two electronic card sorters are used to sort 52,000 IBM cards. If the first sorter operates at 225 cards per minute and the second sorter operates at 175 cards per minute, how long will it take both sorters together to sort all of the cards?

28. Find four consecutive even numbers so that the sum of the first and last is the same as the sum of the second and third. (Be careful!)

Additional Applications

Now that you have had some experience in setting up and solving verbal problems, you are ready to consider a wider variety of applications that lead to first-degree equations. The problems in Exercise 15 are grouped in subject areas. The most difficult problems are double-starred (⋆⋆), moderately difficult problems are single-starred (⋆), and the easier problems are not marked.

Exercise 15

BUSINESS

1. It costs a book publisher $9,000 to prepare a book for publishing (artwork, plates, reviews, etc.); printing costs are $2 per book. If the book is sold to bookstores for $5 a copy, how many copies must be sold for the publisher to break even?

2. A woman borrowed a sum of money from a bank at 6 percent simple interest. At the end of 2.5 years she repaid the bank $575. How much did she borrow from the bank? HINT: $A = P + Prt$, where A is the amount repaid, P is the amount borrowed, r is the interest rate expressed as a decimal, and t is time in years.

⋆ **3.** A variety store sells a particular item costing $4 for $7. If this markup represents the store's pricing policy for all items, what percent markup (on cost) is used? HINT: cost + markup = retail.

⋆ **4.** In a soft-drink bottling plant one machine can fill and cap 20 bottles per minute; another newer machine can do 30 per minute. What will be the total time required to complete a 30,000-bottle order if the older machine is brought on the job 3 hours earlier than the newer machine, and then both machines are used together until the job is finished?

CHEMISTRY

5. In the study of gases there is a simple law called Boyle's law that expresses a relationship between volume and pressure. It states that the product of the pressure and volume, as these quantities change and all other variables are held fixed, remains constant. Stated as a formula, $P_1V_1 = P_2V_2$. If 500 cubic centimeters of air at 70-centimeter pressure were converted to 100-centimeter pressure, what volume would it have?

6. Find P_2 in the preceding problem if $P_1 = 200$ centimeters, $V_1 = 50$ cubic centimeters, and $V_2 = 500$ cubic centimeters.

⋆ **7.** How many liters of pure alcohol must be added to 2 liters of a 40% solution to get a 60% solution?

⋆ **8.** How many gallons of pure alcohol must be added to 3 gallons of a 20% solution to get a 40% solution?

DOMESTIC

9. A car rental company charges $10 per day and 10 cents per mile. If a car was rented for 2 days, how far was it driven if the total rental bill was $50?

10. If you paid \$72 for a pair of skis after receiving a discount of 20 percent, what was the price of the skis before the discount?

11. A friend of yours came out of a post office having spent \$1.32 on thirty 4-cent and 5-cent stamps. How many of each type did he buy?

⋆**12.** The cruising speed of an airplane is 150 miles/hour (relative to ground). You wish to hire the plane for a 3-hour sightseeing trip. You instruct the pilot to fly north as far as possible and still return to the airport at the end of the allotted time.
(*A*) How far north should the pilot fly if there is a 30-mile/hour wind blowing from the north?
(*B*) How far north should the pilot fly if there is no wind blowing?

EARTH SCIENCE

⋆⋆**13.** An earthquake emits a primary wave and a secondary wave. Near the surface of the earth the primary wave travels at about 5 miles/second, and the secondary wave at about 3 miles/second. From the time lag between the two waves arriving at a given seismic station, it is possible to estimate the distance to the quake. (The *epicenter* can be located by getting distance bearings at three or more stations.) Suppose a station measured a time difference of 12 seconds between the arrival of the two waves. How far would the earthquake be from the station?

⋆**14.** As dry air moves upward, it expands and in so doing cools at the rate of about 5.5°F for each 1,000 feet in rise. This ascent is known as the *adiabatic process*. If the ground temperature is 80°F, write an equation that relates temperature T with altitude h (in thousands of feet). How high is an airplane if the pilot observes that the temperature is 25°F?

ECONOMICS

15. Henry Schultz, an economist, formulated a price-demand equation for sugar in the United States as follows: $q = 70.62 - 2.26p$, where p equals wholesale price in cents of 1 pound of sugar, and q equals per capita consumption in pounds of sugar in the United States in any year. At what price per pound would the per capita consumption per year be 25.42 pounds?

⋆**16.** Gross national product (GNP) and net national product (NNP) are both measures of national income and are related as follows:

$$\text{GNP} = \text{NNP} + \text{depreciation}$$

A good estimate for depreciation is $\frac{1}{11}$ of GNP or $\frac{1}{10}$ of NNP.

(*A*) Write a formula for GNP in terms of NNP only.

(*B*) Use part *A* to replace the question marks in the table.

DATE	NNP (BILLIONS OF DOLLARS)	GNP (BILLIONS OF DOLLARS)
1929	96	?
1933	48	?

GEOMETRY

17. The perimeter of a tennis court for singles is 210 feet. Find the dimensions of the court if its length is 3 feet less than 3 times its width.

18. Find the dimensions of a rectangle with perimeter 72 inches if its length is 25 percent longer than its width.

★**19.** A water reed sticks 1 foot out of the water. If it were pulled over to the side until the top just reached the surface, it would be at a point 3 feet from where it originally protruded (Fig. 1). How deep is the water? (Recall that for any right triangle $c^2 = a^2 + b^2$, where c is the length of the longest side.)

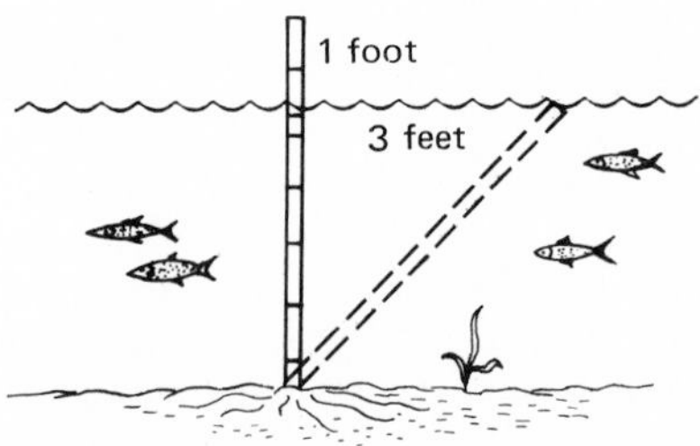

Figure 1

LIFE SCIENCE

20. A fairly good approximation for the normal weight of a person over 60 inches (5 feet) tall is given by the formula $w = 5.5h - 220$, where h is height in inches and w is weight in pounds. How tall should a 121-pound person be?

★**21.** Gregor Mendel (1822), a Bavarian monk and biologist whose name is known to almost everyone today, made discoveries which revolutionized the science of heredity. Out of many experiments in which he crossed peas of one characteristic with those of another, Mendel evolved his now famous laws of heredity. In one experiment he crossed hybrid yellow round peas (which contained green and wrinkled as recessive genes) and obtained 560 peas of the following types: 319 yellow round, 101 yellow wrinkled, 108 green round, and 32 green wrinkled. From his laws of heredity he predicted the ratio 9:3:3:1. Using the ratio, calculate the theoretical expected number of each type of pea from this cross, and compare it with the experimental results.

22. In biology there is an approximate rule, called the bioclimatic rule for temperate climates, that states that in spring and early summer, periodic phenomena, such as blossoming for a given species, appearance of certain insects, and ripening of fruit, usually come about 4 days later for each 500 feet of altitude increase or 1° latitude increase from any given base. In terms of formulas we have

$$d = 4\left(\frac{h}{500}\right) \qquad \text{and} \qquad d = 4L$$

where d = change in days, h = change in altitude in feet, and L = change in latitude in degrees.

What change in altitude would delay pear trees from blossoming for 16 days? What change in latitude would accomplish the same thing?

PHYSICS-ENGINEERING

23. If a small solid object is thrown downward with an initial velocity of 50 feet/second, its velocity after time t is given approximately by

$$v = 50 + 32t$$

How many seconds are required for the object to attain a velocity of 306 feet/second?

★**24.** In 1849, during a celebrated experiment, the French mathematician Fizeau made the first accurate approximation of the speed of light. By using a rotating disc with notches equally spaced on the circumference and a reflecting mirror 5 miles away (Fig. 2), he was able to measure the elapsed time for the light traveling to the mirror and back. Calculate his estimate for the speed of light (in miles per second) if his measurement for the elapsed time was $\frac{1}{20{,}000}$ seconds? ($d = rt$)

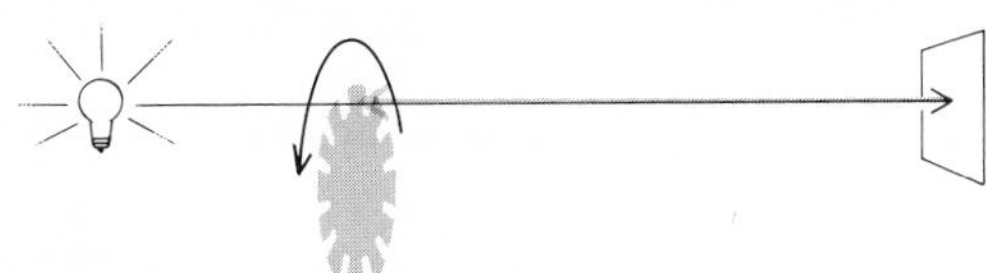

Figure 2

25. A type of physics problem with wide applications is the *lever problem*. For a lever, relative to a fulcrum, to be in static equilibrium (balanced) the sum of the downward forces times their respective distances on one side of the fulcrum must equal the sum of the downward forces times their respective distances on the other side of the fulcrum (Fig. 3).

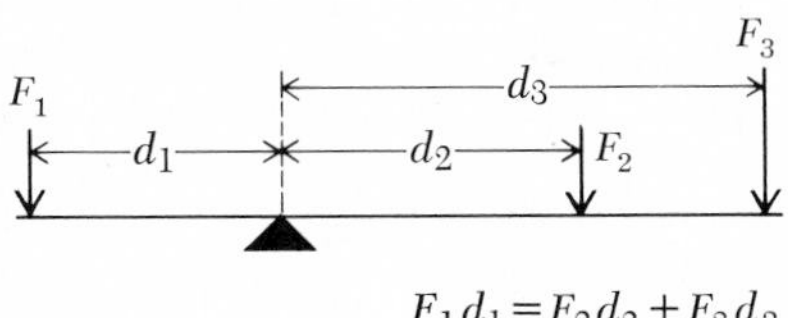

Figure 3

If a person has a 3-foot wrecking bar and places a fulcrum 3 inches from one end, how much can he lift if he applies a force of 50 pounds to the long end?

26. Where would a fulcrum have to be placed to balance an 8-foot bar with a 6-pound weight on one end and a 7-pound weight on the other?

27. Two people decided to move a 1,920-pound rock by use of a 9-foot steel bar (Fig. 4). If they place the fulcrum 1 foot from the rock and one of them applies a force of 150 pounds on the other end, how much force will the second person have to apply 2 feet from that end to lift the rock?

Figure 4

★**28.** If two pulleys are fastened together as in the diagram (Fig. 5), and the radius of the larger pulley is 10 inches and the smaller one 2 inches, how heavy a weight can one lift by exerting an 80-pound pull on the free rope?

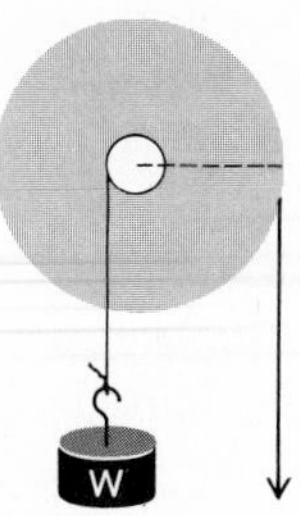

Figure 5

29. In a simple electric circuit, such as that found in a flashlight, the voltage provided by the batteries is related to the resistance and current in the circuit by Ohm's law,

$$E = IR$$

where E = electromotive force, volts

I = current, amperes

R = resistance, ohms

(*A*) If a two-cell battery puts out 3 volts and a current of 0.2 ampere flows through the circuit, what is the total resistance in the circuit?
(*B*) How much current will flow through a five-cell flashlight circuit (putting out 1.5 volts per cell), if the total resistance in the circuit is 25 ohms?

PSYCHOLOGY

30. In 1948 Professor Brown,[†] a psychologist, trained a group of rats (in an experiment on motivation) to run down a narrow passage in a cage to receive food in a goal box. He then put a harness on each rat and connected it to an overhead wire that was attached to a scale (Fig. 6). In this way he could place the rat at different distances from the food and measure the pull (in grams) of the rat toward the food. He found that a relation between motivation (pull) and position was given approximately by the equation

$$p = -\tfrac{1}{5}d + 70 \qquad 30 \le d \le 175$$

where pull p is measured in grams and distance d is measured in centimeters. If the pull registered was 40 grams, how far was the rat from the goal box?

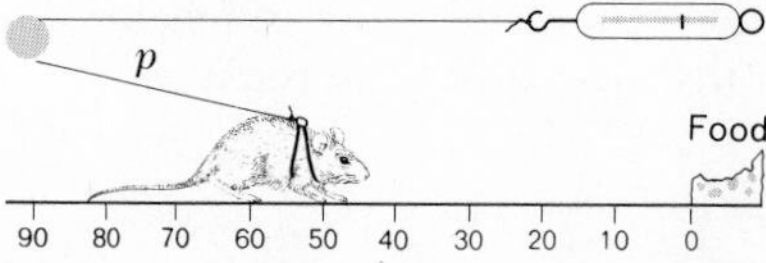

Figure 6

[†]Judson S. Brown, "Gradients of Approach and Avoidance Responses and Their Relation to Level of Motivation," *J. Comp. Physiol. Psychol.*, vol. 41, pp. 450–465, March 1948.

★**31.** Professor Brown performed the same kind of experiment as described in the preceding problem except that he replaced the food in the goal box with a mild electric shock. With the same kind of apparatus, he was able to measure the avoidance strength relative to the distance from the object to be avoided. He found that the avoidance strength a (measured in grams) was related to the distance d that the rat was from the shock (measured in centimeters) approximately by the equation

$$a = -\tfrac{4}{3}d + 230 \qquad 30 \le d \le 175$$

If the same rat was trained as described in this and the last problem, at what distance (to one decimal place) from the goal box would the approach and avoidance strength be the same? (What do you think that the rat would do at this point?)

PUZZLES ★★**32.** After 12:00 noon exactly, what time will the hands of a clock be together again?

★★**33.** A classic problem is the courier problem. If a column of men 3 miles long is marching at 5 miles/hour, how long will it take a courier on a motorcycle traveling at 25 miles/hour to deliver a message from the end of the column to the front and then return?

3.3 First-Degree Inequalities in One Variable

In the preceding chapters we worked with simple inequalities with obvious solutions; for example, $x < 5$, $-3 \le t < 4$, and $m \le -7$. In this section we will consider inequalities that do not have obvious solutions. Can you guess the real number solutions for

$$2x - 3 < 4x + 5$$

By the end of this section you will be able to solve this type of inequality almost as easily as you solved equations of the corresponding type.

The familiar ideas and procedures which apply to equations also apply to inequalities, but with certain important changes. As with equations, we are interested in performing operations on inequalities that will produce simpler *equivalent inequalities* (inequalities that have the same solution set), leading eventually to an inequality with an obvious solution. The operations in Theorem 3 produce equivalent inequalities.

THEOREM 3 If a, b, and c are real numbers and $a > b$, then

(A) $a + c > b + c$

(B) $a - c > b - c$

(C) $ca > cb$ if c is positive — inequality sign stays the same

(D) $ca < cb$ if c is negative — inequality sign reverses

(*E*) $\frac{a}{c} > \frac{b}{c}$ if c is positive inequality sign stays the same

(*F*) $\frac{a}{c} < \frac{b}{c}$ if c is negative inequality sign reverses

(*G*) $b < a$

The same theorem holds if each inequality sign is reversed, or if $>$ is replaced with $\geq$ and $<$ is replaced with $\leq$. Thus, we find that we can perform essentially the same operations on inequality statements that we perform on equations with the exception that *the inequality sign reverses if we multiply or divide both sides of an inequality by a negative number*; otherwise, the inequality sign does not change. For example, let us start with the true statement

$$-2 > -6$$

If we add 3 to both sides, we obtain

$$-2 + 3 > -6 + 3$$
$$1 > -3$$

and the inequality sign must remain the same to have a true statement. If we multiply both sides of $-2 > -6$ by 2, we obtain

$$2(-2) > 2(-6)$$
$$-4 > -12$$

and again the inequality sign stays the same. But if we multiply both sides of $-2 > -6$ by -2, we must change the sense of the inequality sign to get a true statement. Thus,

$$4 < 12$$

Similarly, if we divide both sides of $-2 > -6$ by 2, the inequality sign does not change, but if we divide by -2, it must change.

Let us prove the multiplication properties (parts *C* and *D*) of Theorem 3. Starting with

$$a > b$$

as given and using the definition of $>$ from Chap. 1, we write

$$b < a$$

Now, by definition of $<$, there exists a positive number p such that

$$b + p = a$$

or, subtracting b from both sides of the equation, we can write

$$p = a - b \tag{1}$$

Now if we multiply both sides of (1) by a positive number c, we obtain

$$cp = ca - cb$$

where the left side is still positive (Why?). Adding cb to both sides, we get

$$cb + cp = ca$$

Since cp is positive, then, by definition of $<$,

$$cb < ca$$

or

$$ca > cb$$

Thus, if $a > b$ and c is positive, then $ca > cb$. That is, if we multiply both sides of an inequality by a positive number, the inequality sign does not change.

Now suppose we multiply both sides of (1) above with a negative number c, then we obtain

$$cp = ca - cb \tag{2}$$

but the left side of (2) is now negative (Why?). Subtracting cp from both sides and adding cb to both sides of (2) we get

$$cb = ca - cp$$

or

$$ca - cp = cb$$

But if cp is negative, then $-cp$ is positive, hence, by definition of $<$,

$$ca < cb$$

Thus, if $a > b$ and c is negative, then $ac < ab$. That is, if we multiply both sides of an inequality by a negative number, the inequality sign reverses.

Several examples should make the process of using Theorem 3 in solving first-degree inequalities clear.

EXAMPLE 10 (*A*) Solve and graph: $3x - 4 > x - 2$

SOLUTION

$$3x - 4 > x - 2$$

$$3x > x + 2$$

$$2x > 2$$

$$x > 1 \qquad \text{inequality sign does not reverse (Why?)}$$

(*B*) Solve and graph: $3(x - 1) \leq 5(x + 2) - 5$

SOLUTION

$$3(x-1) \le 5(x+2) - 5$$
$$3x - 3 \le 5x + 10 - 5$$ 
$$3x - 3 \le 5x + 5$$
$$3x \le 5x + 8$$
$$-2x \le 8$$
$$x \ge -4 \quad \text{inequality sign reverses (Why?)}$$

PROBLEM 10 Solve and graph:

(A) $4x - 2 > 2x - 6$ (B) $2(2x+3) \le 6(x-2) + 10$

ANSWER (A) $x > -2$ (B) $x \le 4$

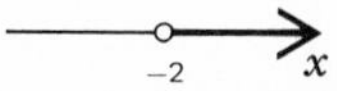

EXAMPLE 11 Solve and graph: $-8 \le 3x - 5 < 7$

SOLUTION We proceed as above, except we try to isolate x in the middle:

$$-8 \le 3x - 5 < 7$$
$$-8 + 5 \le 3x - 5 + 5 < 7 + 5$$
$$-3 \le 3x < 12$$
$$\frac{-3}{3} \le \frac{3x}{3} < \frac{12}{3}$$
$$-1 \le x < 4$$

PROBLEM 11 Solve and graph: $-3 < 2x + 3 \le 9$

ANSWER $-3 < x \le 3$

Exercise 16

A *Solve:*

1. $x - 4 < -1$

2. $x - 2 > 5$

3. $x + 3 > -4$

4. $x + 5 < -2$

5. $3x < 6$

6. $2x > 8$

7. $-3x < 6$

8. $-2x > 8$

9. $\frac{x}{3} < -7$

10. $\frac{x}{5} > -2$

11. $\dfrac{x}{-5} > -2$

12. $\dfrac{x}{-3} < -7$

13. $2x - 3 > 5$

14. $3x + 7 < 13$

Solve and graph:

15. $7x - 8 < 4x + 7$

16. $4x + 8 \geq x - 1$

17. $3 - x \geq 5(3 - x)$

18. $2(x - 3) + 5 < 5 - x$

B **19.** $-7n \geq 21$

20. $-5t < -10$

21. $\dfrac{N}{-2} > 4$

22. $\dfrac{M}{-3} \leq -2$

23. $-4x - 7 > 5$

24. $-2x + 8 < 4$

25. $4x < 7x + 12$

26. $3y > 5y - 2$

27. $3 - m < 4(m - 3)$

28. $2(1 - u) \geq 5u$

29. $3 - x \geq 5(3 - x)$

30. $3 - (2 + x) > -9$

31. $-3 \leq x - 5 \leq 8$

32. $2 < x + 3 < 5$

33. $-24 \leq 6x \leq 6$

34. $-6 \leq 3x \leq 9$

35. $-4 < 5t + 6 \leq 21$

36. $2 \leq 3m - 7 < 14$

37. $-8 \leq 3m - 2 < 7$

38. $1 < 4A + 5 < 25$

For Problems 39 to 48 set up appropriate inequalities and solve.

39. What numbers satisfy the condition: "3 less than twice the number is greater than or equal to -6"?

40. What numbers satisfy the condition: "5 more than twice the number is less than or equal to 7"?

41. What numbers satisfy the condition: "If 15 is diminished by 3 times the number, the result is less than 6"?

42. What numbers satisfy the condition: "5 less than 3 times the number is less than or equal to 4 times the number"?

43. If the area of a rectangle of length 10 inches must be greater than 65 square inches, how large may the width be?

44. If the perimeter of a rectangle with a length of 10 inches must be smaller than 30 inches, how large may the width be?

45. For a business to make a profit it is clear that revenue R must be greater than costs C; in short, a profit will result only if $R > C$. If a company manufactures records and its cost equation for a week is $C = 300 + 1.5x$ and its revenue equation is $R = 2x$, where x is the number of records sold in a week, how many records must be sold for the company to realize a profit?

46. As dry air moves upward it expands and in so doing cools at a rate of about 5.5°F for each 1,000 foot rise up to about 40,000 feet. If the ground temperature is 70°F, then the temperature T at height h is given approximately by $T = 70 - 0.0055h$. For what range in altitude will the temperature be between 26 and -40°F?

C **47.** If the power demands in an 110-volt electrical circuit in a home range is between 220 and 2,750 watts, what is the range of current flowing through the circuit? ($W = EI$, where W = power in watts, E = pressure in volts, and I = current in amperes.)

48. In an 110-volt electrical house circuit a 30-ampere fuse is used. In order to keep the fuse from "blowing," the total wattages in all of the appliances on that circuit must be kept below what figure? ($W = EI$, where W = power in watts, E = pressure in volts, and I = current in amperes.) HINT: $I = W/E$ and $I \leq 30$.

49. If both a and b are positive numbers and b/a is greater than 1, then is $a - b$ positive or negative?

50. If both a and b are negative numbers and b/a is greater than 1, then is $a - b$ positive or negative?

51. Supply the reasons for the proof of Theorem 3A:

PROOF	REASONS
1 $a > b$	1
2 $b < a$	2
3 $b + p = a$ for some $p > 0$	3
4 $(b + c) + p = a + c$	4
5 $b + c < a + c$	5
6 $a + c > b + c$	6

52. Prove Theorem 3*B*. (See the preceding problem.)

53. Let m and n be real numbers with $m > n$. Then there exists a positive real number p such that $m = n + p$. Find the fallacy in the following argument:

$$m = n + p$$

$$(m - n)m = (m - n)(n + p)$$

$$m^2 - mn = mn + mp - n^2 - np$$

$$m^2 - mn - mp = mn - n^2 - np$$

$$m(m - n - p) = n(m - n - p)$$

$$m = n$$

54. Let us assume that m and n are positive real numbers with $m > n$, then

$$mn > n^2$$

$$mn - m^2 > n^2 - m^2$$

$$m(n-m) > (n+m)(n-m)$$
$$m > n+m$$
$$0 > n$$

which says that n is negative, a contradiction to our original assumption. Can you find the error?

3.4 Absolute Value in Equations and Inequalities

In Sec. 1.5 we defined the *absolute value* of a real number x as follows:

$$|x| = \begin{cases} x & \text{if } x \text{ is positive or } 0 \\ -x & \text{if } x \text{ is negative} \end{cases}$$

Thus,

$$|7| = 7 \qquad |0| = 0 \qquad |-7| = 7$$

EQUATIONS INVOLVING ABSOLUTE VALUE

If we want to solve an equation such as

$$|x| = 6$$

then it follows from the definition of absolute value that x can be replaced with either $+6$ or -6. We write the solution in the compact form

$$x = \pm 6$$

Similarly, to solve the slightly more complicated equation

$$|2x - 3| = 5$$

we can replace $2x - 3$ with either $+5$ or -5. Thus, the single equation $|2x - 3| = 5$ is equivalent to the two equations

$$2x - 3 = 5 \qquad \text{or} \qquad 2x - 3 = -5$$

which are often written more compactly as

$$2x - 3 = \pm 5$$

and both are solved simultaneously as follows:

$$2x = 3 \pm 5$$
$$x = \frac{3 \pm 5}{2}$$
$$x = 4, -1$$

CHECK

$x = 4$ $\qquad |2(4) - 3| = |5| = 5$

$x = -1$ $\qquad |2(-1) - 3| = |-5| = 5$

EXAMPLE 12 Solve:

(A) $|x - 7| = 4$

SOLUTION

$$x - 7 = \pm 4$$
$$x = 7 \pm 4$$
$$x = 11, 3$$

(B) $|3x + 8| = 4$

SOLUTION

$$3x + 8 = \pm 4$$
$$3x = -8 \pm 4$$
$$x = \frac{-8 \pm 4}{3}$$
$$x = -\tfrac{4}{3}, -4$$

PROBLEM 12 Solve:

(A) $|m + 3| = 9$ (B) $|2x + 3| = 7$

ANSWER (A) 6, −12 (B) 2, −5

INEQUALITIES INVOLVING ABSOLUTE VALUE

Inequalities such as $|x| < 5$ and $|x - 3| \leq 4$ are encountered frequently and are compact ways of writing double-inequality statements. The following theorem shows how we can transform such statements into double-inequality forms that can then be solved by methods considered in the last section.

THEOREM 4 For p a positive number,

$|x| < p$ is equivalent to $-p < x < p$

Part of the proof of this theorem is included in the exercises.

EXAMPLE 13 Solve and graph:

(A) $|x| < 5$

SOLUTION $-5 < x < 5$

(B) $|2x - 3| \leq 5$

SOLUTION $-5 \leq 2x - 3 \leq 5$

$$-2 \le 2x \le 8$$

$$-1 \le x \le 4$$

PROBLEM 13 Solve and graph:

(A) $|x| \le 3$ (B) $|2x + 1| < 9$

ANSWER (A) $-3 \le x \le 3$

(B) $-5 < x < 4$

Exercise 17

A *Solve and graph:*

1. $|x| = 5$ **2.** $|x| = 7$

3. $|t - 3| = 4$ **4.** $|y - 5| = 3$

5. $|x + 1| = 5$ **6.** $|u + 8| = 3$

7. $|t| \le 5$ **8.** $|x| \le 7$

9. $|t - 3| < 4$ **10.** $|y - 5| < 3$

11. $|x + 1| \le 5$ **12.** $|u + 8| \le 3$

B **13.** $|2x - 3| = 5$ **14.** $|3x + 4| = 8$

15. $|2x - 3| \le 5$ **16.** $|5x - 3| \le 12$

17. $|6m + 9| = 13$ **18.** $|5t - 7| = 11$

19. $|9M - 7| < 15$ **20.** $|7u + 9| < 14$

C *A second part to Theorem 4 is: For p a positive number, $|x| > p$ is equivalent to $x < -p$ or $x > p$. Thus, $|x| > 3$ is equivalent to $x < -3$ or $x > 3$. Graphically*

Solve and graph:

21. $|x| \ge 7$ **22.** $|x| \ge 5$

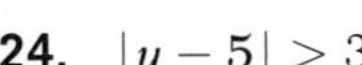

23. $|t - 3| > 4$ **24.** $|y - 5| > 3$

25. $|x + 1| \ge 5$ **26.** $|u + 8| \ge 3$

27. $|3u + 4| > 3$ **28.** $|2y - 8| > 2$

For what values of x does each of the following hold?

29. $|x + 7| = x + 7$ **30.** $|x - 5| = x - 5$

31. $|x - 11| = -(x - 11)$ **32.** $|x + 8| = -(x + 8)$

33. Show that if $|x| < p$, p a positive number, then $-p < x < p$. (Part of Theorem 4.)

Exercise 18 Chapter Review

A **1.** Solve and check: $4x - 9 = x - 15$

2. Find three consecutive numbers whose sum is 159.

Solve and graph:

3. $4x + 9 \leq -3$ **4.** $|x| = 6$

5. $|x| < 6$ **6.** $|y + 9| < 5$

7. What numbers satisfy the condition: "5 less than 5 times the number is less than or equal to 10"?

B **8.** Solve and check: $2x + 3(x - 1) = 5 - (x - 4)$

9. Find the dimensions of a rectangle with perimeter 42 feet if the length is 3 feet less than twice the width.

10. A pile of coins consists of nickels and quarters. How many of each kind are there if the value of the pile is $1.75 and there are 5 more nickels than quarters?

Solve and graph:

11. $3(2 - x) - 2 \leq 2x - 1$ **12.** $-14 \leq 3x - 2 < 7$

13. $|4x - 7| = 5$ **14.** $|4x - 7| \leq 5$

15. Indicate true (T) or false (F):

(*A*) If $x < y$ and $a > 0$, then $ax < ay$.

(*B*) If $x < y$ and $a < 0$, then $ax > ay$.

C **16.** Show that $-b/a$, $a \neq 0$, is a solution to $ax + b = 0$.

17. What is the maximum and minimum number of solutions for an equation of the type $ax + b = 0$, $a \neq 0$?

18. Solve: $3 - 2\{x - 2[3x - 2(x + 1)]\} = 5(x - 2) + 2$

19. How many gallons of pure acid must be mixed with 5 gallons of 40% solution to get a 45% solution?

Solve and graph:

20. $|x - 2| > 3$ **21.** $|3x - 8| > 2$

Chapter 4
FIRST-DEGREE EQUATIONS AND INEQUALITIES IN TWO VARIABLES

4.1 Cartesian Coordinate System

To form a cartesian coordinate system we start with two sets of objects, the set of all points in a plane (a set of geometric objects) and the set of all ordered pairs of real numbers. In a plane select two number lines, one vertical and one horizontal, and let them cross at their respective origins (Fig. 1). Up and to the right are the usual choices for the positive direc-

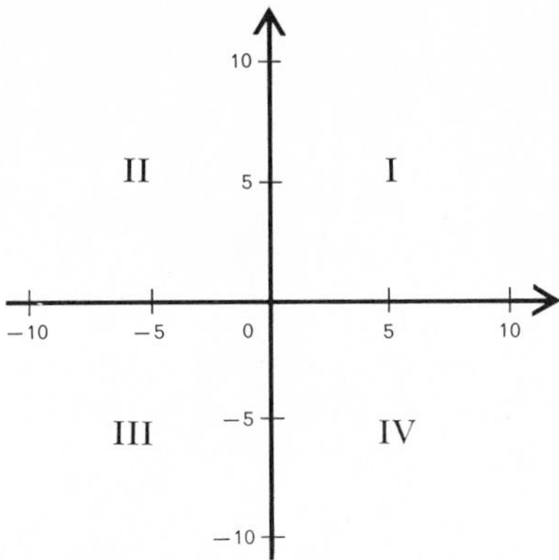

Figure 1

tions. These two number lines are called the *vertical axis* and the *horizontal axis* or (together) the *coordinate axes*. The coordinate axes divide the plane into four parts called *quadrants*. The quadrants are numbered counterclockwise from I to IV. All points in the plane lie in one of the four quadrants except for points on the coordinate axes.

Pick a point P in the plane at random (see Fig. 2). Pass horizontal and vertical lines through the point. The vertical line will intersect the horizontal axis at a point with coordinate a, and the horizontal line will intersect the vertical axis at a point with coordinate b. These two numbers form the *coordinates*

$$(a, b)$$

of the point P in the plane.

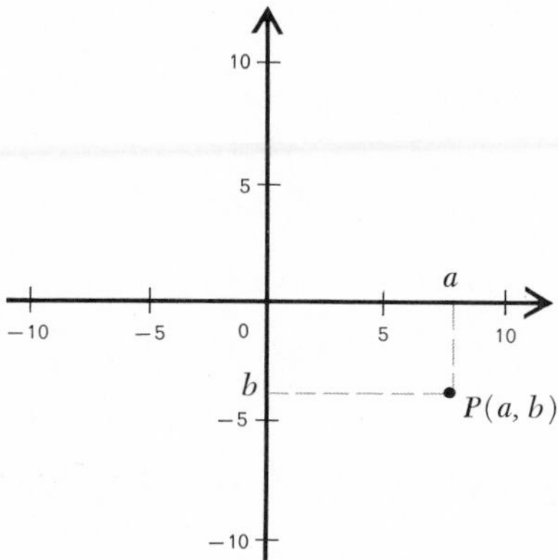

Figure 2

The first number a (called the *abscissa* of P) in the ordered pair (a, b) is the directed distance of the point from the vertical axis (measured on the horizontal scale); the second number b (called the *ordinate* of P) is the directed distance of the point from the horizontal axis (which is measured on the vertical scale). The point with coordinate $(0, 0)$ is called the *origin*.

We know that a and b exist for each point in the plane since every point on each axis has a real number associated with it. Hence, by the procedure described, each point located in the plane can be labeled with a unique pair of real numbers. Conversely, by reversing the process, each pair of real numbers can be associated with a unique point in the plane.

The system that we have just defined is called a *cartesian coordinate system* (sometimes referred to as a rectangular coordinate system).

EXAMPLE 1 Find the coordinates of each of the points A, B, C, and D.

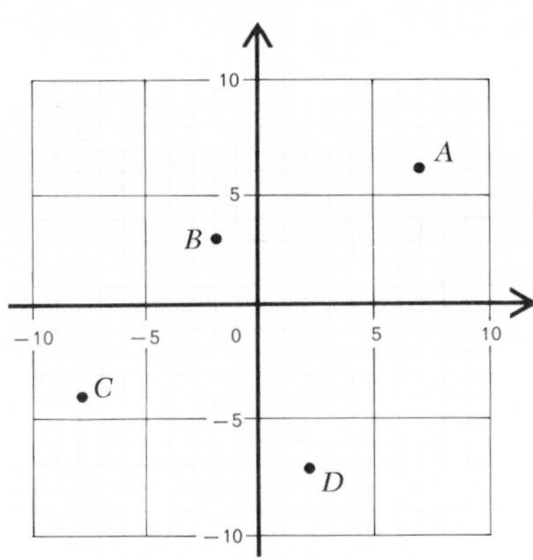

SOLUTION $A(7, 6)$ $B(-2, 3)$ $C(-8, -4)$ $D(2, -7)$

PROBLEM 1 Find the coordinates, using the figure in Example 1, for each of the following points:

(A) 2 units to the right and 1 unit up from A

(B) 2 units to the left and 2 units down from C

(C) 1 unit up and 1 unit to the left of D

(D) 2 units to the right of B

ANSWER (A) $(9, 7)$ (B) $(-10, -6)$ (C) $(1, -6)$ (D) $(0, 3)$

EXAMPLE 2 Graph (associate each ordered pair of numbers with a point in the cartesian coordinate system):

$(2, 7)$, $(7, 2)$, $(-8, 4)$, $(4, -8)$, $(-8, -4)$, $(-4, -8)$

SOLUTION

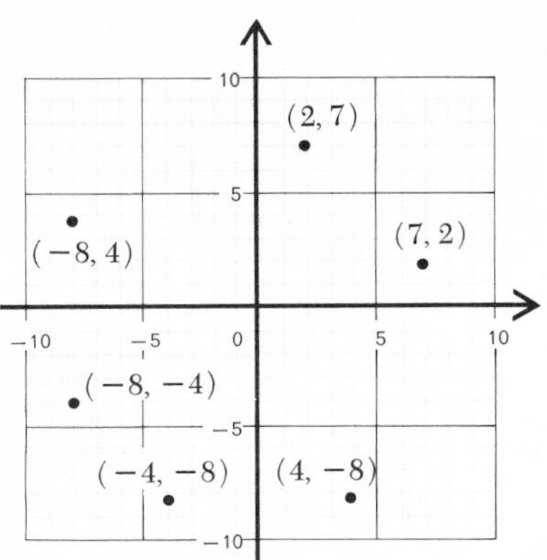

It is very important to note that the ordered pair (2, 7) and the set {2, 7} are not the same thing; $\{2, 7\} = \{7, 2\}$, but $(2, 7) \neq (7, 2)$.

PROBLEM 2 Graph: (3, 4), (−3, 2), (−2, −2), (4, −2), (0, 1), and (−4, 0).

ANSWER

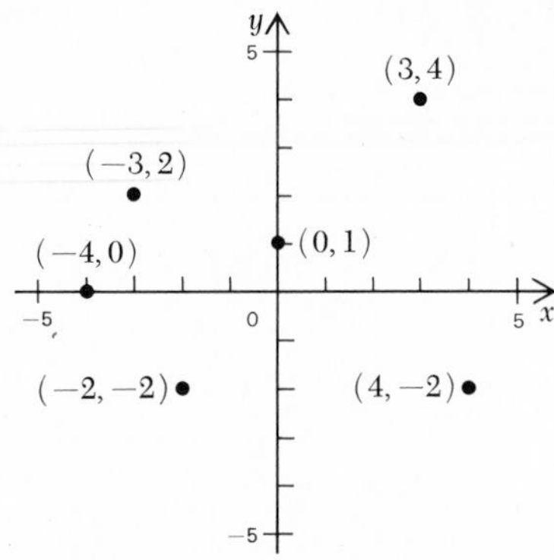

Exercise 19

A *Write down the coordinates of each labeled point.*

1.

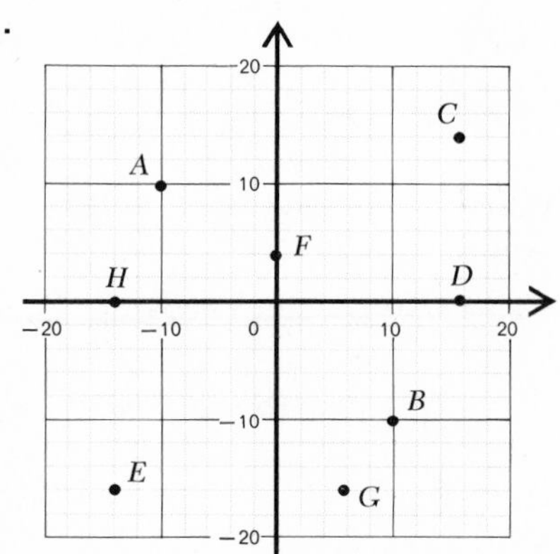

2.

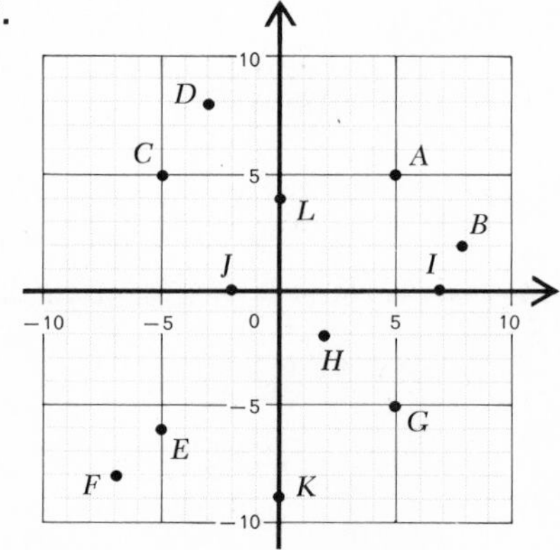

Graph each set of ordered pairs of numbers on the same coordinate system.

3. (4, 4), (−4, 1), (−3, −3), (5, −1), (0, 2), and (−2, 0)

4. (3, 1), (−2, 3), (−5, −1), (2, −1), (4, 0), and (0, −5)

5. (2, 7), (7, 2), (−6, 3), (−4, −7), (2, 3), (0, −8), and (9, 0)

6. (−9, 8), (8, −9), (0, 5), (4, −8), (−3, 0), (7, 7), and (−6, −6)

B *Write down the coordinates of each labeled point to the nearest quarter of a unit.*

7.

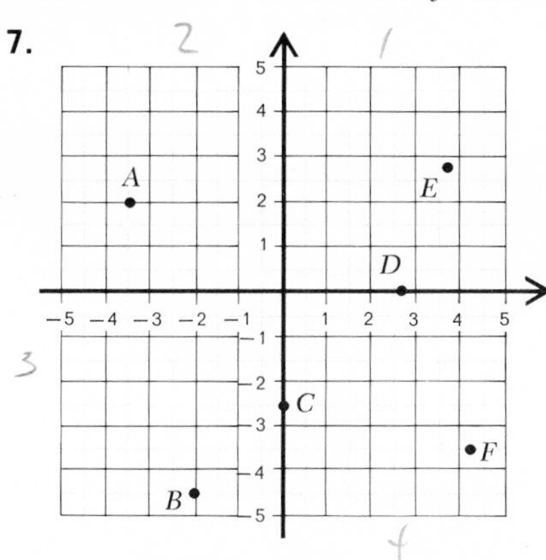

8.

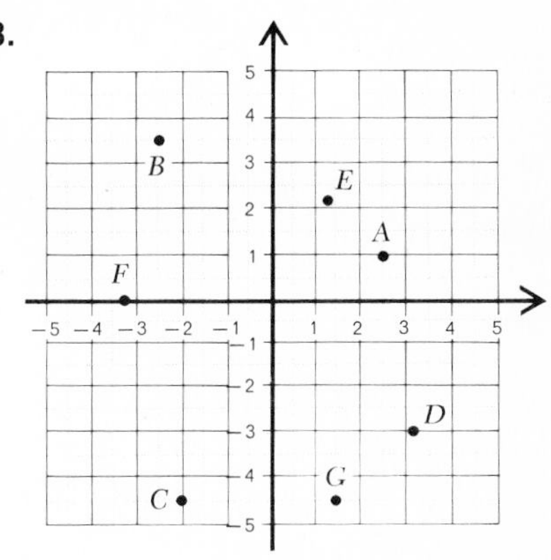

9. Graph the following ordered pairs of numbers on the same coordinate system: $A(3\frac{1}{2}, 2\frac{1}{2})$, $B(-4\frac{1}{2}, 3)$, $C(0, -3\frac{3}{4})$, $D(-2\frac{3}{4}, -3\frac{3}{4})$, and $E(4\frac{1}{4}, -3\frac{3}{4})$

10. Graph the following ordered pairs of numbers on the same coordinate system: $A(1\frac{1}{2}, 3\frac{1}{2})$, $B(-3\frac{1}{4}, 0)$, $C(3, -2\frac{1}{2})$, $D(-4\frac{1}{2}, 1\frac{3}{4})$, and $E(-2\frac{1}{2}, -4\frac{1}{4})$

11. Without graphing, tell which quadrants contain the graph of each of the following ordered pairs (see Fig. 8):

(*A*) $(-23, 403)$ (*B*) $(32\frac{1}{2}, -430)$ (*C*) $(2{,}001, 25)$ (*D*) $(-0.008, -3.2)$

12. Without graphing, tell which quadrants contain the graph of each of the following ordered pairs:

(*A*) $(-20, -4)$ (*B*) $(-3, 22\frac{3}{4})$ (*C*) $(4, 35{,}000)$ (*D*) $(\sqrt{2}, -3)$

13. What is the first coordinate of any point on a vertical axis in a cartesian coordinate system?

14. What is the second coordinate of any point on a horizontal axis in a cartesian coordinate system?

15. In which quadrants do the coordinates of a point have the same sign?

16. In which quadrants do the coordinates of a point have opposite signs?

17. In which quadrants is the abscissa positive?

18. In which quadrants is the ordinate positive?

C **19.** Write the coordinates of the vertices (corners) of a 5 by 3 rectangle lying in the first quadrant with longest side along the horizontal axis and one vertex at the origin.

20. Repeat Problem 19 with the exception that the rectangle is in the third quadrant.

21. If a 12- by 8-inch rectangle is centered at the origin with longest side horizontal, write down the coordinates of each vertex.

22. Repeat Problem 21 with the exception that the longest side is vertical.

4.2 First-Degree Equations and Straight Lines

The invention of the cartesian coordinate system represented a very important advance in mathematics. It was through the use of this system that René Descartes (1596–1650), a French philosopher-mathematician, was able to transform geometric problems requiring long tedious reasoning into algebraic problems which could be solved quite mechanically. This joining of algebra and geometry has now become known as *analytic geometry*.

Two fundamental problems of analytic geometry are:

1 Given an equation, find its graph.
2 Given a geometric figure, such as a straight line, circle, or ellipse, find its equation.

In this and the next section we will consider both of these problems relative to straight lines. Later we will consider other types of graphs. Let us start by trying to find the solution set for

$$y = 2x - 4$$

A *solution of an equation in two variables* is an ordered pair of real numbers that satisfy the equation. If we agree that the first element in the ordered pair will replace x and the second y, then

$$(0, -4)$$

is a solution of $y = 2x - 4$, as can easily be checked by replacing x with 0 and y with -4. How do we find other solutions? The answer is easy: We simply assign to x in $y = 2x - 4$ any convenient value and solve for y. For example, if $x = 3$ then

$$\begin{aligned} y &= 2(3) - 4 \\ &= 2 \end{aligned}$$

Hence,

$$(3, 2)$$

is another solution of $y = 2x - 4$. It is clear that by proceeding in this manner, we can get solutions to this equation without end. Thus, the solution set is infinite. Let us make up a table of some solutions and graph them in a cartesian coordinate system, identifying the horizontal axis with x and the vertical axis with y.

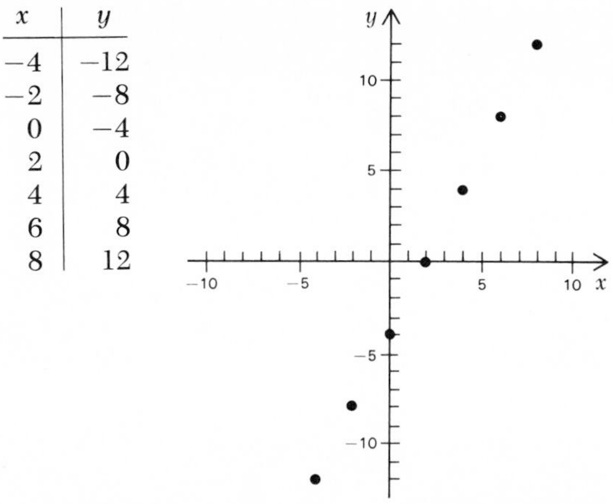

x	y
-4	-12
-2	-8
0	-4
2	0
4	4
6	8
8	12

It appears that the graph of the equation is a straight line. If we knew this for certain, then graphing $y = 2x - 4$ would be easy. We would simply find two solutions of the equation, plot them, and then plot as much of $y = 2x - 4$ as we like by drawing a line through the two points using a straightedge. It turns out that it is true that the graph of $y = 2x - 4$ is a straight line. In fact, we have the following general theorem, which we state without proof.

THEOREM 1 The graph of any equation of the form

$$Ax + By = C$$

where A, B, and C are constants (A and B both not 0), and x and y are variables, is a straight line. Every straight line in a cartesian coordinate system is the graph of an equation of this form.

It immediately follows that any equation of the form

$$y = mx + b$$

where m and b are constants, is also a straight line since it can be written in the form $-mx + y = b$, which is a special case of $Ax + By = C$.

Thus, to graph any equation of the form

$$Ax + By = C$$

or

$$y = mx + b$$

we plot any two points of the solution set and use a straightedge to draw a line through these two points. It is sometimes wise to find a third point as a check point.

It should be obvious that we cannot draw a straight line extending indefinitely in either direction. We will settle for the part of the line in which we are interested—usually the part close to the origin unless otherwise stated.

EXAMPLE 3 (*A*) The graph of $y = 2x - 4$ is

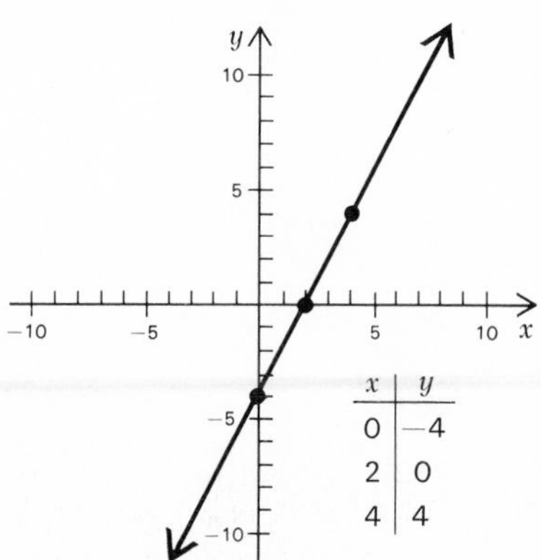

x	y
0	−4
2	0
4	4

(*B*) To graph $x + 3y = 6$, assign to either x or y any convenient value and solve for the other variable. If we let $x = 0$, a convenient value, then

$$0 + 3y = 6$$
$$3y = 6$$
$$y = 2$$

Thus, (0, 2) is a solution.

If we let $y = 0$, another convenient choice, then

$$x + 3(0) = 6$$
$$x + 0 = 6$$
$$x = 6$$

Thus, (6, 0) is a solution.

To find a check point, choose another value for x or y, say $x = -6$, then

$$-6 + 3y = 6$$
$$3y = 12$$
$$y = 4$$

Thus, (−6, 4) is also a solution.

Now plot these three points and draw a line through them. (If a straight line does not pass through all three points, then you have made a mistake and must go back and check your work.)

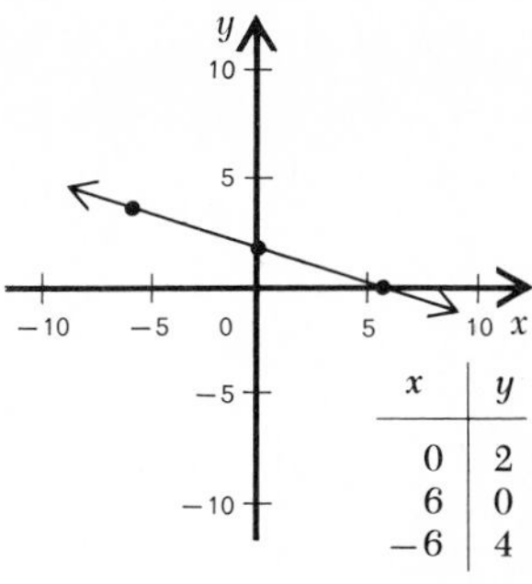

PROBLEM 3 Graph:

(A) $y = \frac{x}{2} - 3$

(B) $2x + 3y = 15$

ANSWER (A)

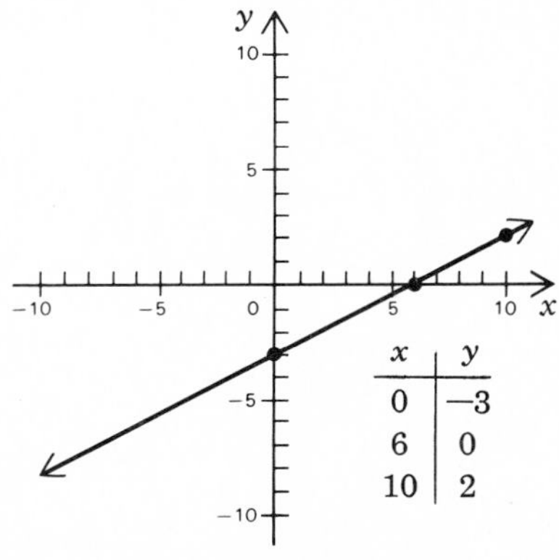

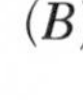

(B)

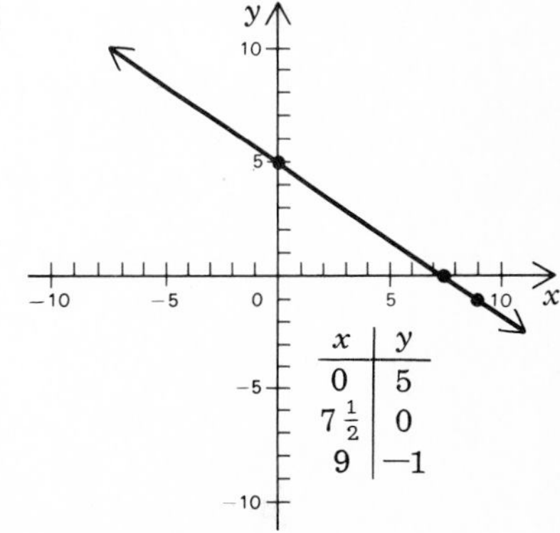

Vertical and horizontal lines in rectangular coordinate systems have particularly simple equations.

EXAMPLE 4 Graph $y = 4$ and $x = 3$ in a rectangular coordinate system.

SOLUTION To graph $y = 4$ or $x = 3$ in a rectangular coordinate system, each equation must be provided with the missing variable (usually done mentally) as follows:

$y = 4$	is equivalent to	$0x + y = 4$
$x = 3$	is equivalent to	$x + 0y = 3$

In the first case, we see that no matter what value is assigned to x, $0x = 0$; thus, as long as $y = 4$, x can assume any value, and the graph of $y = 4$ is a horizontal line crossing the y axis at 4. Similarly, in the second case y can assume any value as long as $x = 3$, and the graph of $x = 3$ is a vertical line crossing the x axis at 3. Thus,

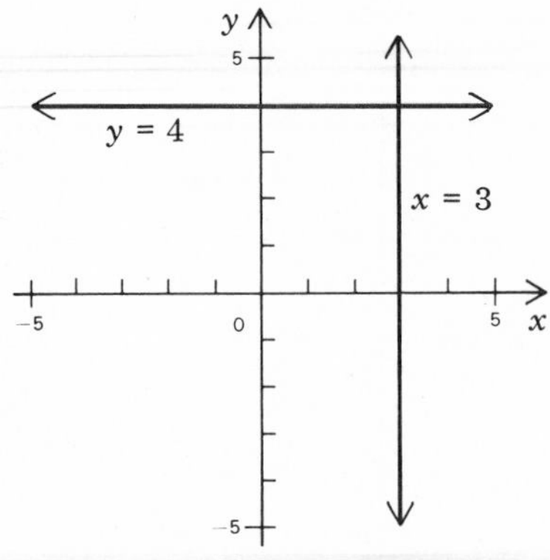

PROBLEM 4 Graph $y = -3$ and $x = -4$ in a rectangular coordinate system.

ANSWER

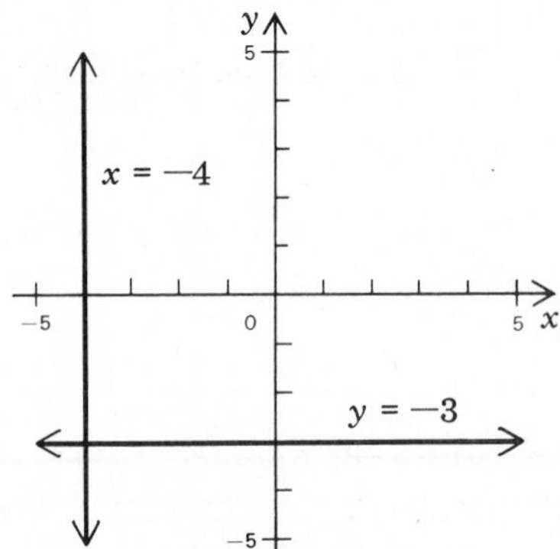

It should now be clear why first-degree equations in two variables are often referred to as *linear equations*.

Exercise 20

Graph in a rectangular coordinate system.

A **1.** $y = 2x$ **2.** $y = x$ **3.** $y = 2x - 3$

4. $y = x - 1$ **5.** $y = \frac{x}{3}$ **6.** $y = \frac{x}{2}$

7. $y = \frac{x}{3} + 2$
8. $y = \frac{x}{2} + 1$
9. $x + y = -4$
10. $x + y = 6$
11. $x - y = 3$
12. $x - y = 5$
13. $3x + 4y = 12$
14. $2x + 3y = 12$
15. $8x - 3y = 24$
16. $3x - 5y = 15$
17. $y = 3$
18. $x = 2$
19. $x = -4$
20. $y = -3$

B

21. $y = \frac{1}{2}x$
22. $y = \frac{1}{4}x$
23. $y = \frac{1}{2}x - 1$
24. $y = \frac{1}{4}x + 1$
25. $y = -x + 2$
26. $y = -2x + 6$
27. $y = \frac{1}{3}x - 1$
28. $y = \frac{1}{2}x + 2$
29. $3x + 2y = 10$
30. $2x + y = 7$
31. $5x - 6y = 15$
32. $7x - 4y = 21$
33. $y = 0$
34. $x = 0$

Write in the form $y = mx + b$ and graph.

35. $x + 6 = 3x + 2 - y$
36. $y - x - 2 = x + 1$

Write in the form $Ax + By = C$ and graph.

37. $y + 8 = 2 - x - y$
38. $6x - 3 + y = 2y + 4x + 5$

Use a different scale on the vertical axis to keep the size of the graph within reason.

39. $I = 6t, 0 \leq t \leq 10$
40. $d = 60t,\ 0 \leq t \leq 10$
41. $v = 10 + 32t,\ 0 \leq t \leq 5$
42. $A = 100 + 10t,\ 0 \leq t \leq 10$
43. Graph $x + y = 3$ and $2x - y = 0$ on the same coordinate system. Determine by inspection the coordinates of the point where the two graphs cross. Show that the coordinates of the point of intersection satisfies both equations.
44. Repeat the preceding problem with the equations

$$2x - 3y = -6 \quad \text{and} \quad x + 2y = 11$$

C

45. Graph $y = mx - 2$ for $m = 2$, $m = \frac{1}{2}$, $m = 0$, $m = -\frac{1}{2}$, and $m = -2$, all on the same coordinate system.
46. Graph $y = -\frac{1}{2}x + b$ for $b = -6$, $b = 0$, and $b = 6$ all on the same coordinate system.
47. Graph $y = |2x|$ and $y = |\frac{1}{2}x|$ on the same coordinate system.
48. Graph $y = |x|$. HINT: Graph $y = x$ for $x \geq 0$, and $y = -x$ for $x < 0$.
49. In 1948 Professor Brown, a psychologist, trained a group of rats (in an experiment on motivation) to run down a narrow passage in a cage to receive food in

a goal box. A harness was put on each rat and the harness was then connected to an overhead wire that was attached to a scale. In this way the rat could be placed at different distances (in centimeters) from the food and Professor Brown could then measure the pull (in grams) of the rat toward the food. It was found that a relation between motivation (pull) and position was given approximately by the equation $p = -\frac{1}{5}d + 70$, $30 \leq d \leq 175$. Graph this equation for the indicated values of d.

50. In a simple electric circuit, such as found in a flashlight, if the resistance is 30 ohms, the current in the circuit I (in amperes) and the electromotive force E (in volts) are related by the equation $E = 30I$. Graph this equation for $0 \leq I \leq 1$.

51. In biology there is an approximate rule, called the bioclimatic rule, for temperate climates that states that in spring and early summer periodic phenomena such as blossoming for a given species, appearance of certain insects, and ripening of fruit usually come about 4 days later for each 500 feet of altitude. Stated as a formula,

$$d = 4\left(\frac{h}{500}\right)$$

where d = change in days and h = change in altitude in feet. Graph the equation for $0 \leq h \leq 4{,}000$.

4.3 Other Standard Equations for Straight Lines

In the last section we considered the problem: Given a linear equation of the form

$$Ax + By = C$$

or

$$y = mx + b$$

find its graph. Now we will consider the reverse problem: Given certain information about a straight line in a rectangular coordinate system, find its equation.

SLOPE-INTERCEPT FORM OF THE EQUATION OF A LINE

As we saw in the last section, any equation of the form $Ax + By = C$, $B \neq 0$, can always be written in the form

$$y = mx + b$$

where m and b are constants. This form has several interesting and useful properties. In particular, the constants m and b have special geometric significance. The significance of b is easily seen by assigning x the value

0 and noting that $y = b$. Thus $(0, b)$ is a solution of the equation, and since $(0, b)$ is the point where the graph crosses the y axis, b is called the *y intercept.*

To determine the geometric significance of m, let us choose two points (x_1, y_1) and (x_2, y_2) on the line $y = mx + b$ (Fig. 3). Since these points lie on the line, their coordinates must satisfy the equation $y = mx + b$.

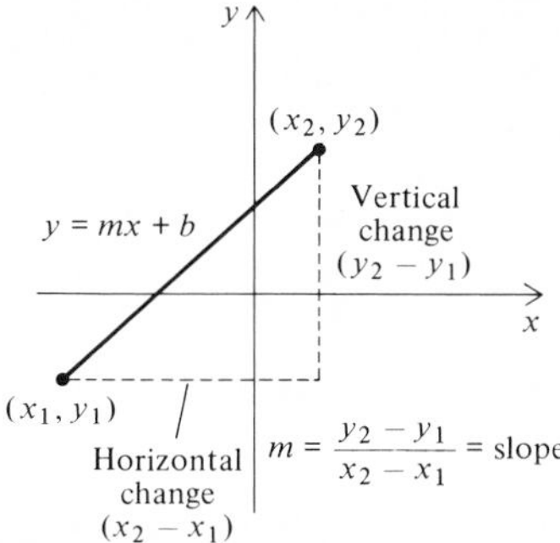

Figure 3

Thus,

$$y_1 = mx_1 + b \qquad \text{and} \qquad y_2 = mx_2 + b$$

Solving both of these equations for b, we obtain

$$b = y_1 - mx_1 \qquad \text{and} \qquad b = y_2 - mx_2$$

Since $y_1 - mx_1$ and $y_2 - mx_2$ are both equal to b, we set them equal to each other and solve for m (a good exercise for the reader). Thus

SLOPE FORMULA

$$m = \frac{y_2 - y_1}{x_2 - x_1} = \frac{\text{vertical change}}{\text{horizontal change}}$$

Interpreting this last equation in terms of Fig. 3, it is seen that m is the ratio of the change in y to the corresponding change in x between any two distinct points on the line. The number m is called the slope of the line, and is a measure of the steepness of the line. Thus if an equation of a line is written in the form $y = mx + b$, the constant m is the slope of the line. If we are given the coordinates of two points (x_1, y_1) and (x_2, y_2) and want to find the slope of the line joining the two points, then we use the formula

$$m = (y_2 - y_1)/(x_2 - x_1)$$

EXAMPLE 5 (A) Find the slope and y intercept of the line $y = \frac{1}{3}x + 2$.

SOLUTION Slope $= m = \frac{1}{3}$; y intercept $= b = 2$.

(*B*) Find the equation of a line with slope -2 and y intercept 3.

SOLUTION $y = -2x + 3$

(*C*) Find the slope of a line passing through (1, 5) and (6, −7).

SOLUTION $m = \frac{y_2 - y_1}{x_2 - x_1} = \frac{-7-5}{6-1} = -\frac{12}{5}$ or $\frac{5-(-7)}{1-6} = -\frac{12}{5}$

NOTE: It does not matter which we call (x_1, y_1) and (x_2, y_2), as long as we stick to the choice once it is made.

PROBLEM 5 (*A*) Find the slope and y intercept of the line $y = (x/2) - 7$.

(*B*) Find the equation of a line with slope $-\frac{1}{3}$ and y intercept 6.

(*C*) Find the slope of a line passing through (−3, 5) and (−1, −3).

ANSWER (*A*) $m = -\frac{1}{2}$, $b = -7$ (*B*) $y = -(x/3) + 6$ (*C*) $m = -4$

The slope of a line may be positive, negative, 0, or not defined. (Which case applies to vertical lines? Which case to horizontal lines?) It should now be clear why $y = mx + b$ is called the *slope-intercept form* of the equation of a line.

POINT-SLOPE FORM OF THE EQUATION OF A LINE

In Example 5 we found the equation of a line given its slope and y intercept. Often it is necessary to find the equation of a line given its slope and the coordinates of a point through which it passes, or to find the equation of a line given the coordinates of two points through which it passes.

Let a line have slope m and pass through the fixed point (x_1, y_1). If (x, y) is to be a point on the line, the slope of the line passing through (x, y) and (x_1, y_1) must be m (see Fig. 4). Thus the equation

$$\frac{y - y_1}{x - x_1} = m$$

restricts a variable point (x, y), so that only those points in the plane lying on the line will have coordinates that satisfy the equation, and vice versa. This equation is usually written in the form

POINT-SLOPE FORM

$$y - y_1 = m(x - x_1)$$

and is referred to as the *point-slope form* of the equation of a line. Using

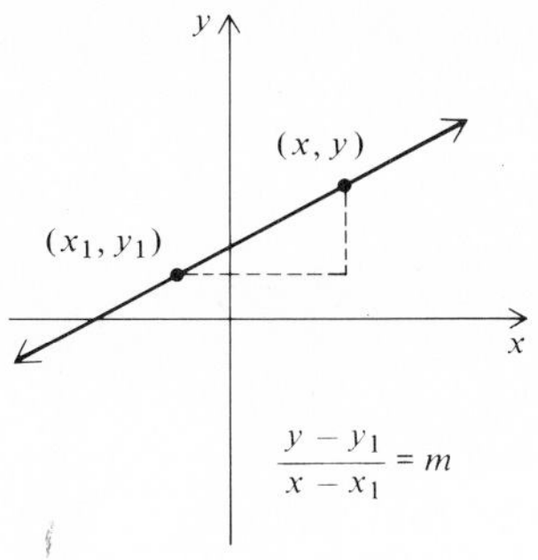

Figure 4

this equation in conjunction with the slope formula, we can also find the equation of a line knowing only the coordinates of two points through which it passes.

EXAMPLE 6 (*A*) Find an equation of a line with slope $-\frac{1}{3}$ that passes through (6, −3). Write the resulting equation in the form $y = mx + b$.

SOLUTION

$$y - y_1 = m(x - x_1)$$
$$y - (-3) = -\tfrac{1}{3}(x - 6)$$
$$y + 3 = -\tfrac{1}{3}(x - 6)$$
$$y + 3 = -\frac{x}{3} + 2$$
$$y = -\tfrac{1}{3}x - 1$$

(*B*) Find an equation of a line that passes through the two points (−2, −6) and (2, 2).

SOLUTION First find the slope of the line using the slope formula, then proceed as in part *A*, using the coordinates of either point for (x_1, y_1).

$$m = \frac{y_2 - y_1}{x_2 - x_1} = \frac{2 - (-6)}{2 - (-2)} = 2$$

$$y - y_1 = m(x - x_1)$$
$$y - (-6) = 2[x - (-2)]$$
$$y + 6 = 2(x + 2)$$
$$y + 6 = 2x + 4$$
$$y = 2x - 2$$

PROBLEM 6 (A) Find the equation of a line with slope $\frac{2}{3}$ that passes through $(-3, 4)$.

(B) Find the equation of a line that passes through the two points $(6, -1)$ and $(-2, 3)$. Transform the equation into the form $y = mx + b$.

ANSWER (A) $y - 4 = \frac{2}{3}(x + 3)$ or $y = \frac{2}{3}x + 6$ (B) $y = -(x/2) + 2$

VERTICAL AND HORIZONTAL LINES

If a line is vertical, it does not have a slope. Since points on a vertical have constant abscissas and arbitrary ordinates, the equation of a vertical line is of the form

$$x + 0y = c$$

or simply

$$x = c \qquad \text{vertical line}$$

where c is the abscissa of each point on the line. Similarly, if a line is horizontal (slope 0), then every point on the line has constant ordinate and arbitrary abscissa. Thus the equation of a horizontal line is of the form

$$0x + y = c$$

or simply

$$y = c \qquad \text{horizontal line}$$

where c is the ordinate of each point on the line.

EXAMPLE 7 The equation of a vertical line through $(-2, -4)$ is $x = -2$, and the equation of a horizontal line through the same point is $y = -4$.

PROBLEM 7 What are the equations of vertical and horizontal lines through $(3, -8)$?

ANSWER $x = 3 \qquad y = -8$

PARALLEL AND PERPENDICULAR LINES

It can be shown that nonvertical parallel lines have the same slope, and that nonvertical perpendicular lines have slopes that are negative reciprocals of each other. (The details are left to another course.) Thus, the lines determined by

$$y = \tfrac{2}{3}x - 5 \qquad \text{and} \qquad y = \tfrac{2}{3}x + 8$$

are parallel since both have the same slope, $\frac{2}{3}$. And the lines that are determined by

$$y = \frac{x}{3} + 5 \qquad \text{and} \qquad y = -3x - 7$$

are perpendicular, since their slopes are negative reciprocals of each other; that is

$$\frac{1}{3} = -\left(\frac{1}{-3}\right)$$

Exercise 21

A *Find the slope, y intercept, and graph each equation.*

1. $y = 2x - 3$ **2.** $y = x + 1$

3. $y = -x + 2$ **4.** $y = -2x + 1$

Write the equation of the line with slope and y intercept as indicated.

5. Slope $= 5$ **6.** Slope $= 3$
y intercept $= -2$ y intercept $= -5$

7. Slope $= -2$ **8.** Slope $= -1$
y intercept $= 4$ y intercept $= 2$

Write the equation of the line that passes through the given point with the indicated slope.

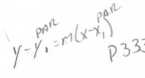

9. $m = 2$, $(5, 4)$ **10.** $m = 3$, $(2, 5)$

11. $m = -2$, $(2, 1)$ **12.** $m = -3$, $(1, 3)$

Find the slope of the line that passes through the given points.

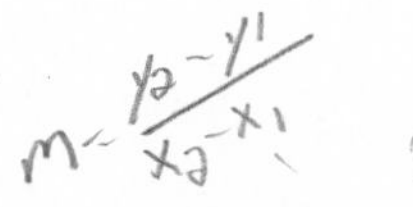

13. $(3, 2)$ and $(5, 6)$ **14.** $(1, 3)$ and $(2, 4)$

15. $(2, 1)$ and $(10, 5)$ **16.** $(1, 3)$ and $(7, 5)$

Write the equation of the line through each indicated pair of points.

17. $(3, 2)$ and $(5, 6)$ **18.** $(1, 3)$ and $(2, 4)$

19. $(2, 1)$ and $(10, 5)$ **20.** $(1, 3)$ and $(7, 5)$

B *Find the slope, y intercept, and graph each equation.*

21. $y = -\frac{x}{3} + 2$ **22.** $y = -\frac{x}{4} - 1$

23. $x + 2y = 4$ **24.** $x - 3y = -6$

25. $2x + 3y = 6$ **26.** $3x + 4y = 12$

Write the equation of the line with slope and y intercept as given.

27. Slope $= -\frac{1}{2}$
y intercept $= -2$

28. Slope $= -\frac{1}{3}$
y intercept $= -5$

29. Slope $= \frac{2}{3}$
y intercept $= \frac{3}{2}$

30. Slope $= -\frac{3}{2}$
y intercept $= \frac{5}{2}$

Write the equation of the line that passes through the given point with the indicated slope. Transform the equation into the form $y = mx + b$.

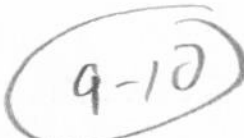

31. $m = -2$, $(-3, 2)$

32. $m = -3$, $(4, -1)$

33. $m = \frac{1}{2}$, $(-4, 3)$

34. $m = \frac{2}{3}$, $(-6, -5)$

Find the slope of the line that passes through the given points.

35. $(3, 7)$ and $(-6, 4)$

36. $(-5, -2)$ and $(-5, -4)$

37. $(4, -2)$ and $(-4, 0)$

38. $(-3, 0)$ and $(3, -2)$

Write the equation of the line through each of the indicated pairs of points. Transform the equation into the form $y = mx + b$.

39. $(3, 7)$ and $(-6, 4)$

40. $(-5, -2)$ and $(5, -4)$

41. $(4, -2)$ and $(-4, 0)$

42. $(-3, 0)$ and $(3, -4)$

Write the equations of the vertical and horizontal lines through each point.

43. $(-3, 5)$

44. $(6, -2)$

45. $(-1, 22)$

46. $(5, 0)$

C *Write the equation of the line through the indicated pairs of points. Transform the equation into the form $Ax + By = C$.*

47. $(3, 5)$ and $(3, -4)$

48. $(-2, 5)$ and $(3, 5)$

49. $(-2, 1)$ and $(2, 3)$

50. $(3, -1)$ and $(-3, -5)$

51. A sporting goods store sells a pair of ski boots costing \$20 for \$33 and a pair of skis costing \$60 for \$93. (*A*) If the markup policy of the store for items costing over \$10 is assumed to be linear and is reflected in the pricing of these two items, write an equation that relates retail price R with cost C. (*B*) Graph this equation for $10 \leq C \leq 300$. (*C*) Use the equation to find the cost of a surfboard retailing for \$240.

52. The management of a company manufacturing ball-point pens estimates costs for running the company to be \$200 per day at zero output, and \$700 per day at an output of 1,000 pens. (*A*) Assuming total cost per day, c, is linearly related to total output per day, x, write an equation relating these two quantities. (*B*) Graph the equation for $0 \leq x \leq 2{,}000$.

4.4 First-Degree Inequalities In Two Variables

We know how to graph first-degree equations such as

$$y = 2x - 3$$

or

$$2x - 3y = 5$$

but how do we graph

$$y \leq 2x - 3$$

or

$$2x - 3y > 5$$

We will find that graphing inequalities is almost as easy as graphing the equations. The following discussion leads to a simple solution to the problem.

A line in a cartesian coordinate system divides the plane into two *half planes*. A vertical line divides the plane into left and right half planes; a nonvertical line divides the plane into upper and lower half planes, according to the location of each point as indicated in Fig. 5. Now let us compare the graphs of

$$y < 2x - 3$$

$$y = 2x - 3$$

and

$$y > 2x - 3$$

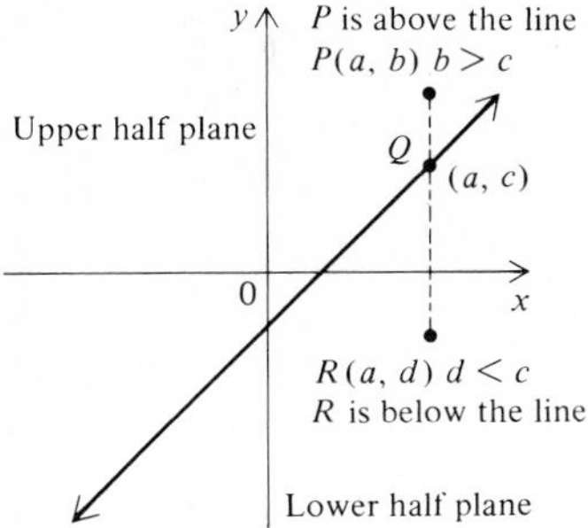

Figure 5

Consider the vertical line $x = x_0$, and ask what the relationship of y is to $2x_0 - 3$ as we move (x_0, y) up and down this vertical line (see Fig. 6). If we are at point Q, then $y = 2x_0 - 3$; if we move up the vertical line to P, the ordinate of (x_0, y) increases and $y > 2x_0 - 3$; if we move down the line to R, the ordinate of (x_0, y) decreases and $y < 2x_0 - 3$. Since the same results are obtained for each point on the x axis, we conclude that the graph of $y > 2x - 3$ is the upper half plane determined by $y = 2x - 3$, and $y < 2x - 3$ is the lower half plane.

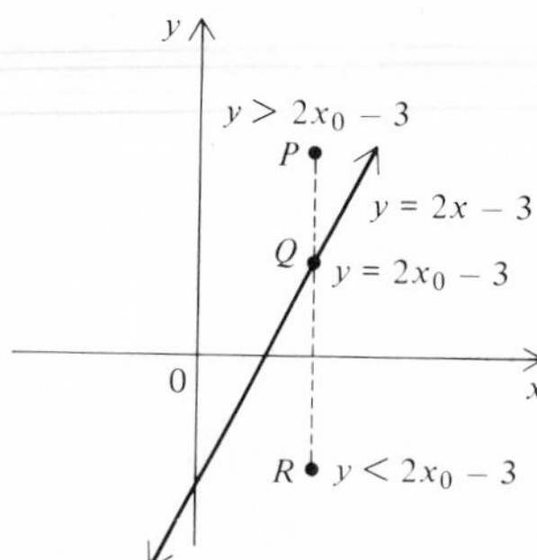

Figure 6

In graphing $y > 2x - 3$, we show the line $y = 2x - 3$ as a broken line, indicating that it is not part of the graph; in graphing $y \geq 2x - 3$, we show the line $y = 2x - 3$ as a solid line, indicating that it is part of the graph. Figure 7 illustrates four typical cases.

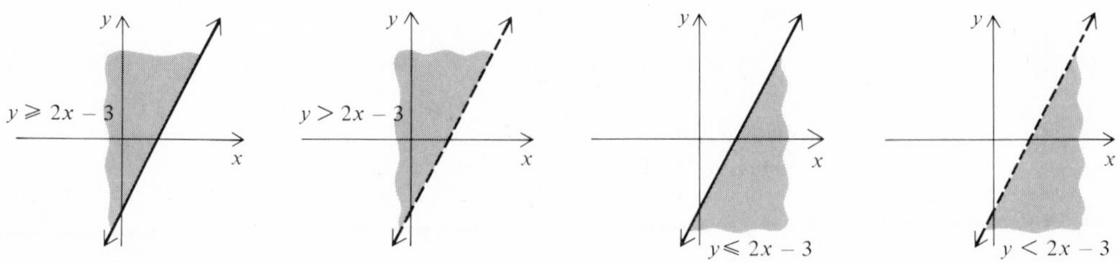

Figure 7

The preceding discussion suggests the following important theorem, which we state without proof.

THEOREM 2 The graph of a linear inequality

$$Ax + By < C$$

or

$$Ax + By > C$$

with $B \neq 0$, is either the upper half plane or the lower half plane (but not both) determined by the line $Ax + By = C$. If $B = 0$, the graph of

$$Ax < C$$

or

$$Ax > C$$

is either the left half plane or the right half plane (but not both) as determined by the line $Ax = C$.

This theorem leads to a simple, fast procedure for graphing linear inequalities in two variables:

1 First, graph $Ax + By = C$:
  as a broken line if equality is not included in original statement
  as a solid line if equality is included in original statement
2 Choose a test point in the plane not on the line—the origin is the best choice if it is not on the line—and substitute the coordinates into the inequality.
3 The graph of the original inequality includes:
  the half plane containing the test point, if the inequality is satisfied by that point, or
  the half plane not containing the test point, if the inequality is not satisfied by that point

EXAMPLE 8 Graph $3x - 4y \leq 12$.

SOLUTION First graph the line $3x - 4y = 12$. Pick a convenient test point above or below the line. The origin requires the least computation. We see that $3 \cdot 0 - 4 \cdot 0 \leq 12$ is true; hence, the graph of the inequality is the upper half plane.

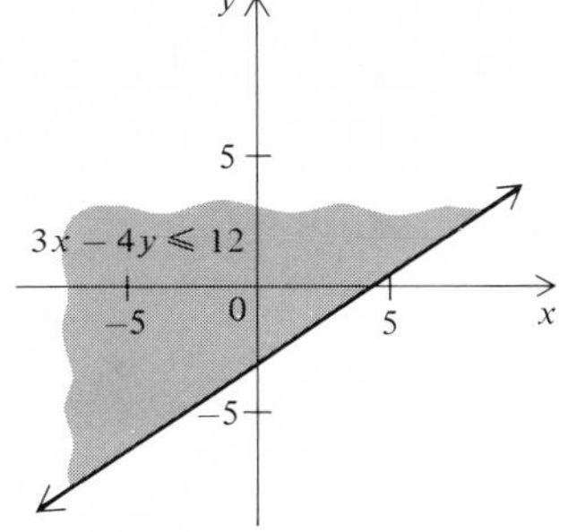

PROBLEM 8 Graph $2x + 3y \leq 6$.

ANSWER

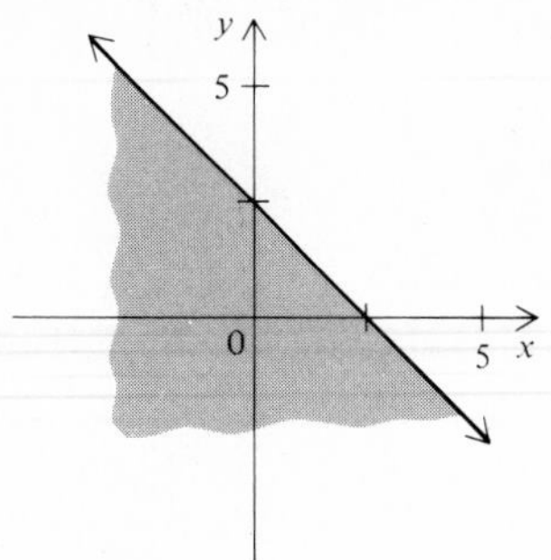

EXAMPLE 9 Graph:

(*A*) $y > -3$

(*B*) $2x < 5$

(*C*) $-2 \leq x \leq 4$

SOLUTION

(*A*)

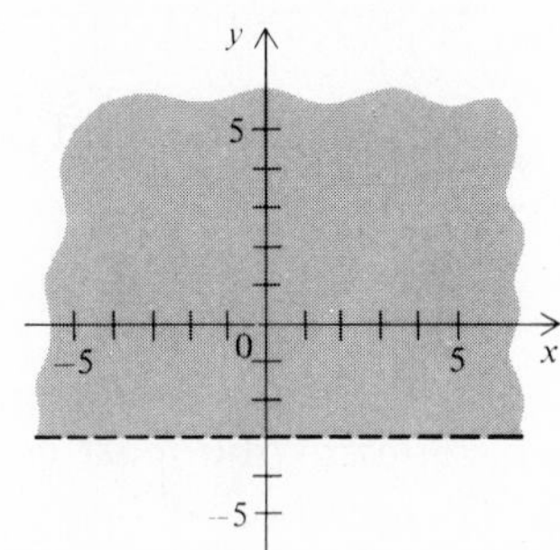

(*B*)

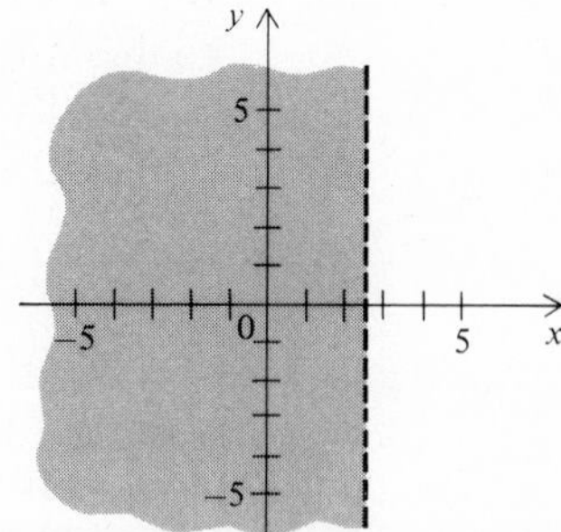

(*C*)

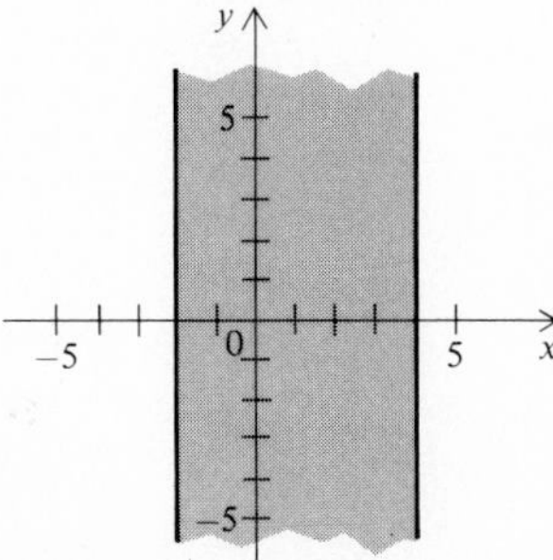

PROBLEM 9 Graph:

(*A*) $y < 2$

(*B*) $3x > -8$

(*C*) $1 \leq y \leq 5$

ANSWER

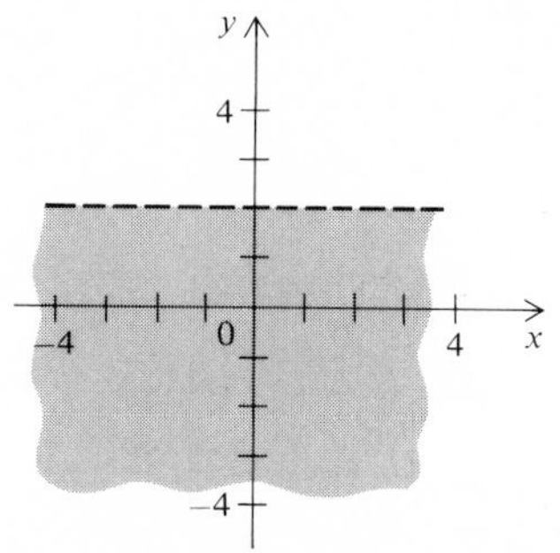

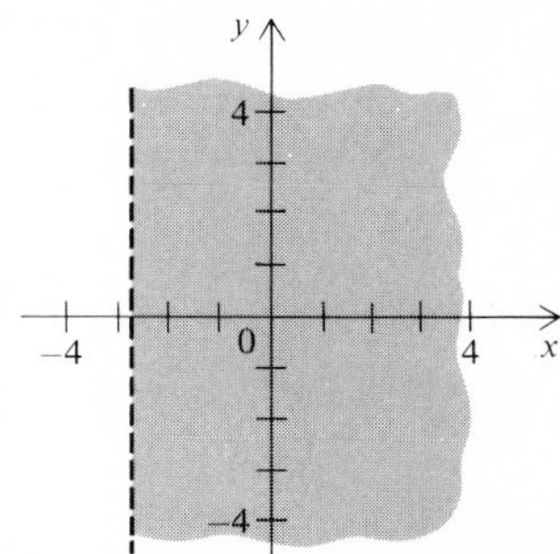

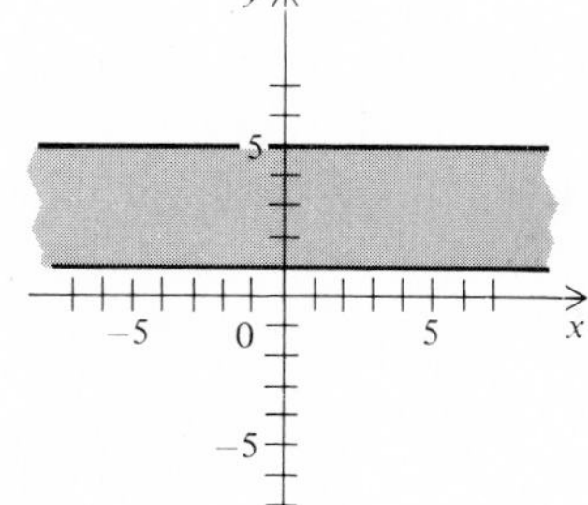

Exercise 22

Graph each inequality:

A

1. $x + y \leq 6$

2. $x + y \geq 4$

3. $x - y > 3$

4. $x - y < 5$

5. $y \geq x - 2$

6. $y \leq x + 1$

B

7. $2x - 3y < 6$

8. $3x + 4y < 12$

9. $3y - 2x \geq 24$

10. $3x + 2y \geq 18$

11. $y \geq \frac{x}{3} - 2$

12. $y \leq \frac{x}{2} - 4$

13. $y \leq \frac{2}{3}x + 5$

14. $y > \frac{x}{3} + 2$

15. $x \geq -5$

16. $y \leq 8$

17. $y < 0$

18. $x > 0$

19. $-1 < x \leq 3$

20. $-3 \leq y < 2$

21. $-2 \leq y \leq 2$

22. $-1 \leq x \leq 4$

C *Graph each set:*

23. $\{(x, y) \mid -3 \leq x \leq 3 \text{ and } 1 \leq y \leq 5\}$

24. $\{(x, y) \mid 1 \leq x \leq 5 \text{ and } -2 \leq y \leq 2\}$

25. $\{(x, y) \mid x \geq 0,\ y \geq 0, \text{ and } 3x + 4y \leq 12\}$

26. $\{(x, y) \mid x \geq 0,\ y \geq 0, \text{ and } 3x + 2y \leq 18\}$

27. A manufacturer of surfboards makes a standard model and a competition model. The pertinent manufacturing data is summarized in the table.

	STANDARD MODEL (WORKHOURS PER BOARD)	COMPETITION MODEL (WORKHOURS PER BOARD)	MAXIMUM WORKHOURS AVAILABLE PER WEEK
Fabricating	6	8	120
Finishing	1	3	30

If x is the number of standard models and y is the number of competition models produced per week, write a system of inequalities that indicate the restrictions on x and y. Graph this system showing the region of permissible values for x and y.

Exercise 23 Chapter Review

A *Graph each in a rectangular coordinate system.*

1. $y = 2x - 3$

2. $2x + y = 6$

3. $x - y \geq 6$

4. $y > x - 1$

5. What is the slope and y intercept for the graph of $y = -2x - 3$?

6. Write the equation of a line that passes through (2, 4) with slope -2.

7. What is the slope of the line that passes through (1, 3) and (3, 7)?

8. Write the equation of the line that passes through (1, 3) and (3, 7).

B *Graph each in a rectangular coordinate system.*

9. $y = \frac{1}{3}x - 2$

10. $3x - 2y = 9$

11. $x = -3$

12. $4x - 5y \leq 20$

13. $y < \dfrac{x}{2} + 1$

14. $x \geq -3$

15. $-4 \leq y < 3$

16. What is the slope and y intercept for the graph of $x + 2y = -6$?

17. Write the equation of a line that passes through $(-3, 2)$ with slope $-\frac{1}{3}$. Write the final answer in the form $y = mx + b$.

18. What is the slope of the line that passes through $(-3, 2)$ and $(3, -2)$?

19. Write the equation of a line that passes through $(-3, 2)$ and $(3, -2)$. Write the final answer in the form $y = mx + b$.

20. Write the equation of a vertical axis through $(-2, 5)$.

21. In which quadrant(s) is the ordinate of a point negative?

22. In which quadrant(s) is the abscissa of a point negative?

C 23. Graph $y = mx - 1$ for $m = \frac{1}{2}$, 0, and 2, all on the same coordinate system.

24. Graph $y = \dfrac{x}{2} + b$ for $b = -4$, 0, and 4, all on the same coordinate system.

25. What is the slope and y intercept for the graph of $3x - 2y = 5$?

26. What is the slope of the line that passes through $(3, 4)$ and $(3, -4)$?

27. Write the equation of a line that passes through $(3, 4)$ and $(3, -4)$.

28. Write the equation of a line that passes through $(-5, 3)$ and $(2, 3)$.

29. Write the equation of a line that passes through $(-2, 5)$ and $(3, -2)$ in the form $Ax + By = C$, where A, B, and C are integers, $A > 0$.

30. Graph $\{(x, y) \mid y \geq 0, 1 \leq x \leq 5, 2x + 3y \leq 18\}$.

Chapter 5
SYSTEMS OF LINEAR EQUATIONS

5.1 Systems of Linear Equations in Two Variables

Many practical problems can be solved conveniently using two-equation–two-unknown methods. For example, if a 12-foot board is cut in two pieces so that one piece is 4 feet longer than the other piece, how long is each piece? We could solve this problem using one-equation–one-unknown methods studied earlier, but we can also proceed as follows using two variables.

Let

$x =$ the length of the longer piece

$y =$ the length of the shorter piece

then

$$x + y = 12$$

$$x - y = 4$$

To solve this system is to find all the ordered pairs of real numbers that satisfy both equations at the same time. In general, we are interested in solving linear systems of the type

$$ax + by = m$$

$$cx + dy = n$$

where a, b, c, d, m, and n are real constants. There are several methods of solving systems of this type. We will consider two that are widely used.

SOLVING BY GRAPHING

We proceed by graphing both equations on the same coordinate system. Then the coordinates of any points that the graphs have in common must be solutions to the system since they must satisfy both equations. (Why?)

EXAMPLE 1 Solve by graphing:

$$x + y = 12$$

$$x - y = 4$$

SOLUTION

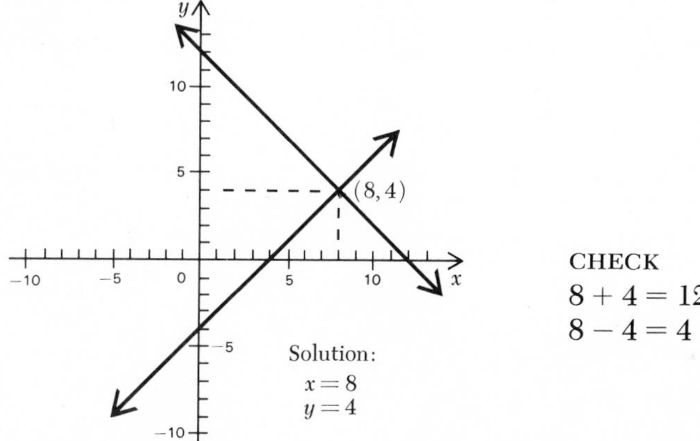

CHECK
$8 + 4 = 12$
$8 - 4 = 4$

PROBLEM 1 Solve by graphing:

$$x + y = 10$$

$$x - y = 6$$

ANSWER $x = 8, \quad y = 2$

It is clear that Example 1 and problem 1 each has exactly one solution since the two lines in each case intersect in exactly one point Let us now look at three cases that illustrate the three possible ways two lines can be related to each other in a rectangular coordinate system. (Since one of the cases involves parallel lines, recall that it can be shown that if the slopes of two lines are the same, then the two lines are parallel.)

EXAMPLE 2 Solve each of the following systems by graphing.

(A) $2x - 3y = 2$ $\quad x + 2y = 8$

(B) $4x + 6y = 12$ $\quad 2x + 3y = -6$

(C) $2x - 3y = -6$ $\quad -x + \frac{3}{2}y = 3$

SOLUTION (A)

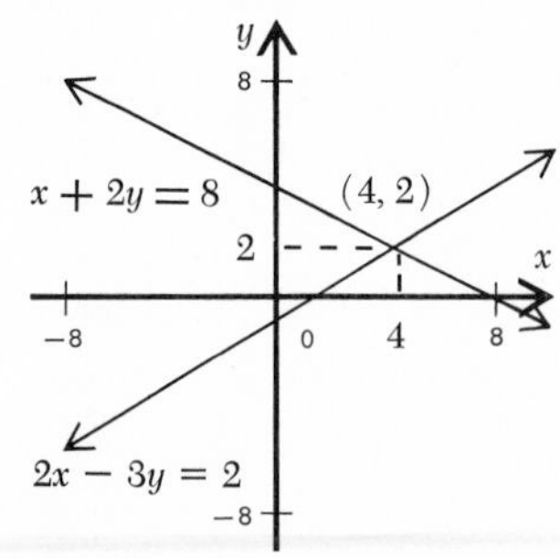

Lines intersect at one point only:
Exactly one solution:
$x = 4,\ y = 2$

(B)

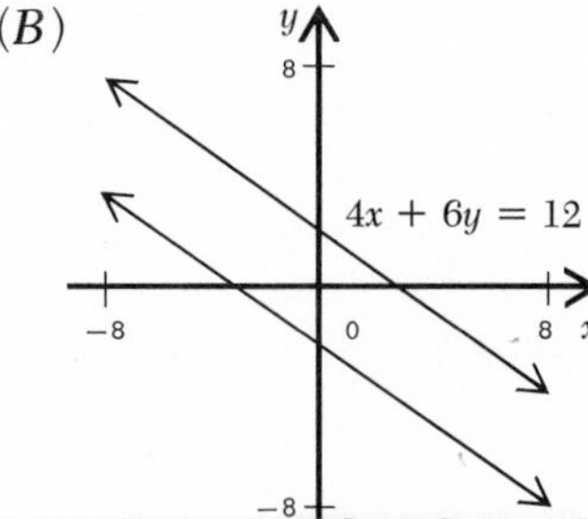

Lines are parallel
(Both have slope $\frac{2}{3}$):
No solution

(C)

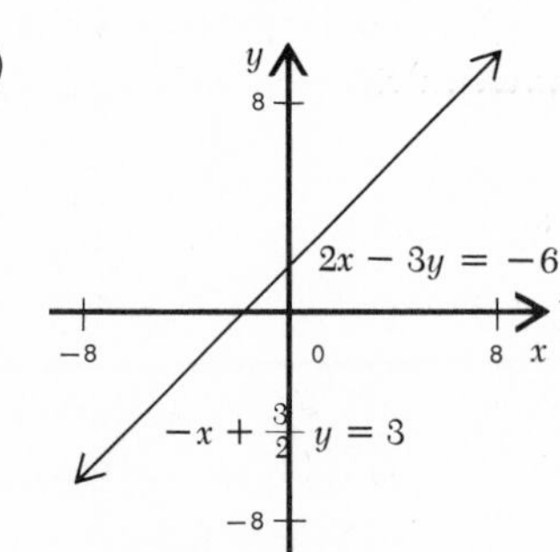

Lines coincide:
Infinite number of solutions

PROBLEM 2 Solve each of the following systems by graphing.

(A) $2x + 3y = 12$ $\quad x - 3y = -3$

(B) $x - 3y = -3$ $\quad -2x + 6y = 12$

(C) $2x - 3y = 12$ $\quad -x + \frac{3}{2}y = -6$

ANSWER (A) $x = 3,\ y = 2$ (B) no solution (C) infinitely many solutions

Now we know exactly what to expect when solving a system of two linear equations in two unknowns:

Exactly one pair of numbers as a solution

No solutions

An infinite number of solutions

In most applications the first case prevails. If we find a pair of numbers that satisfy the system of equations and the slopes of the graphs of the equations are different, then that pair of numbers is the only solution of the system, and we need not look further for others.

The graphical method of solving systems of equations yields considerable information as to what to expect in the way of solutions to a system of two linear equations in two unknowns. In addition, graphs frequently reveal relationships in problems that would otherwise be hidden. On the other hand, if one is interested in solutions with several-decimal-place accuracy, the graphical method is often not practical. The method of elimination, to be considered next, will take care of this deficiency.

SOLVING BY ELIMINATION

The elimination method can produce answers to any decimal accuracy desired. This method has to do with the replacement of systems of equations with simpler equivalent systems (by performing appropriate operations) until we get a system whose solution is obvious. *Equivalent systems* are, as you would expect, systems with the same solution set. What operations on a system produce equivalent systems? The following theorem, stated but not proved, answers this question.

THEOREM 1 Equivalent systems result if

(*A*) Any algebraic expression in an equation is replaced with its equal.

(*B*) Both sides of the equations are multiplied by a nonzero constant.

(*C*) The two equations are combined by addition or subtraction and the result replaces either of the two original equations.

Part *C* follows from the addition and subtraction properties of equality. We add (or subtract) the left side of one equation to (or from) the left side of another and add (or subtract) the right side to (or from) the right side of the other. Solving a system of equations by use of this theorem is best illustrated by examples.

EXAMPLE 3 Solve the system:

$$3x + 2y = 13$$

$$2x - y = 4$$

SOLUTION We use Theorem 1 to eliminate one of the variables, and thus obtain a system whose solution is obvious:

$$\begin{aligned} (A)\quad & 3x + 2y = 13 \\ (B)\quad & 2x - y = 4 \end{aligned}$$

If we multiply equation (*B*) by 2 and add the result to equation (*A*), we can eliminate *y*.

$$\begin{aligned} (A)\quad & 3x + 2y = 13 \\ 2(B)\quad & 4x - 2y = 8 \\ (A) + 2(B)\quad & 7x = 21 \\ & x = 3 \end{aligned}$$

Eliminate *y* by addition.

$$\begin{aligned} 2 \cdot 3 - y &= 4 \\ -y &= -2 \\ y &= 2 \end{aligned}$$

Substitute $x = 3$ back into either equation (*A*) or equation (*B*), the simpler of the two, and solve for *y*.

CHECK

(*A*)	(*B*)
$3x + 2y = 13$	$2x - y = 4$
$3 \cdot 3 + 2 \cdot 2 \stackrel{?}{=} 13$	$2 \cdot 3 - 2 \stackrel{?}{=} 4$
$9 + 4 \stackrel{\checkmark}{=} 13$	$6 - 2 \stackrel{\checkmark}{=} 4$

PROBLEM 3 Solve the system:

$$2x + 3y = 7$$
$$3x - y = 5$$

ANSWER $x = 2$, $y = 1$

EXAMPLE 4 Solve the system:

$$2x + 3y = 1$$
$$5x - 2y = 12$$

SOLUTION

$$\begin{aligned} (A)\quad & 2x + 3y = 1 \\ (B)\quad & 5x - 2y = 12 \end{aligned}$$

If we multiply equation (*A*) by 2 and equation (*B*) by 3 and add, we can eliminate *y*.

$$\begin{aligned} 4x + 6y &= 2 \\ 15x - 6y &= 36 \\ 19x &= 38 \\ x &= 2 \end{aligned}$$

$$\begin{aligned} 2 \cdot 2 + 3y &= 1 \\ 3y &= -3 \\ y &= -1 \end{aligned}$$

Substitute $x = 2$ back into either equation (*A*) or equation (*B*).

CHECK

(A)

$$2x + 3y = 1$$
$$2 \cdot 2 + 3(-1) \stackrel{?}{=} 1$$
$$4 - 3 \stackrel{\checkmark}{=} 1$$

(B)

$$5x - 2y = 12$$
$$5 \cdot 2 - 2(-1) \stackrel{?}{=} 12$$
$$10 + 2 \stackrel{\checkmark}{=} 12$$

PROBLEM 4 Solve the system:

$$3x - 2y = 8$$
$$2x + 5y = -1$$

ANSWER $x = 2$, $y = -1$

EXAMPLE 5 Solve the system:

$$x + 3y = 2$$
$$2x + 6y = -3$$

SOLUTION

$$(A) \quad x + 3y = 2$$
$$(B) \quad 2x + 6y = -3$$

$$\begin{array}{r} 2x + 6y = 4 \\ 2x + 6y = -3 \\ \hline 0 = 7 \end{array}$$

A contradiction!

Hence, no solution.

Our assumption that there are values for x and y that satisfy equation (A) and equation (B) simultaneously must be false (otherwise, we have proved that $0 = 7$); thus the system has no solutions. Systems of this type are said to be *inconsistent*—conditions have been placed on the unknowns x and y that are impossible to meet. Geometrically, what does this mean? It means that the graphs of the two equations are parallel; otherwise, there would have to be a solution.

PROBLEM 5 Solve the system:

$$2x - y = 2$$
$$-4x + 2y = 1$$

ANSWER No solution

EXAMPLE 6 Solve the system:

$$-2x + y = -8$$
$$x - \tfrac{1}{2}y = 4$$

SOLUTION

$$\begin{aligned} (A)\quad -2x + y &= -8 \\ (B)\quad x - \tfrac{1}{2}y &= 4 \end{aligned}$$

$$\begin{aligned} -2x + y &= -8 \\ 2x - y &= 8 \\ \hline 0 &= 0 \end{aligned}$$

Both unknowns have been eliminated! Actually, if we had multiplied equation (B) by -2, we would have obtained equation (A). When one equation is a constant multiple of the other, the system is said to be *dependent*, and their graphs will coincide. There are infinitely many solutions to the system—any solution of one equation will be a solution to the other. What is happening geometrically? The graphs of the two equations actually coincide.

PROBLEM 6 Solve the system:

$$\begin{aligned} 4x - 2y &= 3 \\ -2x + y &= -\tfrac{3}{2} \end{aligned}$$

ANSWER An infinite number of solutions—any solution of one equation is a solution of the other.

Exercise 24

A *Solve by graphing and check:*

1. $x + y = 6$, $x - y = 2$

2. $x + y = 5$, $x - y = 1$

3. $x - 2y = -4$, $2x - y = 10$

4. $x - 3y = 3$, $x - y = 7$

Solve by elimination and check:

5. $-x + y = 1$, $x - 2y = -5$

6. $x + 3y = 13$, $-x + y = 3$

7. $3x - y = -3$, $5x + 3y = -19$

8. $2x + 3y = 1$, $3x - y = 7$

9. $3x + y = -8$, $-5x + 3y = 4$

10. $2x + 4y = 6$, $-3x + y = 5$

B *Solve by graphing:*

11. $x - 2y = 2$
$2x + y = 9$

12. $3x - y = 2$
$x + 2y = 10$

13. $3x - 2y = 12$
$7x + 2y = 8$

14. $-2x + 3y = 12$
$2x - y = 4$

Solve by elimination:

15. $3x - 11y = -7$
$4x + 3y = 26$

16. $11x + 2y = 1$
$9x - 3y = 24$

17. $x + 2y = 4$
$2x + 4y = -9$

18. $3x + 5y = 15$
$6x + 10y = -5$

19. $5m - 3n = 7$
$7m + 12n = -1$

20. $3p + 8q = 4$
$15p + 10q = -10$

21. $6x - 2y = 18$
$-3x + y = -9$

22. $\frac{1}{2}x - y = -3$
$-x + 2y = 6$

C *Solve by graphing:*

23. $3x + 5y = 15$
$6x + 10y = -30$

24. $x + 2y = 4$
$2x + 4y = -8$

25. $3x - 5y = 15$
$x - \frac{5}{3}y = 5$

26. $\frac{1}{2}x - y = -3$
$-x + 2y = 6$

Write in the standard form

$\text{ax} + by = c$

$dx + eg = f$

and solve by elimination.

27. $y = 3x - 3$
$6x = 8 + 3y$

28. $y = 2x - 1$
$6x - 3y = -1$

29. $2x + 3y = 2y - 2$
$3x + 2y = 2x + 2$

30. $2u - 3v = 1 - 3u$
$4v = 7u - 2$

Solve by elimination:

31. $0.5m + 0.2n = 0.54$
$0.3m - 0.6n = 0.18$

32. $0.2x - 0.5y = 0.07$
$0.8x - 0.3y = 0.79$

In the next two problems, first solve for 1/x and 1/y, then solve for x and y.

33. $\frac{2}{x} - \frac{4}{y} = -2$

$\frac{3}{x} + \frac{5}{y} = -\frac{1}{4}$

34. $\frac{5}{x} - \frac{2}{y} = -\frac{7}{2}$

$-\frac{3}{x} + \frac{4}{y} = \frac{7}{2}$

35. Show that the lines in Problem 24 are parallel.

36. Show that the lines in Problem 23 are parallel.

37. Show that the ordered pair of numbers (a, b) that satisfies the system

$Ax + By + C = 0$

$Dx + Ey + F = 0$

will also satisfy the equation

$m(Ax + By + C) + n(Dx + Ey + F) = 0$

for any real numbers m and n.

5.2 Applications

This section contains a wide variety of applications grouped by subject areas similar to those found in Sec. 3.2. All problems should be solved using a two-equation–two-unknown method discussed in the preceding section.

EXAMPLE 7 If you have 25 dimes and quarters in your pocket worth $4, how many of each do you have?

SOLUTION Let $x =$ the number of dimes

$y =$ the number of quarters

then

$x + y = 25$

$10x + 25y = 400$

Multiply the top equation by -10 and add:

$$\begin{aligned} -10x - 10y &= -250 \\ 10x + 25y &= 400 \\ \hline 15y &= 150 \\ y &= 10 \quad \text{quarters} \\ x + 10 &= 25 \\ x &= 15 \quad \text{dimes} \end{aligned}$$

CHECK: $10 + 15 = 25$ coins; $10(25) + 15(10) = 250 + 150 = 400$ cents or \$4

PROBLEM 7 If you have 25 nickels and quarters in your pocket worth \$2.25, how many of each do you have?

ANSWER 20 nickels, 5 quarters

EXAMPLE 8 The population of a town is 30,000, and it is decreasing at the rate of 550 per year. Another town has a population of 18,000 which is increasing at the rate of 1,450 per year. In how many years will both towns be the same size, and what will their population be at that time?

SOLUTION Let x be the population of a town after t years, then

$$x = 30{,}000 - 550t \qquad \text{first town}$$

$$x = 18{,}000 + 1{,}450t \qquad \text{second town}$$

Subtract to eliminate x (or substitute the right member of the second equation into the first member of the first equation):

$$0 = 12{,}000 - 2{,}000t$$

$$2{,}000t = 12{,}000$$

$$t = 6 \text{ years} \qquad \text{(Both towns will be the same size in 6 years.)}$$

Use either equation to find the size of the towns after 6 years.

$$\begin{aligned} x &= 30{,}000 - 550(6) \\ &= 30{,}000 - 3{,}300 \\ &= 26{,}700 \text{ people} \end{aligned}$$

PROBLEM 8 Repeat Example 15 with the first town starting with a population of 150,000 and decreasing at 1,900 per year, and the second town starting with a population of 100,000 and increasing at 3,100 per year.

ANSWER 10 years; 131,000 people

EXAMPLE 9

A jeweler has two bars of gold alloy in stock, one 12 karat and the other 18 karat (24-karat gold is pure gold, 12-karat gold is $\frac{12}{24}$ pure, 18-karat gold is $\frac{18}{24}$ pure, and so on). How many grams of each alloy must be mixed to obtain 10 grams of 14-karat gold?

SOLUTION Let

$x =$ number of grams of 12-karat gold used

$y =$ number of grams of 18-karat gold used

$x + y = 10$ Amount of new alloy.

$\frac{12}{24}x + \frac{18}{24}y = \frac{14}{24}(10)$ Pure gold present before mixing equals pure gold present after mixing.

$$x + y = 10$$
$$6x + 9y = 70$$

Multiply second equation by $\frac{24}{2}$ to simplify, and then solve using methods described above. (We use elimination here.)

$$\begin{aligned} -6x - 6y &= -60 \\ 6x + 9y &= 70 \\ \hline 3y &= 10 \\ y &= 3\tfrac{1}{3} \end{aligned}$$

grams of 18-karat alloy

$$\begin{aligned} x + 3\tfrac{1}{3} &= 10 \\ x &= 6\tfrac{2}{3} \end{aligned}$$

grams of 12-karat alloy

The checking of solutions is left to the reader.

PROBLEM 9

Repeat Example 9 using the fact that the jeweler has only 10-karat and pure gold in stock.

ANSWER $2\frac{6}{7}$ grams of pure gold and $7\frac{1}{7}$ grams of 10-karat gold

Exercise 25

The problems in this exercise are grouped in the following subject areas: business, chemistry, earth sciences, economics, geometry, domestic, life science, music, physics-engineering, psychology, and puzzles. The most difficult problems are double-starred (★★), moderately difficult problems are single-starred (★), and the easier problems are not marked.

BUSINESS

1. If 3 limes and 12 lemons cost 81 cents, and 2 limes and 5 lemons cost 42 cents, what is the cost of 1 lime and 1 lemon?

2. Find the capacity of each of 2 trucks if 3 trips of the larger and 4 trips of the smaller results in a total haul of 41 tons, and if 4 trips of the larger and 3 trips of the smaller results in a total haul of 43 tons.

★ **3.** Two companies have offered you a sales position. Both jobs are essentially the same, but one company pays a straight 8 percent commission and the other pays $51 per week plus 5 percent commission. The best salesmen with either company rarely have sales greater than $4,000 in any one week. Before accepting either offer, it would be helpful to know at what level of sales both companies pay the same and which of the companies pays more on either side of this level. Solve graphically and algebraically.

★ **4.** Solve Problem 3 with the straight-commission company paying 7 percent and the salary-plus-commission company paying $75 per week plus 4 percent commission.

CHEMISTRY

★ **5.** A chemist has two concentrations of hydrochloric acid in stock, a 50% solution and an 80% solution. How much of each should be taken to get 100 cubic centimeters of a 68% solution?

★ **6.** Repeat Problem 5 with the chemist starting with a 40% solution and a 90% solution.

DOMESTIC

7. A school put on a musical comedy and sold 1,000 tickets for a total of $650. If tickets were sold to students for 50 cents and to adults for $1, how many of each type were sold?

8. Repeat Problem 7 with student tickets costing 65 cents and adult tickets costing $1.30.

★ **9.** Wishing to log some flying time, you have rented an airplane for 2 hours. You decide to fly due east until you have to turn around in order to be back at the airport at the end of the 2 hours. The cruising speed of the plane is 120 miles/hour in still air.

(*A*) If there is a 30-miles/hour wind blowing from the east, how long should you head east before you turn around, and how long will it take you to get back?
(*B*) How far from the airport were you when you turned back?
(*C*) Answer (*A*) and (*B*) with the assumption that no wind is blowing.

★**10.** Repeat Problem 9 for a rental period of 5 hours.

EARTH SCIENCE

★★**11.** An earthquake emits a primary wave and secondary wave. Near the surface of the earth the primary wave travels at about 5 miles/second, and the secondary wave at about 3 miles/second. From the time lag between the two waves arriving at a given station, it is possible to estimate the distance to the quake. (The *epicenter* can be located by obtaining distance bearings at three or more stations.) Suppose a station measured a time difference of 16 seconds between the arrival of the two waves: how long did each wave travel, and how far was the earthquake from the station?

★★**12.** A ship using sound-sensing devices above and below water recorded a surface explosion 6 seconds sooner by its underwater device than its above-water device. Sound travels in air at about 1,100 feet/second and in sea water at about

5,000 feet/second. (*A*) How long did it take each sound wave to reach the ship? (*B*) How far was the explosion from the ship?

ECONOMICS

13. BREAKEVEN ANALYSIS It costs a book publisher \$12,000 to prepare a book for publication (artwork, plates, reviews, etc.); printing costs are \$3 per book. (*A*) If the book is sold to bookstores for \$7 a copy, how many copies must be sold to break even, and what are the cost and revenue for this number? HINT: Solve the system

$$C = 12{,}000 + 3n$$

$$R = 7n$$

$$R = C$$

(*B*) Graph the first two equations on the same coordinate system for $0 \le n \le 20{,}000$. Interpret the regions between the lines to the left and to the right of the breakeven point.

★**14.** BREAKEVEN ANALYSIS The management of a small plant that manufactures high-performance surfboards estimates fixed costs (rent, labor, insurance, etc.) to be \$120 per day, and variable costs (materials, packaging, etc.) to be \$60 per board. (*A*) If the company sells each board for \$90, how many must be sold each day to break even, and what are the costs and returns for this number?
(*B*) Graph the cost equation and the return equation on the same coordinate system, assuming a plant capacity of 12 boards per day, and interpret the regions between the lines to the left of the breakeven point and to the right of the breakeven point.

15. SUPPLY AND DEMAND An important problem in economic studies has to do with supply and demand. The quantity of a product that people are willing to buy on a given day generally depends on its price; similarly, the quantity of a product that a supplier is willing to sell on a given day also depends on the price the supplier is able to get for the product.

Let us assume that in a small town on a particular day the demand (in pounds) for hamburger is given by

$$d = 2{,}400 - 1{,}200p \qquad \$0.50 \le p \le \$1.75$$

and the supply by

$$s = -900 + 1{,}800p \qquad \$0.50 \le p \le \$1.75$$

Using these equations, we see that at \$1.50 per pound the people in the town would only be willing to buy 600 pounds of hamburger on that day, whereas the suppliers would be willing to supply 1,800 pounds. Hence, the supply would exceed the demand and force prices down. On the other hand, if the price were 75 cents per pound, the people would then be willing to buy 1,500 pounds of hamburger on that day, but the supplier would only be willing to sell 450 pounds. Thus the demand would exceed the supply, and the prices would go up. At what price would hamburger stabilize for the day; that is, at what price would the supply actually equal the demand ($s = d$)?
(*A*) Solve graphically by graphing the supply-and-demand equations on the same

coordinate system. (The point of intersection of the two graphs is called the *equilibrium point.*)
(*B*) Solve algebraically.
(*C*) Interpret the graph to the left and to the right of the equilibrium point.

16. SUPPLY AND DEMAND In a particular city the weekly supply s and demand d for popular stereo records, relative to average price per record p, is given by the equations

$$d = 5000 - 1{,}000p \qquad \$1.00 \le p \le \$4.00$$

$$s = -3000 + 3{,}000p \qquad \$1.00 \le p \le \$4.00$$

At what price would the supply equal the demand (equilibrium point)? (*A*) Solve graphically. (*B*) Solve algebraically.

GEOMETRY

17. If the sum of two angles in a right triangle is 90° and their difference is 14°, find the two angles.

18. Find the dimensions of a rectangle with perimeter 72 inches if its length is 4 inches longer than its width.

LIFE SCIENCE

19. Suppose we have a sample of 1,236 children who have brown-eyed parents and suppose that all of these brown-eyed parents carry the recessive gene for blue eyes. According to Mendel's laws of heredity, we would expect there to be 3 times as many brown-eyed children as blue-eyed. How many offspring with each eye color would you expect in the 1,236 samples?

20. Repeat Problem 19 with a sample of 2,140.

⋆**21.** A biologist, in a nutrition experiment, wants to prepare a special diet for her experimental animals. She requires a food mixture that contains, among other things, 20 ounces of protein and 6 ounces of fat. She is able to purchase food mixes of the following compositions:

MIX	PROTEIN	FAT
A	20%	2%
B	10%	6%

How many ounces of each mix should she use to get the diet mix? Solve graphically and algebraically.

⋆**22.** An experimental farm placed an order with a chemical company for a chemical fertilizer that would contain, among other things, 120 pounds of nitrogen and 90 pounds of phosphoric acid. The company had two mixtures on hand with the following compositions:

MIXTURE	NITROGEN	PHOSPHORIC ACID
A	20%	10%
B	6%	6%

How many pounds of each mixture should the chemist mix to fill the order?

MUSIC

23. If a taut guitar string is divided into two parts so that 4 times one part is equal to 5 times the other part, a major third will result. How should a 36-inch string be divided to produce a major third?

24. If a taut guitar string is divided into two parts so that 5 times one part is equal to 8 times the other part, a minor sixth will result. How would you divide a 39-inch string to produce a minor sixth.

PHYSICS-ENGINEERING

⋆⋆**25.** In a Gemini-Agena rendezvous flight preparatory to placing people on the moon, the Agena passed over a tracking station in Carnoarvon, Australia, 6 minutes (0.1 hour) before the pursuing, astronaut-carrying Gemini. If the Agena was traveling at 17,000 miles/hour and the Gemini at 18,700 miles/hour, how long did it take the Gemini (after passing the tracking station) to rendezvous with the Agena, and how far from the tracking station, in the direction of motion, did this take place?

PSYCHOLOGY

⋆**26.** A psychologist trained a group of rats (in an experiment on motivation and avoidance) to run down a narrow passage in a cage to receive food in a goal box. He put a harness on each rat, and connected it to an overhead wire that was attached to a scale. In this way he could place the rat at different distances from the food and measure the pull (in grams) of the rat toward the food. He found that a relation between motivation and distance was given approximately by the equation $p = -\frac{1}{5}d + 70$, $30 \le d \le 175$, where p is pull in grams, and d is distance from goal box in centimeters.

The psychologist then replaced the food with a mild electric shock, and with the same apparatus he was able to measure the avoidance strength relative to the distance from the object to be avoided. He found that the avoidance strength was given approximately by $a = -\frac{4}{3}d + 230$, $30 \le d \le 175$, where a is avoidance measured in grams, and d is distance from goal box in centimeters.

If the rat was trained in both experiments, at what distance from the goal box would the approach and avoidance strength be the same? Solve algebraically and graphically. What do you predict that the rat would do if placed to the right of this point (assume goal box is on right)? To the left of this point?

PUZZLES

27. A bank gave you $1.50 in change consisting of only nickels and dimes. If there were 22 coins in all, how many of each type of coin did you receive?

28. A friend of yours came out of a post office having spent $1.32 on thirty 4- and 5-cent stamps. How many of each type were bought?

⋆**29.** A packing carton contains 144 small packages, some weighing 1/4 pound each and the others 1/2 pound each. How many of each type are in the carton if the total contents of the carton weighs 51 pounds?

⋆**30.** If 1 chemical flask and 4 mixing dishes balance 12 test tubes and 2 mixing dishes, and if 2 flasks balance 4 test tubes and 6 mixing dishes, then how many test tubes will balance 1 flask, and how many test tubes will balance 1 mixing dish?

5.3 Systems of Linear Equations in Three Variables

Having learned how to solve systems of linear equations in two variables, there is no reason to stop there. Systems of the form

$$
\begin{aligned}
a_1x + b_1y + c_1z &= k_1 \\
a_2x + b_2y + c_2z &= k_2 \\
a_3x + b_3y + c_3z &= k_3
\end{aligned} \tag{1}
$$

as well as higher order systems are encountered frequently and are worth studying. A triplet of numbers $x = x_0$, $y = y_0$, and $z = z_0$ [also written as an ordered triplet (x_0, y_0, z_0)] is a *solution* of system (1) if each equation is satisfied by this triplet. The set of all such ordered triplets of numbers is called the *solution set* of the system. Two systems are said to be *equivalent* if they have the same solution set.

A particularly easy form of (1) to solve is called a triangular system. Any system (1) is a *triangular system* if

1. One of the equations in the systems yields the value of one of the unknowns directly.
2. After substituting this value into a second equation, the value of a second unknown is determined.
3. After substituting these two values into the third equation, the third unknown is determined.

EXAMPLE 10 Solve the triangular system:

$$
\begin{aligned}
3y &= -3 \\
2y - 3z &= 4 \\
2x + y - z &= 4
\end{aligned}
$$

SOLUTION Solve for y in the first equation.

$$
\begin{aligned}
3y &= -3 \\
y &= -1
\end{aligned}
$$

Substitute this into the second equation to solve for z.

$$
\begin{aligned}
2(-1) - 3z &= 4 \\
-2 - 3z &= 4 \\
z &= -2
\end{aligned}
$$

Now substitute $y = -1$ and $z = -2$ into the third equation to find x:

$$2x + (-1) - (-2) = 5$$

$$2x = 4$$

$$x = 2$$

The solution of the system is $x = 2$, $y = -1$, $z = -2$.

PROBLEM 10 Solve the triangular system:

$$2x = -6$$

$$x - 4z = 5$$

$$3x - y + 2z = -3$$

ANSWER $x = -3$, $y = -10$, $z = -2$

Unfortunately, most of the systems one encounters are not triangular systems. By the use of suitable transformations, however, one is often able to change a complicated system into an equivalent triangular system whose solution is easily determined. The next theorem provides the means to carry out these transformations.

THEOREM 2 A system of equations is transformed into an equivalent system if:

(*A*) The position of any two equations is interchanged.

(*B*) Any equation in the system is multiplied by a nonzero constant.

(*C*) Any equation is replaced by the sum of it and a nonzero constant multiple of another equation in the system.

How is this theorem used to produce equivalent triangular systems? First, we eliminate the same variable from two equations, then from these two equations we eliminate one of the two remaining variables from one of the equations. An example will clarify procedures.

EXAMPLE 11 Solve by transforming into an equivalent triangular system.

$$2x - 3y + z = 10$$

$$x + 2y + 4z = 12$$

$$3x - y - 2z = 1$$

SOLUTION Looking down the columns of coefficients of each of the variables, we find that each column includes a coefficient of either 1 or -1. This means, in this case, we can easily eliminate any one of the variables from two equations. Let us use the first equation to eliminate z from the second and third equations. Multiply (mentally) the first equation by 2 and add to the third equation. This changes the third equation but not the first (Theorem 2*C*).

$$\begin{aligned} 2x - 3y + z &= 10 \\ x + 2y + 4z &= 12 \\ 7x - 7y &= 21 \end{aligned}$$

Now, multiply the first equation by -4 and add to the second equation to eliminate z from the second equation. Also, to simplify the third equation, multiply through by $\frac{1}{7}$ (Theorem 2*B*).

$$\begin{aligned} 2x - 3y + z &= 10 \\ -7x + 14y &= -28 \\ x - y &= 3 \end{aligned}$$

Simplify the second equation by multiplying through by $-\frac{1}{7}$.

$$\begin{aligned} 2x - 3y + z &= 10 \\ x - 2y &= 4 \\ x - y &= 3 \end{aligned}$$

We are almost there. We now eliminate either x or y from one of the last two equations. Let us use the third equation to eliminate x from the second equation. Multiply the third equation by -1 and add to the second equation to obtain

$$\begin{aligned} 2x - 3y + z &= 10 \\ -y &= 1 \\ x - y &= 3 \end{aligned}$$

We now have a triangular system that is equivalent to the original system, and whose solution is easily determined. (The system even looks like a triangle if you interchange the last two equations.) Solving, we obtain

$$x = 2 \quad y = -1 \qquad \text{and} \qquad z = 3$$

PROBLEM 11 Solve by transforming into an equivalent triangular system.

$$\begin{aligned} 2x - y + z &= 0 \\ x + 2y - 2z &= 5 \\ 3x - 4y - 3z &= -5 \end{aligned}$$

ANSWER $x = 1,\ y = 2,\ z = 0$

If we encounter, in the process of transforming a system of equations into an equivalent triangular system, an equation that states a contradiction, such as $0 = -2$, then we must conclude that the system has no solution (that is, the system is inconsistent). If, on the other hand, one of the equations turns out to be $0 = 0$, then the system either has infinitely many solutions or it has none. It is generally easy to determine which at this stage of the solution process.

Exercise 26

A *Solve each triangular system:*

1.
$$\begin{aligned} 2x \qquad &= -6 \\ -x + 4y &= -1 \end{aligned}$$

2.
$$\begin{aligned} 5x - 2y &= -3 \\ 3y &= 12 \end{aligned}$$

3.
$$\begin{aligned} -2x \qquad\qquad &= 2 \\ x - 3y \qquad &= 2 \\ -x + 2y + 3z &= -7 \end{aligned}$$

4.
$$\begin{aligned} 2v + w &= -4 \\ u - 3v + 2w &= 9 \\ -v \qquad &= 3 \end{aligned}$$

Solve by first transforming into an equivalent triangular system using Theorem 2.

5.
$$\begin{aligned} u - 2v &= -4 \\ 5u + 6v &= -4 \end{aligned}$$

6.
$$\begin{aligned} 7s - 2t &= -1 \\ -s + 3t &= -8 \end{aligned}$$

7.
$$\begin{aligned} x - 4y \qquad &= 11 \\ 2x + y \qquad &= 4 \\ -x - 3y + z &= 3 \end{aligned}$$

8.
$$\begin{aligned} 2y - z &= 6 \\ y + 3z &= -4 \\ 2x - y + z &= -6 \end{aligned}$$

9.
$$\begin{aligned} 4s + t &= -3 \\ 3s + 2t &= 4 \\ 6r - 5s - 2t &= 0 \end{aligned}$$

10.
$$\begin{aligned} 2u \qquad + w &= -5 \\ u \qquad - 3w &= -6 \\ 4u + 2v - w &= -9 \end{aligned}$$

B **11.**
$$\begin{aligned} 2x + \qquad - z &= 5 \\ x - 2y - 2z &= 4 \\ 3x + 4y + 3z &= 3 \end{aligned}$$

12.
$$\begin{aligned} x - 3y + z &= 4 \\ -x + 4y - 4z &= 1 \\ 2x - y + 5z &= -3 \end{aligned}$$

13.
$$\begin{aligned} 2a + 4b + 3c &= 6 \\ a - 3b + 2c &= -7 \\ -a + 2b - c &= 5 \end{aligned}$$

14.
$$\begin{aligned} 3u - 2v + 3w &= 11 \\ 2u + 3v - 2w &= -5 \\ u + 4v - w &= -5 \end{aligned}$$

15.
$$\begin{aligned} 2x - 3y + 3z &= -15 \\ 3x + 2y - 5z &= 19 \\ 5x - 4y - 2z &= -2 \end{aligned}$$

16.
$$\begin{aligned} 3x - 2y - 4z &= -8 \\ 4x + 3y - 5z &= -5 \\ 6x - 5y + 2z &= -17 \end{aligned}$$

17. A circle in a rectangular coordinate system can be written in the form $x^2 + y^2 + Dx + Ey + F = 0$. Find D, E, and F so that the circle passes through $(-2, -1)$, $(-1, -2)$, and $(6, -1)$.

C *Solve by first transforming into an equivalent triangular system using Theorem 2.*

19.
$$\begin{aligned} x - 8y + 2z &= -1 \\ x - 3y + z &= 1 \\ 2x - 11y + 3z &= 2 \end{aligned}$$

20.
$$\begin{aligned} -x + 2y - z &= -4 \\ 4x + y - 2z &= 1 \\ x + y - z &= -1 \end{aligned}$$

In the next two problems solve for $1/x$, $1/y$, and $1/z$ first, then solve for x, y, and z.

21.
$$\begin{aligned} \frac{2}{x} + \frac{1}{y} + \frac{3}{z} &= -9 \\ \frac{1}{x} - \frac{4}{y} - \frac{2}{z} &= 15 \\ \frac{1}{x} + \frac{6}{y} + \frac{1}{z} &= -14 \end{aligned}$$

22.
$$\begin{aligned} \frac{1}{x} + \frac{3}{y} + \frac{1}{z} &= 9 \\ \frac{2}{x} - \frac{1}{y} - \frac{4}{z} &= 9 \\ -\frac{1}{x} - \frac{1}{y} + \frac{5}{z} &= -15 \end{aligned}$$

23.
$$\begin{aligned} 4w - x \qquad\qquad &= 5 \\ -3w + 2x - y \qquad &= -5 \\ 2w - 5x + 4y + 3z &= 13 \\ 2w + 2x - 2y - z &= -2 \end{aligned}$$

24.
$$\begin{aligned} 2r - s + 2t - u &= 5 \\ r - 2s + t + u &= 1 \\ -r + s - 3t - u &= -1 \\ -r - 2s + t + 2u &= -4 \end{aligned}$$

25. A zoologist, in an experiment involving mice, finds he needs a food mix that contains, among other things, 23 grams of protein, 6.2 grams of fat, and 16 grams of moisture. He has on hand mixes of the following compositions:

MIX	PROTEIN	FAT	MOISTURE
A	20%	2%	15%
B	10%	6%	10%
C	15%	5%	5%

How many grams of each mix should he use to get the desired diet mix?

26. A newspaper firm uses three printing presses, of different ages and capacities, to print the evening paper. With all three presses running, the paper can be printed in 2 hours. If the newest press breaks down, then the older two presses can print the paper in 4 hours; if the middle press breaks down, the newest and oldest together can print the paper in 3 hours. How long would it take each press alone to print the paper? HINT: Use $2/x + 2/y + 2/z = 1$ as one of the equations.

5.4 Determinants

We digress in this section from the central theme of this chapter—systems of linear equations—to introduce the useful mathematical notion of determinant. Determinants arise quite naturally in many areas in mathematics, including the solving of linear systems. We will consider a few of their uses in this and the next section.

SECOND-ORDER DETERMINANTS

A square array of four real numbers, such as

$$\begin{vmatrix} 2 & -3 \\ 5 & 1 \end{vmatrix}$$

is called a determinant of order 2. (It is important to note that the array of numbers is between parallel lines and not square brackets. If square brackets are used, then the symbol has another meaning.) The above determinant has two rows and two columns—rows are across and columns are up and down. Each number in the determinant is called an *element* of the determinant.

In general, we can symbolize a *second-order determinant* as follows:

$$\begin{vmatrix} a_{11} & a_{12} \\ a_{21} & a_{22} \end{vmatrix}$$

where we use a single letter with a double subscript to facilitate generalization to higher-order determinants. The first number indicates the row in which the element lies, and the second number indicates the column. Thus a_{21} is the element in the second row and first column, and a_{12} is the element in the first row and second column. Each second-order determinant represents a real number given by the formula:

$$\boxed{\begin{vmatrix} a_{11} & a_{12} \\ a_{21} & a_{22} \end{vmatrix} = a_{11}a_{22} - a_{21}a_{12}}$$

EXAMPLE 12

$$\begin{vmatrix} -1 & 2 \\ -3 & -4 \end{vmatrix} = (-1)(-4) - (-3)(2) = 4 - (-6) = 10$$

PROBLEM 12 Find $\begin{vmatrix} 3 & -5 \\ 4 & -2 \end{vmatrix}$

ANSWER 14

THIRD-ORDER DETERMINANTS

A determinant of order 3 is a square array of 9 elements, and represents a real number given by the formula

$$\begin{vmatrix} a_{11} & a_{12} & a_{13} \\ a_{21} & a_{22} & a_{23} \\ a_{31} & a_{32} & a_{33} \end{vmatrix} = a_{11}a_{22}a_{33} - a_{11}a_{32}a_{23} + a_{21}a_{32}a_{13} - a_{21}a_{12}a_{22} + a_{31}a_{12}a_{23} - a_{31}a_{22}a_{13} \tag{2}$$

Note that each term in the expansion on the right of (2) contains exactly one element from each row and each column. Don't panic! You do not need to memorize formula (2). After we introduce the ideas of "minor" and "cofactor," we will state a theorem that can be used to obtain the same result with much less memory strain.

The *minor of an element* in a third-order determinant is a second-order determinant obtained by deleting the row and column that contains the element. For example, in the determinant in formula (2)

$$\text{The minor of } a_{23} = \begin{vmatrix} a_{11} & a_{12} & a_{13} \\ a_{21} & a_{22} & a_{23} \\ a_{31} & a_{32} & a_{33} \end{vmatrix} = \begin{vmatrix} a_{11} & a_{12} \\ a_{31} & a_{32} \end{vmatrix}$$

$$\text{The minor of } a_{32} = \begin{vmatrix} a_{11} & a_{12} & a_{13} \\ a_{21} & a_{22} & a_{23} \\ a_{31} & a_{32} & a_{33} \end{vmatrix} = \begin{vmatrix} a_{11} & a_{13} \\ a_{21} & a_{23} \end{vmatrix}$$

A quantity closely associated with the minor of an element is the cofactor of an element. The *cofactor of an element* a_{ij} (from the ith row and jth column) is the product of the minor of a_{ij} and $(-1)^{i+j}$. That is,

$$\text{Cofactor of } a_{ij} = (-1)^{i+j}\,(\text{minor of } a_{ij})$$

Thus, a cofactor of an element is nothing more than a signed minor. The sign is determined by raising -1 to a power that is the sum of the numbers indicating the row and column in which the element lies. Note that $(-1)^{i+j}$ is -1 if $i+j$ is odd and 1 if $i+j$ is even. Referring again to the determinant in formula (2),

$$\text{The cofactor of } a_{23} = (-1)^{2+3}\begin{vmatrix} a_{11} & a_{12} \\ a_{31} & a_{32} \end{vmatrix} = -\begin{vmatrix} a_{11} & a_{12} \\ a_{31} & a_{32} \end{vmatrix}$$

$$\text{The cofactor of } a_{11} = (-1)^{1+1}\begin{vmatrix} a_{22} & a_{23} \\ a_{32} & a_{33} \end{vmatrix} = \begin{vmatrix} a_{22} & a_{23} \\ a_{32} & a_{33} \end{vmatrix}$$

EXAMPLE 13 Find the cofactor of -2 and 5 in the determinant

$$\begin{vmatrix} -2 & 0 & 3 \\ 1 & -6 & 5 \\ -1 & 2 & 0 \end{vmatrix}$$

SOLUTION The cofactor of $-2 = (-1)^{1+1} \begin{vmatrix} -6 & 5 \\ 2 & 0 \end{vmatrix} = \begin{vmatrix} -6 & 5 \\ 2 & 0 \end{vmatrix}$

$$= (-6)(0) - (2)(5) = -10$$

The cofactor of $5 = (-1)^{2+3} \begin{vmatrix} -2 & 0 \\ -1 & 2 \end{vmatrix} = - \begin{vmatrix} -2 & 0 \\ -1 & 2 \end{vmatrix}$

$$= -[(-2)(2) - (-1)(0)] = 4$$

PROBLEM 13 Find the cofactors of 2 and 3 in the determinant in Example 13.

ANSWER 13, -4

NOTE: The sign in front of the minor, $(-1)^{i+j}$, can be determined rather mechanically by using a checkerboard pattern of $+$ and $-$ signs over the determinant, starting with $+$ in the upper left-hand corner:

$$\begin{matrix} + & - & + \\ - & + & - \\ + & - & + \end{matrix}$$

Use either the checkerboard or the exponent method, whichever is easier for you, to determine the sign in front of the minor.

Now we are ready for the central theorem of this section. It will provide us with an efficient means of evaluating third-order determinants. In addition, it is worth noting that the theorem generalizes completely to include determinants of arbitrary order.

THEOREM 3 The value of a determinant of order 3 is the sum of three products obtained by multiplying each element of any one row (or each element of any one column) by its cofactor.

To prove this theorem we must show that the expansions indicated by the theorem for any row or any column (six cases) produce the expression on the right of formula (2) above. Proofs of special cases of this theorem are left to the C exercises.

EXAMPLE 14 Evaluate by expanding by (A) the first row and (B) the second column:

$$\begin{vmatrix} 2 & -2 & 0 \\ -3 & 1 & 2 \\ 1 & -3 & -1 \end{vmatrix}$$

SOLUTION (A)

$$\begin{vmatrix} 2 & -2 & 0 \\ -3 & 1 & 2 \\ 1 & -3 & -1 \end{vmatrix} = a_{11}\begin{pmatrix}\text{cofactor} \\ \text{of } a_{11}\end{pmatrix} + a_{12}\begin{pmatrix}\text{cofactor} \\ \text{of } a_{12}\end{pmatrix} + a_{13}\begin{pmatrix}\text{cofactor} \\ \text{of } a_{13}\end{pmatrix}$$

$$= 2\left((-1)^{1+1}\begin{vmatrix} 1 & 2 \\ -3 & -1 \end{vmatrix}\right) + (-2)\left((-1)^{1+2}\begin{vmatrix} -3 & 2 \\ 1 & -1 \end{vmatrix}\right) + 0$$

$$= (2)(1)[(1)(-1) - (-3)(2)] + (-2)(-1)[(-3)(-1) - (1)(2)]$$

$$= (2)(5) + (2)(1) = 12$$

(B)

$$\begin{vmatrix} 2 & -2 & 0 \\ -3 & 1 & 2 \\ 1 & -3 & -1 \end{vmatrix} = a_{12}\begin{pmatrix}\text{cofactor} \\ \text{of } a_{12}\end{pmatrix} + a_{22}\begin{pmatrix}\text{cofactor} \\ \text{of } a_{22}\end{pmatrix} + a_{32}\begin{pmatrix}\text{cofactor} \\ \text{of } a_{32}\end{pmatrix}$$

$$= (-2)\left((-1)^{1+2}\begin{vmatrix} -3 & 2 \\ 1 & -1 \end{vmatrix}\right) + (1)\left((-1)^{2+2}\begin{vmatrix} 2 & 0 \\ 1 & -1 \end{vmatrix}\right)$$

$$+ (-3)\left((-1)^{3+2}\begin{vmatrix} 2 & 0 \\ -3 & 2 \end{vmatrix}\right)$$

$$= (-2)(-1)[(-3)(-1) - (1)(2)] + (1)(1)[(2)(-1) - (1)(0)]$$

$$+ (-3)(-1)[(2)(2) - (-3)(0)]$$

$$= (2)(1) + (1)(-2) + (3)(4)$$

$$= 12$$

PROBLEM 14 Evaluate by expanding by (A) the first row and (B) the third column:

$$\begin{vmatrix} 2 & 1 & -1 \\ -2 & -3 & 0 \\ -1 & 2 & 1 \end{vmatrix}$$

ANSWER (A) 3 (B) 3

It should now be clear that we can greatly reduce the work involved in evaluating a determinant by choosing to expand by a row or column with the greatest number of zeros.

Where are determinants used? Many equations and formulas have particularly simple and compact representations in determinant form that are easily remembered. See, for example, Problems 39 to 42 in Exercise 27, and Cramer's rule in the next section.

Exercise 27

A *Evaluate each second-order determinant.*

1. $\begin{vmatrix} 2 & 4 \\ 3 & -1 \end{vmatrix}$

2. $\begin{vmatrix} 2 & 2 \\ -3 & 1 \end{vmatrix}$

3. $\begin{vmatrix} 5 & -4 \\ -2 & 2 \end{vmatrix}$

4. $\begin{vmatrix} 6 & -2 \\ -1 & -3 \end{vmatrix}$

5. $\begin{vmatrix} 3 & -3.1 \\ -2 & 1.2 \end{vmatrix}$

6. $\begin{vmatrix} -1.4 & 3 \\ -0.5 & -2 \end{vmatrix}$

Given the determinant

$$\begin{vmatrix} a_{11} & a_{12} & a_{13} \\ a_{21} & a_{22} & a_{23} \\ a_{31} & a_{32} & a_{33} \end{vmatrix}$$

write the minor of each of the following elements.

7. a_{11}

8. a_{33}

9. a_{23}

10. a_{22}

Write the cofactor of each of the following elements.

11. a_{11}

12. a_{33}

13. a_{23}

14. a_{22}

Given the determinant

$$\begin{vmatrix} -2 & 3 & 0 \\ 5 & 1 & -2 \\ 7 & -4 & 8 \end{vmatrix}$$

write the minor of each of the following elements. (Leave answer in determinant form.)

15. a_{11}

16. a_{22}

17. a_{32}

18. a_{21}

Write the cofactor of each of the following elements and evaluate each.

19. a_{11} **20.** a_{22} **21.** a_{32} **22.** a_{21}

B *Evaluate each of the following determinants using cofactors.*

23. $\begin{vmatrix} 1 & 0 & 0 \\ -2 & 4 & 3 \\ 5 & -2 & 1 \end{vmatrix}$

24. $\begin{vmatrix} 2 & -3 & 5 \\ 0 & -3 & 1 \\ 0 & 6 & 2 \end{vmatrix}$

25. $\begin{vmatrix} 0 & 1 & 5 \\ 3 & -7 & 6 \\ 0 & -2 & -3 \end{vmatrix}$

26. $\begin{vmatrix} 4 & -2 & 0 \\ 9 & 5 & 4 \\ 1 & 2 & 0 \end{vmatrix}$

27. $\begin{vmatrix} 4 & -4 & 6 \\ 2 & 8 & -3 \\ 0 & -5 & 0 \end{vmatrix}$

28. $\begin{vmatrix} 3 & -2 & -8 \\ -2 & 0 & -3 \\ 1 & 0 & -4 \end{vmatrix}$

29. $\begin{vmatrix} -1 & 2 & -3 \\ -2 & 0 & -6 \\ 4 & -3 & 2 \end{vmatrix}$

30. $\begin{vmatrix} 0 & 2 & -1 \\ -6 & 3 & 1 \\ 7 & -9 & -2 \end{vmatrix}$

31. $\begin{vmatrix} 1 & 4 & 1 \\ 1 & 1 & -2 \\ 2 & 1 & -1 \end{vmatrix}$

32. $\begin{vmatrix} 3 & 2 & 1 \\ -1 & 5 & 1 \\ 2 & 3 & 1 \end{vmatrix}$

33. $\begin{vmatrix} 1 & 4 & 3 \\ 2 & 1 & 6 \\ 3 & -2 & 9 \end{vmatrix}$

34. $\begin{vmatrix} 4 & -6 & 3 \\ -1 & 4 & 1 \\ 5 & -6 & 3 \end{vmatrix}$

C *Assuming Theorem 3 applies to determinants of arbitrary order, use it to evaluate the following fourth- and fifth-order determinants.*

35. $\begin{vmatrix} 0 & 1 & 0 & 1 \\ 2 & 4 & 7 & 6 \\ 0 & 3 & 0 & 1 \\ 0 & 6 & 2 & 5 \end{vmatrix}$

36. $\begin{vmatrix} 2 & 6 & 1 & 7 \\ 0 & 3 & 0 & 0 \\ 3 & 4 & 2 & 5 \\ 0 & 9 & 0 & 2 \end{vmatrix}$

37. $\begin{vmatrix} 2 & 0 & 0 & 0 & 0 \\ 0 & 3 & 0 & 0 & 0 \\ 0 & 0 & 2 & 0 & 0 \\ 0 & 0 & 0 & 1 & 0 \\ 0 & 0 & 0 & 0 & 4 \end{vmatrix}$

38. $\begin{vmatrix} -2 & 0 & 0 & 0 & 0 \\ 9 & -1 & 0 & 0 & 0 \\ 2 & 1 & 3 & 0 & 0 \\ -1 & 4 & 2 & 2 & 0 \\ 7 & -2 & 3 & 5 & 5 \end{vmatrix}$

39. Show that $\begin{vmatrix} x & y & 1 \\ 2 & 3 & 1 \\ -1 & 2 & 1 \end{vmatrix} = 0$ is the equation of a line that passes through (2, 3) and (−1, 2).

40. Show that $\begin{vmatrix} x & y & 1 \\ x_1 & y_1 & 1 \\ x_2 & y_2 & 1 \end{vmatrix} = 0$ is the equation of a line that passes through (x_1, y_1) and (x_2, y_2).

41. In analytic geometry it is shown that the area of a triangle with vertices (x_1, y_1), (x_2, y_2), and (x_3, y_3) is the absolute value of

$$\frac{1}{2}\begin{vmatrix} x_1 & y_1 & 1 \\ x_2 & y_2 & 1 \\ x_3 & y_3 & 1 \end{vmatrix}$$

Use this result to find the area of a triangle with vertices (−1, 4), (4, 8), (1, 1).

42. Find the area of a triangle with vertices (−1, 2), (2, 5), and (6, −3). (See the preceding problem.)

43. Prove one case of Theorem 3 by expanding the left side of formula (2) using the first row and cofactors to obtain the right side.

44. Prove one case of Theorem 3 by expanding the left side of formula (2) using the second column and cofactors to obtain the right side.

5.5 Cramer's Rule

Now let us see how determinants arise rather naturally in the process of solving systems of linear equations. We will start by investigating two equations and two unknowns, and then extend any results to three equations and three unknowns.

Instead of thinking of each system of linear equations in two unknowns as a different problem, let us see what happens when we attempt to solve the general system

$$a_{11}x + a_{12}y = k_1 \qquad (1A)$$

$$a_{21}x + a_{22}y = k_2 \qquad (1B)$$

once and for all in terms of the unspecified real constants a_{11}, a_{12}, a_{21}, a_{22}, k_1, and k_2.

We proceed by multiplying equations (1*A*) and (1*B*) by suitable constants so that when the resulting equations are added, left side to left side and right side to right side, one of the variables drops out. Suppose we choose to eliminate y, what constants should we use to make the coefficients of y the same except for the signs? Multiply (1*A*) by a_{22} and equation (1*B*) by $-a_{12}$, then add.

$$\begin{array}{lr} a_{22}(1A): & a_{11}a_{22}x + a_{12}a_{22}y = k_1a_{22} \\ -a_{12}(1B): & -a_{21}a_{12}x - a_{12}a_{22}y = -k_2a_{12} \\ \hline & a_{11}a_{22}x - a_{21}a_{12}x + 0y = k_1a_{22} - k_2a_{12} \\ & (a_{11}a_{22} - a_{21}a_{12})x = k_1a_{22} - k_2a_{12} \\ & x = \dfrac{k_1a_{22} - k_2a_{12}}{a_{11}a_{22} - a_{21}a_{12}} \qquad a_{11}a_{22} - a_{21}a_{12} \neq 0 \end{array}$$

What do the numerator and denominator remind you of? From your experience with determinants in the last section you should recognize these expressions as

$$x = \frac{\begin{vmatrix} k_1 & a_{12} \\ k_2 & a_{22} \end{vmatrix}}{\begin{vmatrix} a_{11} & a_{12} \\ a_{21} & a_{22} \end{vmatrix}}$$

Similarly, starting with system (1) and eliminating x (this is left as an exercise), we obtain

$$y = \frac{\begin{vmatrix} a_{11} & k_1 \\ a_{21} & k_2 \end{vmatrix}}{\begin{vmatrix} a_{11} & a_{12} \\ a_{21} & a_{22} \end{vmatrix}}$$

These results are summarized in the following theorem, which is named after the Swiss mathematician, G. Cramer (1704–1752):

THEOREM 4

(Cramer's rule for two equations and two unknowns.) Given the system

$$a_{11}x + a_{12}y = k_1$$
$$a_{21}x + a_{22}y = k_2$$

with

$$D = \begin{vmatrix} a_{11} & a_{12} \\ a_{21} & a_{22} \end{vmatrix} \neq 0,$$

then

$$x = \frac{\begin{vmatrix} k_1 & a_{12} \\ k_2 & a_{22} \end{vmatrix}}{D} \quad \text{and} \quad y = \frac{\begin{vmatrix} a_{12} & k_1 \\ a_{21} & k_2 \end{vmatrix}}{D}$$

It is easy to remember these determinant formulas for x and y if one observes the following:

1. Determinant D is formed from the coefficients of x and y, keeping the same relative position in the determinant as found in the system.
2. Determinant D appears in the denominator for x and in the denominator for y.
3. The numerator for x can be obtained from D by replacing the coefficients of x, a_{11}, and a_{21}, with the constants k_1 and k_2.
4. The numerator for y can be obtained from D by replacing the coefficients of y, a_{12}, and a_{22}, with the constants k_1 and k_2.

The determinant D is called the *coefficient determinant*. If $D \neq 0$, then the system has exactly one solution, which is given by Cramer's rule. If, on the other hand, $D = 0$, then it can be shown that the system is either inconsistent or dependent; that is, the system either has no solutions or has an infinite number of solutions.

EXAMPLE 15

Solve using Cramer's rule:

$$2x - 3y = 7$$
$$-3x + y = -7$$

SOLUTION

$$D = \begin{vmatrix} 2 & -3 \\ -3 & 1 \end{vmatrix} = -7$$

$$x = \frac{\begin{vmatrix} 7 & -3 \\ -7 & 1 \end{vmatrix}}{-7} = \frac{-14}{-7} = 2 \qquad y = \frac{\begin{vmatrix} 2 & 7 \\ -3 & -7 \end{vmatrix}}{-7} = \frac{7}{-7} = -1$$

PROBLEM 15

Solve using Cramer's rule:

$$3x + 2y = -3$$
$$-4x + 3y = -13$$

ANSWER $x = 1,\ y = -3$

Cramer's rule generalizes completely for any size linear system that has the same number of unknowns as equations. We state without proof the rule for three equations and three unknowns.

THEOREM 5

(*Cramer's rule for three equations and three unknowns.*) Given the system

$$a_{11}x + a_{12}y + a_{13}z = k_1$$
$$a_{21}x + a_{22}y + a_{23}z = k_2$$
$$a_{31}x + a_{32}y + a_{33}z = k_3$$

with

$$D = \begin{vmatrix} a_{11} & a_{12} & a_{13} \\ a_{21} & a_{22} & a_{23} \\ a_{31} & a_{32} & a_{33} \end{vmatrix} \neq 0$$

then

$$x = \frac{\begin{vmatrix} k_1 & a_{12} & a_{13} \\ k_2 & a_{22} & a_{23} \\ k_3 & a_{32} & a_{33} \end{vmatrix}}{D} \qquad y = \frac{\begin{vmatrix} a_{11} & k_1 & a_{13} \\ a_{21} & k_2 & a_{23} \\ a_{31} & k_3 & a_{33} \end{vmatrix}}{D} \qquad z = \frac{\begin{vmatrix} a_{11} & a_{12} & k_1 \\ a_{21} & a_{22} & k_2 \\ a_{31} & a_{32} & k_3 \end{vmatrix}}{D}$$

Once again, notice how the numerators are related to the coefficient determinant D. In each case the k's replace the coefficients of the variable being solved.

EXAMPLE 16

Use Cramer's rule to solve

$$\begin{aligned} x + y \quad &= 1 \\ 3y - z &= -4 \\ x \quad + z &= 3 \end{aligned}$$

SOLUTION

$$D = \begin{vmatrix} 1 & 1 & 0 \\ 0 & 3 & -1 \\ 1 & 0 & 1 \end{vmatrix} = 2$$

$$x = \frac{\begin{vmatrix} 1 & 1 & 0 \\ -4 & 3 & -1 \\ 3 & 0 & 1 \end{vmatrix}}{2} = \frac{4}{2} = 2 \qquad y = \frac{\begin{vmatrix} 1 & 1 & 0 \\ 0 & -4 & -1 \\ 1 & 3 & 1 \end{vmatrix}}{2} = \frac{-2}{2} = -1$$

$$z = \frac{\begin{vmatrix} 1 & 1 & 1 \\ 0 & 3 & -4 \\ 1 & 0 & 3 \end{vmatrix}}{2} = \frac{2}{2} = 1$$

PROBLEM 16 Use Cramer's rule to solve

$$\begin{aligned} 3x \phantom{{}-y} - z &= 5 \\ x - y + z &= 0 \\ x + y \phantom{{}+z} &= 0 \end{aligned}$$

ANSWER $x = 1,\ y = -1,\ z = -2$

In practice, Cramer's rule is rarely used to solve systems of order higher than 2 or 3; more efficient methods are available, including the triangular method discussed in Sec. 5.3. Cramer's rule is, however, a valuable tool in theoretical mathematics.

Exercise 28

Solve, using Cramer's rule:

A

1. $x + 2y = 1$
$x + 3y = -1$

2. $x + 2y = 3$
$x + 3y = 5$

3. $2x + y = 1$
$5x + 3y = 2$

4. $x + 3y = 1$
$2x + 8y = 0$

5. $2x - y = -3$
$-x + 3y = 4$

6. $2x + y = 1$
$5x + 3y = 2$

B

7. $x + y = 0$
$2y + z = -5$
$-x + z = -3$

8. $x + y = -4$
$2y + z = 0$
$-x + z = 5$

9.
$$\begin{aligned} x + y \quad &= 1 \\ 2y + z &= 0 \\ -x + \quad z &= 0 \end{aligned}$$

10.
$$\begin{aligned} x + y \quad &= -4 \\ 2y + z &= 3 \\ -x + \quad z &= 7 \end{aligned}$$

11.
$$\begin{aligned} y + z &= -4 \\ x + \quad 2z &= 0 \\ x - y \quad &= 5 \end{aligned}$$

12.
$$\begin{aligned} x \quad - z &= 2 \\ 2x - y \quad &= 8 \\ x + y + z &= 2 \end{aligned}$$

13.
$$\begin{aligned} 2y - z &= -4 \\ x - y - z &= 0 \\ x - y + 2z &= 6 \end{aligned}$$

14.
$$\begin{aligned} 2x + y \quad &= 2 \\ x - y + z &= -1 \\ x + y + z &= -1 \end{aligned}$$

C *It is clear that $x = 0$, $y = 0$, $z = 0$ is a solution to each of the following systems. Use Cramer's rule to determine if this solution is unique.* (HINT: *If $D \neq 0$ what can you conclude? If $D = 0$ what can you conclude?*)

15.
$$\begin{aligned} x - 4y + 9z &= 0 \\ 4x - y + 6z &= 0 \\ x - y + 3z &= 0 \end{aligned}$$

16.
$$\begin{aligned} 3x - y + 3z &= 0 \\ 5x + 5y - 9z &= 0 \\ -2x + y - 3z &= 0 \end{aligned}$$

17. Prove Theorem 4 for y.

Exercise 29 Chapter Review

A **1.** Solve graphically:
$$\begin{aligned} x - y &= 5 \\ x + y &= 7 \end{aligned}$$

2. Solve by elimination method:
$$\begin{aligned} 2x + 3y &= 7 \\ 3x - y &= 5 \end{aligned}$$

3. $\begin{vmatrix} 2 & -3 \\ -5 & -1 \end{vmatrix} = ?$

4. $\begin{vmatrix} 2 & 3 & -4 \\ 0 & 5 & 0 \\ 1 & -4 & -2 \end{vmatrix} = ?$

5. Solve Problem 2 using Cramer's rule.

6. Solve, using any method:
$$\begin{aligned} 2x - 3y &= -3 \\ 3x + y &= 12 \end{aligned}$$

7. Solve, using two-equation–two-unknown methods: If you have 30 nickels and dimes in your pocket worth \$2.30, how many of each do you have?

B

8. Solve $\begin{aligned} 3x - 2y &= 6 \\ x + 4y &= 16 \end{aligned}$ graphically.

9. Solve $\begin{aligned} 5m - 3n &= 4 \\ -2m + 4n &= -10 \end{aligned}$ by elimination method.

10. $$\frac{a_1k_2 - a_2k_1}{a_1b_2 - a_2b_1} = \frac{\begin{vmatrix} ? & K_1 \\ ? & K_2 \end{vmatrix}}{\begin{vmatrix} a_1 & ? \\ a_2 & ? \end{vmatrix}}$$

11. Solving Problem 9 using Cramer's rule.

12. Solve, using any method: $\begin{aligned} 3x - 2y &= -1 \\ -6x + 4y &= 3 \end{aligned}$

13. $\begin{vmatrix} 1 & 2 & 3 \\ 2 & 0 & 1 \\ -1 & -3 & 4 \end{vmatrix} = ?$

14. $\begin{vmatrix} 2 & -1 & 1 \\ -3 & 5 & 2 \\ 1 & -2 & 4 \end{vmatrix} = ?$

15. Solve: $\begin{aligned} 3x - 2y - 7z &= -6 \\ -x + 3y + 2z &= -1 \\ x + 5y + 3z &= 3 \end{aligned}$

by transforming the system into an equivalent triangular system.

16. Solve: $\begin{aligned} y + 2z &= 4 \\ x - z &= -2 \\ x + y &= 1 \end{aligned}$ using Cramer's rule.

17. Six thousand dollars is to be invested, part at 10 percent and the rest at 6 percent. How much should be invested at each rate if the total annual return from both investments is to be $440? Solve, using two-equation–two-unknown methods.

C

18. Solve: $\begin{aligned} 2x - 6y &= -3 \\ \tfrac{2}{3}x + 2y &= 1 \end{aligned}$ graphically.

19. Solve: $\begin{aligned} x - 4y &= 12 \\ -\frac{x}{4} + y &= 4 \end{aligned}$ by elimination method.

20. Solve Problem 15 using Cramer's rule.

21. Does the following system have solutions other than the trivial solution $x = 0$, $y = 0$, $z = 0$? Explain.

$$2x - y + z = 0$$
$$x + y - z = 0$$
$$4x + y - z = 0$$

22. Solve, using two-equation–two-unknown methods: A chemist has two concentrations of acid in stock, a 40% and a 70% solution. How much of each should she take to get 100 grams of a 49% solution?

23. A container contains 120 packages. Some of the packages weigh $\frac{1}{2}$ pound each and the rest weigh $\frac{1}{3}$ pound each. If the total contents of the container weighs 48 pounds, how many are there of each type package? Solve, using two-equation–two-unknown methods.

Chapter 6

ALGEBRAIC FRACTIONS

6.1 Multiplication and Division

Quotients of polynomials are called *rational expressions*. For example,

$$\frac{1}{x} \qquad \frac{1}{y-3} \qquad \frac{x-2}{x^2-2x+5} \qquad \frac{x^2-3xy+y^2}{3x^3y^4}$$

are all rational expressions. (Recall that a nonzero constant is a polynomial of degree 0.)

Each rational expression involving polynomials with real coefficients names a real number for real number replacements of the variables, division by 0 excluded. Hence, all properties of the real numbers apply to these expressions.

MULTIPLICATION OF RATIONAL EXPRESSIONS

In arithmetic we learned to multiply fractions by multiplying their numerators and their denominators. This is exactly what we do with rational expressions in general.

If P, Q, R, and S represent polynomials, then

$$\frac{P}{Q} \cdot \frac{R}{S} = \frac{PR}{QS}$$

where Q and S are not 0.[†]

EXAMPLE 1

$$\frac{x}{2y} \cdot \frac{x-3}{3y^2} = \frac{x(x-3)}{2y(3y^2)} = \frac{x^2-3x}{6y^3}$$

PROBLEM 1

Multiply: $\frac{3m}{m-2} \cdot \frac{5mn}{m+3}$

ANSWER $\frac{15m^2n}{m^2+m-6}$

Now to one of the basic theorems for rational expressions.

THEOREM 1

(*Fundamental principle of fractions.*) For each polynomial P, Q, and K

$$\frac{PK}{QK} = \frac{P}{Q}$$

where Q and K are not 0.

To prove this theorem, we have only to use the definition of multiplication above, going from right to left:

$$\frac{PK}{QK} = \frac{P}{Q} \cdot \frac{K}{K} = \frac{P}{Q} \cdot 1 = \frac{P}{Q}$$

In words Theorem 1 states that we may multiply the numerator and denominator of a rational form by a nonzero polynomial or divide the numerator and denominator by a nonzero polynomial. This theorem is behind all canceling used to reduce fractional forms to lower terms (canceling common factors from numerator and denominator). And, used from right to left, it is used to raise rational forms to higher terms (multiplying numerator and denominator by common factors). The latter operation is fundamental to the processes of addition and subtraction of rational forms.

[†]When we use equality here, we mean that for all real replacements of the variables involved in the polynomials, the left and right sides of the equation name the same number, except for replacements that lead to zero denominators.

EXAMPLE 2 (A) $\frac{36}{27} = \frac{\overset{1}{\cancel{9}} \cdot 4}{\underset{1}{\cancel{9}} \cdot 3} = \frac{4}{3}$ lower terms

(B) $\frac{6x^3}{9x} = \frac{\overset{1}{\cancel{(3x)}}(2x^2)}{\underset{1}{\cancel{(3x)}}\,3} = \frac{2x^2}{3}$ lower terms

(C) $\frac{8x(x-5)}{12(x+3)(x-5)} = \frac{\overset{1}{\cancel{4(x-5)}}2x}{\underset{1}{\cancel{4(x-5)}}3(x+3)} = \frac{2x}{3(x+3)}$ lower terms

(D) $\frac{3}{4} = \frac{3 \cdot 3}{4 \cdot 3} = \frac{9}{12}$ higher terms

(E) $\frac{2}{5x} = \frac{2(2xy^2)}{5x(2xy^2)} = \frac{4xy^2}{10x^2y^2}$ higher terms

(F) $\frac{2}{5(x+7)} = \frac{2(x-2)}{5(x+7)(x-2)}$ higher terms

PROBLEM 2 Replace question marks with appropriate symbols:

(A) $\frac{24}{32} = \frac{?}{4}$ (B) $\frac{8m^2}{12m^4} = \frac{2}{?}$

(C) $\frac{9y(y+2)}{6y^2(y-2)(y+2)} = \frac{3}{?}$ (D) $\frac{2}{3} = \frac{?}{12}$

(E) $\frac{7}{4x} = \frac{14xy}{?}$ (F) $\frac{3m}{4(m-3)} = \frac{?}{8m(m-3)(m+2)}$

ANSWER (A) 3 (B) $3m^2$ (C) $2y(y-2)$ (D) 8 (E) $8x^2y$
(F) $6m^2(m+2)$

Now let us consider a few slightly more complicated examples that involve multiplication and reducing to lowest terms. By reducing a rational form to *lowest terms* we mean dividing out (canceling) all common polynomial factors with integer coefficients from the numerator and denominator.

EXAMPLE 3 (A) $\frac{3a^2b}{4c^2d} \cdot \frac{8c^2d^3}{9ab^2} = \frac{(3a^2b) \cdot (8c^2d^3)}{(4c^2d) \cdot (9ab^2)} = \frac{24a^2bc^2d^3}{36ab^2c^2d} = \frac{(2ad^2)(\cancel{12abc^2d})}{(3b)(\cancel{12abc^2d})} = \frac{2ad^2}{3b}$

This process is easily shortened to the following when it is realized that, in effect, any factor in either numerator may "cancel" any like factor in either denominator. Thus,

$$\frac{\overset{1\cdot a\cdot 1}{\cancel{3}\cancel{a^2}\cancel{b}}}{\underset{1\cdot 1\cdot 1}{\cancel{4}\cancel{c^2}\cancel{d}}}\cdot\frac{\overset{2\cdot 1\cdot d^2}{\cancel{8}\cancel{c^2}\cancel{d^3}}}{\underset{3\cdot 1\cdot b}{\cancel{9}\cancel{d}\cancel{b^2}}}=\frac{2ad^2}{3b}$$

(B) $$(x^2-4)\cdot\frac{2x-3}{x+2}=\frac{\overset{1}{\cancel{(x+2)}}(x-2)}{1}\cdot\frac{(2x-3)}{\underset{1}{\cancel{(x+2)}}}=(x-2)(2x-3)$$

(C) $$\frac{4a^2-9b^2}{4a^2+12ab+9b^2}\cdot\frac{6a^2b}{8a^2b^2-12ab^3}=\frac{\overset{1}{\cancel{(2a-3b)}}\overset{1}{\cancel{(2a+3b)}}}{\underset{(2a+3b)}{\cancel{(2a+3b)^2}}}\cdot\frac{\overset{3a}{\cancel{6a^2b}}}{\underset{2b\quad 1}{\cancel{4ab^2}\cancel{(2a-3b)}}}$$

$$=\frac{3a}{2b(2a+3b)}$$

PROBLEM 3 Multiply and reduce to lowest terms:

(A) $\dfrac{4x^2y^3}{9w^2z}\cdot\dfrac{3wz^2}{2xy^4}$ (B) $\dfrac{x+5}{x^2-9}\cdot(x+3)$

(C) $\dfrac{x^2-9y^2}{x^2-6xy+9y^2}\cdot\dfrac{6x^2y}{2x^2+6xy}$

ANSWER (A) $\dfrac{2xz}{3wy}$ (B) $\dfrac{x+5}{x-3}$ (C) $\dfrac{3xy}{x-3y}$

DIVISION OF RATIONAL EXPRESSIONS

The following theorem on division follows directly from the general definition of division for real numbers (recall: $A \div B = Q$ if and only if $A = BQ$).

THEOREM 2 If P, Q, R, and S represent polynomials, then

$$\boxed{\frac{P}{Q}\div\frac{R}{S}=\frac{P}{Q}\cdot\frac{S}{R}}$$

where Q, S, and R are not 0.

In words, the theorem states that to divide one rational expression by another, invert the divisor and multiply. To prove the theorem we need only show that the product of $\dfrac{R}{S}$ and $\dfrac{P}{Q}\cdot\dfrac{S}{R}$ is $\dfrac{P}{Q}$, a problem left for the exercises.

EXAMPLE 4

(A) $\dfrac{6a^2b^3}{5cd} \div \dfrac{3a^2c}{10bd} = \dfrac{6a^2b^3}{5cd} \cdot \dfrac{10bd}{3a^2c} = \dfrac{4b^4}{c^2}$

(B) $(x+4) \div \dfrac{2x^2-32}{6xy} = \dfrac{x+4}{1} \cdot \dfrac{6xy}{2(x-4)(x+4)} = \dfrac{3xy}{x-4}$

(C) $\dfrac{10x^3y}{3xy+9y} \div \dfrac{4x^2-12x}{x^2-9} = \dfrac{10x^3y}{3y(x+3)} \cdot \dfrac{(x+3)(x-3)}{4x(x-3)} = \dfrac{5x^2}{6}$

PROBLEM 4 Divide and reduce to lowest terms:

(A) $\dfrac{8w^2z^2}{9x^2y} \div \dfrac{4wz}{6xy^2}$ (B) $\dfrac{2x^2-8}{4x} \div (x+2)$

(C) $\dfrac{x^2-4x+4}{4x^2y-8xy} \div \dfrac{x^2+x-6}{6x^2+18x}$

ANSWER (A) $\dfrac{4wz}{3x}$ (B) $\dfrac{x-2}{2x}$ (C) $\dfrac{3}{2y}$

Exercise 30

In answers do not change improper fractions to mixed fractions; that is, write $\frac{7}{2}$, *not* $3\frac{1}{2}$.

A *Replace question marks with appropriate symbols.*

1. $\dfrac{36y}{54y} = \dfrac{2}{?}$ **2.** $\dfrac{21x}{28x} = \dfrac{?}{4}$ **3.** $\dfrac{4}{5} = \dfrac{28m^3}{?}$

4. $\dfrac{3}{7} = \dfrac{?}{21x^2}$ **5.** $\dfrac{9xy}{12y^2} = \dfrac{3x}{?}$ **6.** $\dfrac{6x^3}{4xy} = \dfrac{?}{2y}$

Multiply and reduce to lowest terms.

7. $\dfrac{10}{9} \cdot \dfrac{12}{15}$ **8.** $\dfrac{3}{7} \cdot \dfrac{14}{9}$ **9.** $\dfrac{2a}{3bc} \cdot \dfrac{9c}{a}$

10. $\dfrac{2x}{3yz} \cdot \dfrac{6y}{4x}$ **11.** $\dfrac{3x^2}{4} \cdot \dfrac{16y}{12x^3}$ **12.** $\dfrac{2x^2}{3y^2} \cdot \dfrac{9y}{4x}$

Divide and reduce to lowest terms.

13. $\dfrac{9m}{8n} \div \dfrac{3m}{4n}$ **14.** $\dfrac{6x}{5y} \div \dfrac{3x}{10y}$ **15.** $\dfrac{a}{4c} \div \dfrac{a^2}{12c^2}$

16. $\dfrac{2x}{3y} \div \dfrac{4x}{6y^2}$ **17.** $\dfrac{x}{3y} \div 3y$ **18.** $2xy \div \dfrac{x}{y}$

B *Perform the indicated operations and reduce to lowest terms.*

19. $\dfrac{8x^2}{3xy} \cdot \dfrac{12y^3}{6y}$

20. $\dfrac{6a^2}{7c} \cdot \dfrac{21cd}{12ac}$

21. $\dfrac{21x^2y^2}{12cd} \div \dfrac{14xy}{9d}$

22. $\dfrac{3uv^2}{5w} \div \dfrac{6u^2v}{15w}$

23. $\dfrac{9u^4}{4v^3} \div \dfrac{-12u^2}{15v}$

24. $\dfrac{-6x^3}{5y^2} \div \dfrac{18x}{10y}$

25. $\dfrac{3c^2d}{a^3b^3} \div \dfrac{3a^3b^3}{cd}$

26. $\dfrac{uvw}{5xyz} \div \dfrac{5vy}{uwxz}$

27. $\dfrac{3x^2y}{x-y} \cdot \dfrac{x-y}{6xy}$

28. $\dfrac{x+3}{2x^2} \cdot \dfrac{4x}{x+3}$

29. $\dfrac{x+3}{x^3+3x^2} \cdot \dfrac{x^3}{x-3}$

30. $\dfrac{a^2-a}{a-1} \cdot \dfrac{a+1}{a}$

31. $\dfrac{x-2}{4y} \div \dfrac{x^2+x-6}{12y^2}$

32. $\dfrac{4x}{x-4} \div \dfrac{8x^2}{x^2-6x+8}$

33. $\dfrac{6x^2}{4x^2y-12xy} \cdot \dfrac{x^2+x-12}{3x^2+12x}$

34. $\dfrac{2x^2+4x}{12x^2y} \cdot \dfrac{6x}{x^2+6x+8}$

35. $(t^2-t-12) \div \dfrac{t^2-9}{t^2-3t}$

36. $\dfrac{2y^2+7y+3}{4y^2-1} \div (y+3)$

37. $\dfrac{m+n}{m^2-n^2} \div \dfrac{m^2-mn}{m^2-2mn+n^2}$

38. $\dfrac{x^2-6x+9}{x^2-x-6} \div \dfrac{x^2+2x-15}{x^2+2x}$

39. $-(x^2-3x) \cdot \dfrac{x-2}{x-3}$

40. $-(x^2-4) \cdot \dfrac{3}{x+2}$

C

41. $\left(\dfrac{d^5}{3a} \div \dfrac{d^2}{6a^2}\right) \cdot \dfrac{a}{4d^3}$

42. $\dfrac{d^5}{3a} \div \left(\dfrac{d^2}{6a^2} \cdot \dfrac{a}{4d^3}\right)$

43. $\dfrac{2x^2}{3y^2} \cdot \dfrac{-6yz}{2x} \cdot \dfrac{y}{-xz}$

44. $\dfrac{-a}{-b} \cdot \dfrac{12b^2c}{15ac} \cdot \dfrac{-10}{4b}$

45. $\dfrac{9-x^2}{x^2+5x+6} \cdot \dfrac{x+2}{x-3}$

46. $\dfrac{2-m}{2m+m^2} \cdot \dfrac{m^2+4m+4}{m^2-4}$

47. $\dfrac{x^2-xy}{xy+y^2} \div \left(\dfrac{x^2-y^2}{x^2+2xy+y^2} \div \dfrac{x^2-2xy+y^2}{x^2y+xy^2}\right)$

48. $\left(\dfrac{x^2-xy}{xy+y^2} \div \dfrac{x^2-y^2}{x^2+2xy+y^2}\right) \div \dfrac{x^2-2xy+y^2}{x^2y+xy^2}$

49. $(x^2-x-6)/(x-3) = x+2$, except for what values of x?

50. $(x^2 - 1)/(x - 1)$ and $x + 1$ name the same real number for (*all, all but one, no*) replacements of x by real numbers.

51. Prove Theorem 2.

6.2 Addition and Subtraction

Addition and subtraction of rational expressions are based on the corresponding properties of real number fractions. Thus,

> If D, P, Q, and K represent polynomials, then
>
> $$\frac{P}{D} + \frac{Q}{D} = \frac{P + Q}{D} \qquad (1)$$
>
> $$\frac{P}{D} - \frac{Q}{D} = \frac{P - Q}{D} \qquad (2)$$
>
> $$\frac{P}{D} = \frac{PK}{DK} \qquad (3)$$
>
> where D and K are not 0.

Verbally, if the denominators of two rational expressions are the same, we may either add or subtract the expressions by adding or subtracting the numerators and placing the result over the common denominator; if the denominators are not the same, we use property (3) to change the form of each fraction so they have a common denominator, and then use either (1) or (2).

Even though any common denominator will do, the problem will generally become less involved if the least common denominator (lcd) is used. If the lcd is not obvious (often it is), then we factor each denominator completely, including numerical coefficients. *The lcd should then contain each different factor in the denominators to the highest power it occurs in any one denominator.*

EXAMPLE 5 (*A*)

$$\frac{x-2}{x-3} - \frac{1}{x-3} = \frac{x-2-1}{x-3} = \frac{x-3}{x-3} = 1$$

(*B*)

$$\frac{2}{2y} + \frac{1}{4y^2} - 1 = \frac{1(2y)}{(2y)(2y)} + \frac{1}{4y^2} - \frac{4y^2}{4y^2} = \frac{2y + 1 - 4y^2}{4y^2}$$

$$= \frac{1 + 2y - 4y^2}{4y^2}$$

Note that the lcd is $4y^2$.

(C) $$\frac{4}{x^2-4}-\frac{3}{x^2-x-2}=\frac{4}{(x-2)(x+2)}-\frac{3}{(x-2)(x+1)}$$

We see that the lcd is $(x-2)(x+2)(x+1)$. Then

$$\frac{4(x+1)}{(x-2)(x+2)(x+1)}-\frac{3(x+2)}{(x-2)(x+2)(x+1)}=\frac{4(x+1)-3(x+2)}{(x-2)(x+2)(x+1)}$$

$$=\frac{4x+4-3x-6}{(x-2)(x+2)(x+1)}=\frac{(x-2)}{(x-2)(x+2)(x+1)}=\frac{1}{(x+2)(x+1)}$$

PROBLEM 5 Combine into single fractions and simplify:

(A) $\frac{5(x-4)}{2(x+2)}-\frac{3(x-2)}{2(x+2)}$ (B) $\frac{1}{3y^2}-\frac{1}{6y}+1$

(C) $\frac{3}{x^2-1}-\frac{2}{x^2+2x+1}$

ANSWER (A) $\frac{x-7}{x+2}$ (B) $\frac{2-y+6y^2}{6y^2}$ (C) $\frac{x+5}{(x-1)(x+1)^2}$

Exercise 31

A *Replace each question mark with an appropriate polynomial.*

1. $\frac{3x}{7y}=\frac{?}{14y^2}$

2. $\frac{5m}{3n}=\frac{?}{12mn^2}$

3. $\frac{4}{(2x-3)}=\frac{?}{(2x-3)(x+2)}$

4. $\frac{3m}{(m-n)(m+n)}=\frac{?}{(m-n)(m+n)^2}$

5. $x-3=\frac{?}{x+2}$

6. $2x=\frac{?}{3(x-3)}$

Combine into single fractions and simplify.

7. $\frac{2x}{3y}+\frac{5}{3y}$

8. $\frac{5y}{4x}+\frac{3}{4x}$

9. $\frac{7x}{5x^2} - \frac{2}{5x^2}$

10. $\frac{3m}{2m^2} - \frac{1}{2m^2}$

11. $\frac{3x}{y} + \frac{1}{4}$

12. $\frac{2}{x} - \frac{1}{3}$

13. $\frac{2}{x} + 1$

14. $x + \frac{1}{x}$

15. $\frac{u}{v^2} - \frac{1}{v} + \frac{u^2}{v^3}$

16. $\frac{1}{x} - \frac{y}{x^2} + \frac{y^2}{x^3}$

17. $\frac{4x}{2x-1} - \frac{2}{2x-1}$

18. $\frac{5a}{a-1} - \frac{5}{a-1}$

19. $\frac{2x}{4x^2-9} - \frac{3}{4x^2-9}$

20. $\frac{y}{y^2-9} - \frac{3}{y^2-9}$

21. $\frac{2}{x+1} + \frac{3}{x-2}$

22. $\frac{1}{x-2} + \frac{1}{x+3}$

23. $\frac{2}{3y} - \frac{3}{y+3}$

24. $\frac{3}{2x} - \frac{2}{x+2}$

B *Reduce to lowest terms.*

25. $\frac{12m^3n^5}{9m^3n^8}$

26. $\frac{12x^5y^2}{6x^7y}$

27. $\frac{8m^2n^2 + 8mn^3}{4m^3n^3}$

28. $\frac{6x^3y - 6x^2y^2}{8x^3y^2}$

29. $\frac{a^2 - 16b^2}{ab - 4b^2}$

30. $\frac{2x^2 + 3xy}{4x^2 - 9y^2}$

Combine into single fractions and simplify.

31. $\frac{5}{3k} - \frac{6x-4}{3k}$

32. $\frac{1}{2a^2} - \frac{2b-1}{2a^2}$

33. $\frac{4t-3}{18t^3} + \frac{3}{4t} - \frac{2t-1}{6t^2}$

34. $\frac{3y+8}{4y^2} - \frac{2y-1}{y^3} - \frac{5}{8y}$

35. $\frac{3}{x+3} - \frac{3x+1}{(x-1)(x+3)}$

36. $\frac{4}{2x-3} - \frac{2x+1}{(2x-3)(x+2)}$

37. $\frac{1}{m^2-n^2} + \frac{1}{m^2+2mn+n^2}$

38. $\frac{3x-1}{2x^2+x-3} - \frac{2}{x-1}$

39. $2 + \frac{x+1}{x-3}$

40. $\frac{t+1}{t-1} - 1$

41. $x - 3 - \frac{x-1}{x-2}$

42. $\frac{x^2-2x}{x-2} + x - 3$

43. $5 + \frac{a}{a+1} - \frac{a}{a-1}$

44. $\frac{1}{y+2} + 3 - \frac{2}{y-2}$

45. $\frac{2x}{x^2-y^2} + \frac{1}{x+y} - \frac{1}{x-y}$

46. $\frac{1}{x+3} - \frac{1}{x-3} + \frac{2x}{x^2-9}$

47. $\frac{3s}{3s^2-12} + \frac{1}{2s^2+4s}$

48. $\frac{2t}{3t^2-48} + \frac{t}{4t+t^2}$

49. $\frac{1}{m^2-n^2} - \frac{1}{m^2-2mn+n^2}$

50. $\frac{3}{x^2-1} - \frac{2}{x^2-2x+1}$

C 51. $\frac{5}{y-3} - \frac{2}{3-y}$

52. $\frac{3}{x-1} + \frac{2}{1-x}$

53. $\frac{x+7}{ax-bx} + \frac{y+9}{by-ay}$

54. $\frac{1}{5x-5} - \frac{1}{3x-3} + \frac{1}{1-x}$

55. $\frac{x}{x^2-x-2} - \frac{1}{x^2+5x-14} - \frac{2}{x^2+8x+7}$

56. $\frac{m^2}{m^2+2m+1} + \frac{1}{3m+3} - \frac{1}{6}$

57. $\frac{xy^2}{x^3-y^3} - \frac{y}{x^2+xy+y^2}$

58. $\frac{x}{x^2-xy+y^2} - \frac{xy}{x^3+y^3}$

Replace question marks with appropriate expressions.

59. $-\frac{1}{3-x} = \frac{?}{x-3}$

60. $-\frac{-3}{3-x} = -\frac{?}{x-3}$

61. $\frac{y-x}{3-x} = \frac{?}{x-3}$

62. $-\frac{a-b}{b-a} = ?$

6.3 Ratio and Proportion

You have no doubt been using ratios for many years to compare two quantities and will recall that the *ratio of one quantity to another* is simply the quotient of the two quantities—the first divided by the second.

EXAMPLE 6 If 12 male and 16 female students are in a freshman English class, then the ratio of males to females is 12/16 or 3/4, which is often written

$3:4$ (read "3 to 4")

The ratio of females to males is 16/12 or 4/3 or

$4:3$ (read "4 to 3")

PROBLEM 6 If there are 1,000 male and 800 female students in a small college, what is the ratio of

(*A*) males to females? (*B*) females to males?

ANSWER (*A*) 5 : 4 (*B*) 4 : 5

In addition to comparing known quantities, another reason why we want to know something about ratios is that they often lead to a simple way of finding unknown quantities.

EXAMPLE 7 Suppose you are told that in a school the ratio of girls to boys is 3 : 5 and that there are 1,450 boys. How many girls are in the school?

SOLUTION Let $x =$ the number of girls in the school, then

$$\frac{x}{1{,}450} = \frac{3}{5}$$

$$x = 1{,}450\ (3/5)$$

$$= 870 \text{ girls}$$

PROBLEM 7 If in a school the ratio of boys to girls is 2 : 3, and there are 1,200 girls, how many boys are in the school?

ANSWER 800

The statement of equality between two ratios is called a *proportion*. Knowing that various pairs of quantities are proportional leads to simple solutions of many types of problems. For example, to find the height of a tree, the distance across a lake, or many other inaccessible distances, one can use the proportional property of similar triangles (two triangles are similar if their corresponding angles are equal): If two triangles are similar, the ratios of corresponding sides are equal. Thus, in the similar triangles in the figure

$$\frac{a}{d} = \frac{b}{e} \qquad \frac{a}{d} = \frac{c}{f} \qquad \frac{e}{b} = \frac{f}{c}$$

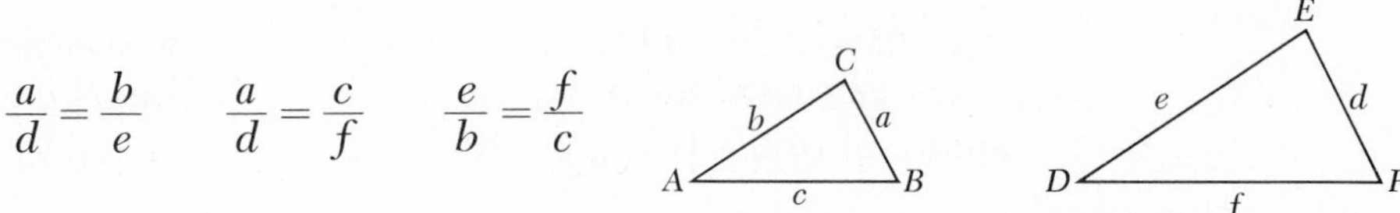

EXAMPLE 8 To measure the length of a lake, you can proceed as indicated in Fig. 1. Place stakes *B* and *C* at either end of the lake, then place a stake *A* away from the lake so that either *AC* or *AB* can easily be measured. Place a stake *D* on *AB*, then locate a stake *E* on *AC* so that angle *ADE* is the same as angle *ABC*. Thus, triangle *ADE* and triangle *ABC* are similar (Why?).

Measure AB, AD, and DE, and determine the length of the lake, BC, from the proportion

$$\frac{BC}{DE} = \frac{AB}{AD}$$

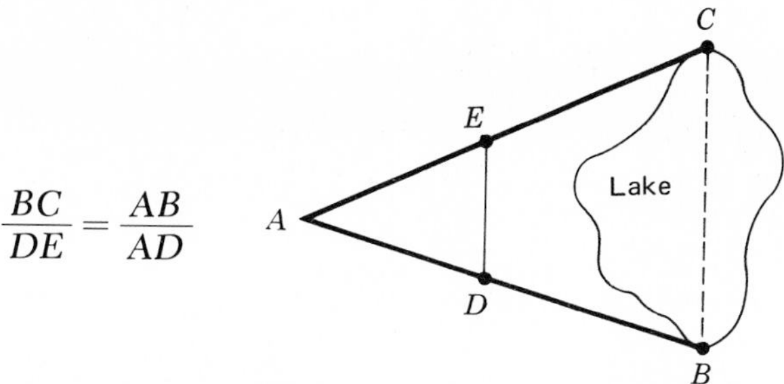

Figure 1

suppose $AB = 1{,}200$ feet, $DE = 300$ feet, $AD = 400$ feet, and $x = BC$, then

$$\frac{x}{300} = \frac{1{,}200}{400}$$

$$x = \frac{(1{,}200)(300)}{400}$$

$$= 900 \text{ feet}$$

PROBLEM 8 Find the length of the lake in Example 8 if $AC = 2{,}400$ feet, $AE = 800$ feet and $DE = 600$ feet.

ANSWER 1,800 feet

Exercise 32

A *Write as a ratio:*

1. 64 girls to 16 boys

2. 16 boys to 64 girls

3. 30 inches to 10 inches

4. 25 feet to 5 feet

5. 25 square feet to 100 square feet

6. 30 square inches to 90 square inches

Solve each proportion:

7. $\frac{y}{16} = \frac{5}{4}$

8. $\frac{x}{12} = \frac{2}{3}$

9. $\frac{y}{13} = \frac{21}{39}$

10. $\frac{d}{12} = \frac{27}{18}$ **11.** $\frac{35}{56} = \frac{x}{32}$ **12.** $\frac{18}{27} = \frac{h}{6}$

13. If the ratio of girls to boys is 7 : 9 and there are 630 boys, how many girls are there?

14. If in a school the ratio of boys to girls is 5 : 7 and there are 840 girls, how many boys are there?

15. Find the length of the lake in Fig. 1 if $AB = 1{,}800$ feet, $AE = 600$ feet, and $DE = 340$ feet?

16. Find the length of the lake in Fig. 1 if $AC = 1{,}500$ feet, $AD = 375$ feet, and $DE = 420$ feet.

B **17.** PHOTOGRAPHY If you enlarge a 6- by 3-inch picture so that the longer side is 8 inches, how wide will the enlargement be?

18. SCALE DRAWINGS An architect wishes to make a scale drawing of a 48- by 30-foot rectangular building. If his drawing of the building is 6 inches long, how wide is it?

19. COMPUTERS If an IBM electronic card sorter can sort 1,250 cards in 5 minutes, how long will it take the sorter to sort 11,250 cards?

20. INTELLIGENCE The IQ (Intelligence Quotient) is found by dividing the mental age, as indicated by standard tests, by the chronological age and multiplying by 100. For example, if a girl had a mental age of 12 and a chronological age of 10, her IQ would be 120. If an 11-year-old has an IQ of 132 (superior intelligence), compute her mental age.

21. OPTICS (MAGNIFICATION) In Fig. 2 triangles CBA and CDE are similar; hence, corresponding parts are proportional. If the object is 0.4 inch, $AC = 1.4$ inches, and CD = 4.9 inches, what is the size of the image?

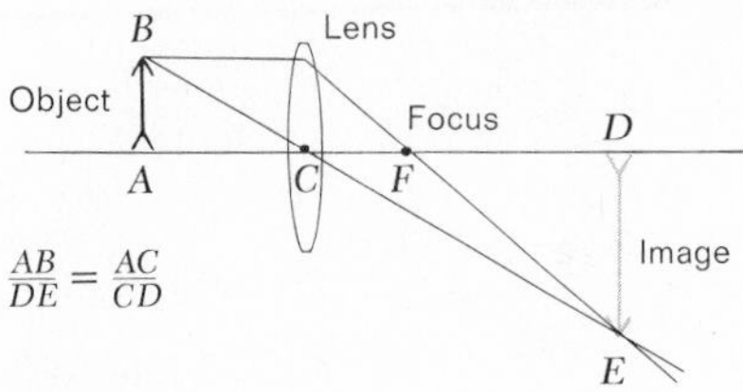

Figure 2

22. COMMISSIONS If you were charged a commission of \$57 on the purchase of 300 shares of stock, what would be the proportionate commission on 500 shares?

C **23.** HYDRAULIC LIFTS If in Fig. 3 the diameter of the smaller pipe is $\frac{1}{2}$ inch and the diameter of the larger pipe is 10 inches, how much force would be required to lift a 3,000-pound car? (Neglect weight of lift equipment.)

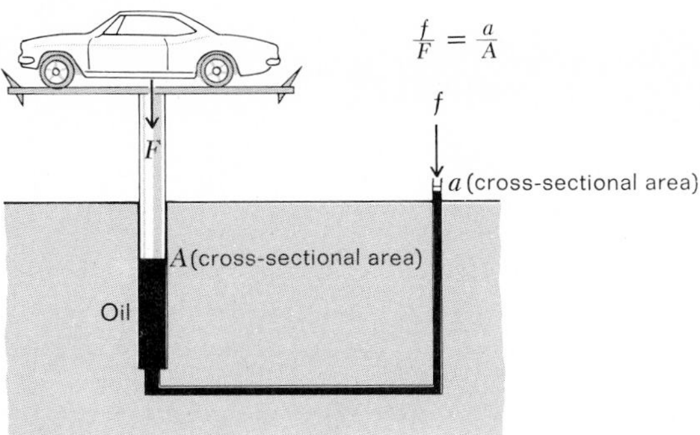

Figure 3

24. POPULATION SAMPLING Zoologists Green and Evans (1940) estimated the total population of snowshoe hares in the Lake Alexander area of Minnesota as follows: They captured and banded 948 hares, and then released them. After an appropriate period for mixing, they again captured a sample of 421 and found 167 of these marked. Set up an appropriate proportion and estimate the total hare population in the region.

25. ASTRONOMY Do you have any idea how one might measure the circumference of the earth? In 240 B.C. Eratosthenes measured the size of the earth from its curvature. At Syene, Egypt (lying on the Tropic of Cancer) the sun was directly overhead at noon on June 21. At the same time in Alexandria, a town 500 miles directly north, the sun's rays fell at an angle of 7.5° to the vertical. Using this information and a little knowledge of geometry (see Fig. 4), Eratosthenes was able to approximate the circumference of the earth using the following proportion: 7.5 is to 360 as 500 is to the circumference of the earth. Compute Eratosthenes' estimate.

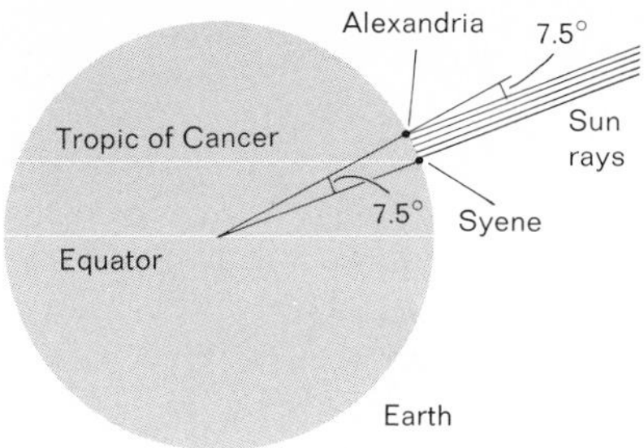

Figure 4

6.4 Equations and Inequalities Involving Fractions

We are now ready to consider a wider variety of equations and inequalities involving fractional forms. Earlier methods will now be streamlined a bit to produce results rather quickly. Several examples should make the process clear.

EXAMPLE 9 What operation can we perform on the equation

$$\frac{x+1}{3} - \frac{x}{4} = \frac{1}{2}$$

to eliminate the denominators? If we can find a number that is exactly divisble by each denominator, then we can use the multiplication property of equality to clear the denominators. The least common denominator of the fractions is exactly what we are looking for! Thus, we multiply both members of the equation by 12:

$$12\left(\frac{x+1}{3} - \frac{x}{4}\right) = 12\left(\frac{1}{2}\right)$$

$$\overset{4}{\cancel{12}} \cdot \frac{(x+1)}{\underset{1}{\cancel{3}}} - \overset{3}{\cancel{12}} \cdot \frac{x}{\underset{1}{\cancel{4}}} = \overset{6}{\cancel{12}} \cdot \frac{1}{\underset{1}{\cancel{2}}}$$

$$4(x+1) - 3x = 6$$

$$4x + 4 - 3x = 6$$

$$x = 2$$

CHECK

$$\frac{2+1}{3} - \frac{2}{4} = 1 - \frac{1}{2} = \frac{1}{2}$$

PROBLEM 9 Solve and check:

$$\frac{x+2}{2} - \frac{x}{3} = 5$$

ANSWER $x = 24$

EXAMPLE 10 Equations will often have rational coefficients written as decimal fractions. Some equations of this type are more easily solved if they are first cleared of decimals. The following is a case in point:

$$0.2x + 0.3(x - 5) = 13$$

$$10(0.2x) + 10[0.3(x - 5)] = 10 \cdot 13$$

$$2x + 3(x - 5) = 130$$

$$5x - 15 = 130$$
$$5x = 145$$
$$x = 29$$

CHECK

$$0.2(29) + 0.3(29 - 5) = 5.8 + 7.2$$
$$= 13$$

PROBLEM 10 Solve and check:

$$0.3(x + 2) + 0.5x = 3$$

ANSWER $x = 3$

EXAMPLE 11 Solve:

$$\frac{2x - 3}{4} + 6 > 2 + \frac{4x}{3}$$

SOLUTION

$$12\frac{2x - 3}{4} + 12 \cdot 6 > 12 \cdot 2 + 12\frac{4x}{3}$$
$$6x - 9 + 72 > 24 + 16x$$
$$6x + 63 > 24 + 16x$$
$$-10x > -39$$
$$x < 3.9$$

PROBLEM 11 Solve:

$$\frac{4x - 3}{3} + 8 > 6 + \frac{3x}{2}$$

ANSWER $x < 6$

If an equation involves variables in one or more denominators, such as

$$\frac{3}{x} - \frac{1}{2} = \frac{4}{x}$$

we may proceed in essentially the same way as above, as long as we stay away from any value that makes a denominator 0. In this case

$$x \neq 0$$

EXAMPLE 12 Solve and check:

$$\frac{3}{x} - \frac{1}{2} = \frac{4}{x}$$

SOLUTION

$$\frac{3}{x} - \frac{1}{2} = \frac{4}{x} \qquad x \neq 0$$

$$(2x)\frac{3}{x} - (2x)\frac{1}{2} = (2x)\frac{4}{x}$$

$$6 - x = 8$$

$$-x = 2$$

$$x = -2$$

CHECK

$$\frac{3}{-2} - \frac{1}{2} \stackrel{?}{=} \frac{4}{-2}$$

$$-\frac{4}{2} \stackrel{\checkmark}{=} -\frac{4}{2}$$

PROBLEM 12 Solve and check:

$$\frac{2}{3} - \frac{2}{x} = \frac{4}{x}$$

ANSWER $x = 9$

EXAMPLE 13 Solve and check:

$$\frac{3x}{x-2} - 4 = \frac{14-4x}{x-2}$$

SOLUTION

$$\frac{3x}{x-2} - 4 = \frac{14-4x}{x-2}$$

$x \neq 2$, since division by 0 is not defined

$$(x-2)\frac{3x}{x-2} - 4(x-2) = (x-2)\frac{14-4x}{x-2}$$

$$3x - 4x + 8 = 14 - 4x$$

$$3x = 6$$

$$x = 2$$

Since x cannot equal 2 (see above), the original equation has no solution.

PROBLEM 13 Solve and check:

$$\frac{2x}{x-1} - 3 = \frac{7-3x}{x-1}$$

ANSWER $x = 2$

Exercise 33

Solve each equation and inequality, and check each equation.

A **1.** $\frac{x}{5} - 2 = \frac{3}{5}$ **2.** $\frac{x}{7} - 1 = \frac{1}{7}$

3. $\frac{x}{3} + \frac{x}{6} = 4$

4. $\frac{y}{4} + \frac{y}{2} = 9$

5. $\frac{m}{4} - \frac{m}{3} = \frac{1}{2}$

6. $\frac{n}{5} - \frac{n}{6} = \frac{\text{[illegible]}}{5}$

7. $x - \frac{2}{3} > \frac{x}{3} + 2$

8. $\frac{x}{5} - 3 < \frac{3}{5} - x$

9. $\frac{5}{12} - \frac{m}{3} = \frac{4}{9}$

10. $\frac{2}{3} - \frac{x}{8} = \frac{5}{6}$

11. $x > \frac{x}{3} - \frac{1}{2}$

12. $m - \frac{1}{2} = \frac{8}{3}$

13. $0.7x + 0.9x = 32$

14. $0.3x + 0.5x = 24$

15. $\frac{1}{2} - \frac{2}{x} = \frac{3}{x}$

16. $\frac{2}{x} - \frac{1}{3} = \frac{5}{x}$

17. $\frac{1}{m} - \frac{1}{9} = \frac{4}{9} - \frac{2}{3m}$

18. $\frac{1}{2t} + \frac{1}{8} = \frac{2}{t} - \frac{1}{4}$

B

19. $\frac{x-2}{3} + 1 = \frac{x}{7}$

20. $\frac{x+3}{2} - \frac{x}{3} = 4$

21. $\frac{x-3}{2} - 1 > \frac{x}{4}$

22. $-2 - \frac{x}{4} < \frac{1+x}{3}$

23. $-2 - \frac{B}{4} \leq \frac{1+B}{3}$

24. $\frac{y-3}{4} - 1 > \frac{y}{2}$

25. $0.1(x-7) + 0.05x = 0.8$

26. $0.4(x+5) - 0.3x = 17$

27. $\frac{2x-3}{9} - \frac{x+5}{6} = \frac{3-x}{2} - 1$

28. $\frac{3x+4}{3} - \frac{x-2}{5} = \frac{2-x}{15} - 1$

29. $\frac{p}{3} - \frac{p-2}{2} \leq \frac{p}{4} - 4$

30. $\frac{3q}{7} - \frac{q-4}{3} > 4 + \frac{2q}{7}$

31. $-4 \leq \frac{9}{5}C + 32 \leq 68$

32. $-1 \leq \frac{2}{3}m + 5 \leq 11$

33. $\frac{7}{y-2} - \frac{1}{2} = 3$

34. $\frac{9}{L+1} - 1 = \frac{12}{L+1}$

35. $\frac{3}{2x-1} + 4 = \frac{6x}{2x-1}$

36. $\frac{5x}{x+5} = 2 - \frac{25}{x+5}$

37. $\frac{2E}{E-1} = 2 + \frac{5}{2E}$

38. $\frac{3N}{N-2} - \frac{9}{4N} = 3$

39. $\frac{n-5}{6n-6} = \frac{1}{9} - \frac{n-3}{4n-4}$

40. $\frac{1}{3} - \frac{s-2}{2s+4} = \frac{s+2}{3s+6}$

41. $\frac{D^2+2}{D^2-4} = \frac{D}{D-2}$

42. $\frac{5}{x-3} = \frac{33-x}{x^2-6x+9}$

C **43.** $\frac{3x}{24} - \frac{2 - x}{10} = \frac{5 + x}{40} - \frac{1}{15}$

44. $\frac{2x}{10} - \frac{3 - x}{14} = \frac{2 + x}{5} - \frac{1}{2}$

45. $\frac{2x}{5} - \frac{1}{2}(x - 3) \leq \frac{2x}{3} - \frac{3}{10}(x + 2)$

46. $\frac{2}{3}(x + 7) - \frac{x}{4} > \frac{1}{2}(3 - x) + \frac{x}{6}$

47. $-5 \leq \frac{5}{9}(F - 32) \leq 10$

48. $-10 \leq \frac{5}{9}(F - 32) \leq 25$

49. $\frac{5t - 22}{t^2 - 6t + 9} - \frac{11}{t^2 - 3t} - \frac{5}{t} = 0$

50. $\frac{5}{x - 3} = \frac{33 - x}{x^2 - 6x + 9}$

51. $5 - \frac{2x}{3 - x} = \frac{6}{x - 3}$

52. $\frac{3x}{2 - x} + \frac{6}{x - 2} = 3$

53. $\frac{1}{c^2 - c - 2} - \frac{3}{c^2 - 2c - 3} = \frac{1}{c^2 - 5c + 6}$

54. $\frac{5t - 22}{t^2 - 6t + 9} - \frac{11}{t^2 - 3t} - \frac{5}{t} = 0$

6.5 Applications

This section contains a wide variety of applications that give rise to equations or inequalities involving fractional forms. These are realistic problems both from the point of view of the subject matter and the kinds of numbers involved.

EXAMPLE 14 Five individuals formed a glider club and decided to share the cost of a glider equally. They found, however, that if they let three more join the club, the share for each of the original five would be reduced by \$120. What was the total cost of the glider?

SOLUTION Let C = the total cost of the glider.

$$\text{Cost per share for 5 members} - \text{cost per share for 8 members} = \text{reduction in cost for each of the original five}$$

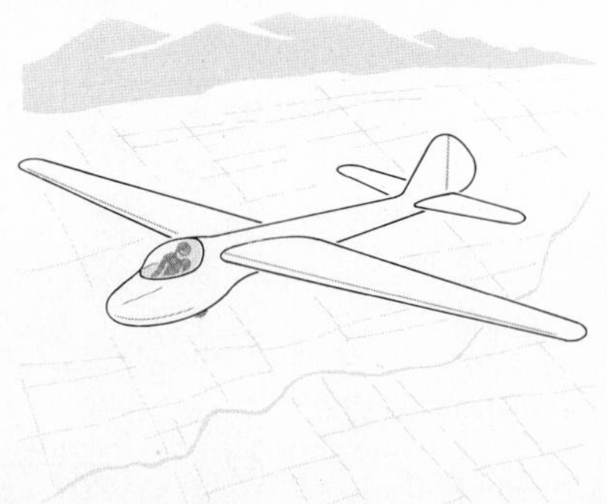

$$\frac{C}{5} - \frac{C}{8} = 120$$

$$40 \cdot \frac{C}{5} - 40 \cdot \frac{C}{8} = 40 \cdot 120$$

$$8C - 5C = 4{,}800$$

$$3C = 4{,}800$$

$$C = \$1{,}600$$

total cost of glider

PROBLEM 14 Three individuals bought a sail boat together. If they had taken in a fourth person, the cost for each would have been reduced \$200. What was the total cost of the boat?

ANSWER \$2,400

EXAMPLE 15 If the temperature for a 24-hour period in the Antarctica ranged between $-49°F$ and $14°F$ (that is, $-49 \le F \le 14$), what was the range in Celsius degrees? (Recall $F = \frac{9}{5}C + 32$)

SOLUTION Since $F = \frac{9}{5}C + 32$, we replace F in $-49 \le F \le 14$ with $\frac{9}{5}C + 32$ and solve the double inequality:

$$-49 \le \tfrac{9}{5}C + 32 \le 14$$

$$-49 - 32 \le \tfrac{9}{5}C + 32 - 32 \le 14 - 32$$

$$-81 \le \tfrac{9}{5}C \le -18$$

$$(\tfrac{5}{9})(-81) \le (\tfrac{5}{9})(\tfrac{9}{5}C) \le (\tfrac{5}{9})(-18)$$

$$-45 \le C \le -10$$

PROBLEM 15 Repeat Example 4 for $-31 \le F \le 5$.

ANSWER $-35 \le C \le -15$

EXAMPLE 16 A speedboat takes 1.5 times longer to go 120 miles up a river than to return. If the boat cruises at 25 miles/hour in still water, what is the rate of the current.

SOLUTION Let x = rate of current in miles/hour

$25 - x$ = rate of boat upstream

$25 + x$ = rate of boat downstream

Time upstream = 1.5 time downstream

$$\frac{\text{Distance upstream}}{\text{Rate upstream}} = 1.5\,\frac{\text{distance downstream}}{\text{rate downstream}}$$ Recall $t = d/r$ from $d = rt$

$$\frac{120}{25 - x} = (1.5)\frac{120}{25 + x}$$

$$\frac{120}{25 - x} = \frac{180}{25 + x}$$

$$(25 + x)120 = (25 - x)180$$

$$3{,}000 + 120x = 4{,}500 - 180x$$

$$300x = 1{,}500$$

$$x = 5 \text{ miles/hour} \qquad \text{rate of current}$$

CHECK

$$\text{Time upstream} = \frac{120}{20} = 6 \text{ hours}$$

$$\text{Time downstream} = \frac{120}{30} = 4 \text{ hours}$$

Therefore, time upstream is 1.5 times longer than time downstream.

PROBLEM 16 A fishing boat takes twice as long to go 24 miles up a river than to return. If the boat cruises at 9 miles/hour in still water, what is the rate of the current?

ANSWER 3 miles/hour

EXAMPLE 17 In an electronic computer center a card-sorter operator is given the job of alphabetizing a given quantity of IBM cards. He knows that an older sorter can do the job by itself in 6 hours. With the help of a newer machine the job is completed in 2 hours. How long would it take the new machine to do the job alone?

SOLUTION Let $x =$ time for new machine to do job alone
Quantity of work = rate × time
Rate of old machine = $\frac{1}{6}$ job per hour
Rate of new machine = $1/x$ job per hour

$$\begin{pmatrix}\text{Quantity of}\\ \text{work done by}\\ \text{first machine}\\ \text{in 2 hours}\end{pmatrix} + \begin{pmatrix}\text{Quantity of}\\ \text{work done by}\\ \text{second machine}\\ \text{in 2 hours}\end{pmatrix} = 1 \text{ whole job}$$

$$(\text{rate})(\text{time}) + (\text{rate})(\text{time}) = 1$$

$$\frac{1}{6}(2) + \frac{1}{x}(2) = 1$$

$$\frac{1}{3} + \frac{2}{x} = 1$$

$$x + 6 = 3x$$

$$2x = 6$$

$$x = 3 \text{ hours}$$

CHECK

$$\tfrac{1}{6}(2) + \tfrac{1}{3}(2) = \tfrac{1}{3} + \tfrac{2}{3} = 1 \text{ whole job}$$

PROBLEM 17 At a family cabin water is pumped and stored in a large water tank. Two pumps are used for this purpose. One can fill the tank by itself in 6 hours, and the other can do the job in 9 hours. How long will it take both pumps operating together to fill the tank?

ANSWER $3\frac{3}{5}$ hours

Exercise 34

*These problems are not grouped from easy (***A***) to difficult theoretical (***C***). They are grouped somewhat according to type. Some are easy and some are difficult. The most difficult problems are double-starred (⋆⋆), moderately difficult problems are single-starred (⋆), and the easier problems are not marked.*

ANTHROPOLOGY

1. Anthropologists, in their study of race and human genetic groupings, use ratios called indices. One widely used index is the cephalic index, the ratio of the breadth of the head to its length expressed as a percent (looking down from above). Thus

$$C = \frac{100B}{L}$$

(long-headed, $C < 75$; intermediate, $75 \leq C \leq 80$; round-headed, $C > 80$.) If an Indian tribe in Baja, California had a cephalic index of 66 and the average breadth of their heads was 6.6 inches, what was the average length of their heads?

BUSINESS

⋆⋆ **2.** If a stock that you bought on Monday went up 10 percent on Tuesday and fell 10 percent on Wednesday, how much did you pay for the stock on Monday if you sold it on Wednesday for \$99?

3. For a business to make a profit it is clear that revenue R must be greater than costs C; in short, a profit will result only if $R > C$. If a company manufactures records and its cost equation for a week is $C = 300 + 1.5x$, where x is the number of records manufactured in a week, and its revenue equation is $R = 2x$, where x is the number of records sold in a week, how many records must be sold for the company to realize a profit? Solve using inequality methods.

⋆ **4.** In an electronic computer center a card-sorter operator is given the job of alphabetizing a given quantity of IBM cards. The operator knows that an older sorter can do the job by itself in 3 hours. With the help of a newer machine the job is completed in 1 hour. How long would it take the new machine to do the job alone?

CHEMISTRY

5. In a chemistry experiment the solution of hydrochloric acid is to be kept between 30 and 35°C. What would the range of temperature be in Fahrenheit degrees $[C = \frac{5}{9}(F - 32)]$? Solve using inequality methods.

⋆ **6.** How many gallons of pure alcohol must be added to 3 gallons of a 20% solution to get a 40% solution?

DOMESTIC

★ **7.** You are at a river resort and rent a motor boat for 5 hours at 7 A.M. You are told that the boat will travel at 8 miles/hour upstream and 12 miles/hour returning. You decide that you would like to go as far up the river as you can and still be back at noon. At what time should you turn back, and how far from the resort will you be at that time?

★ **8.** A travel agent, arranging a round-trip charter flight from New York to Europe, tells 100 signed-up members of a group that they can save $120 each if they can get 50 more people to sign up for the trip to fill the plane. What is the total round-trip charter cost of the jet transport? What is the cost per person if 100 go? What is the cost per person if 150 go?

9. To be eligible for a certain university, a student must have a total average of not below 70 on three entrance examinations. If a student received a 55 and a 73 on the first two examinations, what must he receive on the third examination to be eligible to enter the university? Solve using inequality methods.

EARTH SCIENCE

★**10.** An explosion is set off on the surface of the water 11,000 feet from a ship. If the sound reaches the ship through the water 7.77 seconds before it arrives through the air and if sound travels through water 4.5 times faster than through air, how fast (to the nearest foot) does sound travel in air and in water?

11. A scuba diver (Fig. 5) knows that 1 atmosphere of pressure is the weight of a column of air 1 square inch extending straight up from the surface of the earth without end (14.7 pounds/square inch). Also, the water pressure below the surface increases 1 atmosphere for each 33 feet of depth. In terms of a formula

$$P = 1 + \frac{D}{33}$$

where P is pressure in atmospheres and D is depth in feet. At what depth will the pressure be 3.6 atmospheres?

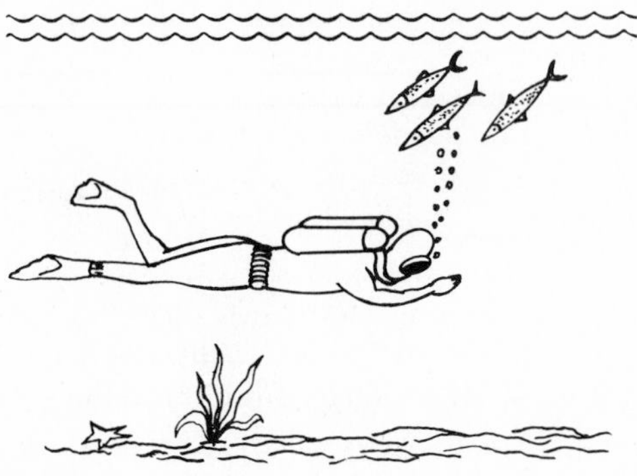

Figure 5

ECONOMICS

12. (*A*) What would a net monthly salary have to be in 1940 to have the same purchasing power of a net monthly salary of $550 in 1965? HINT: Use the table on page 177 and an appropriate proportion. (*B*) Answer the same question for 1950 and 1960.

YEAR	COST-OF-LIVING INDEX
1940	48
1945	62
1950	82
1955	93
1960	103
1965	110

LIFE SCIENCE

⋆**13.** The naturalists in Yosemite National Park decided to estimate the number of bears in the most popular part of the park, Yosemite Valley. They used live traps (Fig. 6) to capture 50 bears; these bears were marked and released. A week later 50 more were captured, and it was observed that 10 of these were marked. Use this information to estimate the total number of bears in the valley.

Figure 6

⋆**14.** Gregor Mendel (1822), a Bavarian monk and biologist whose name is known to almost everyone today, made discoveries which revolutionalized the science of heredity. Out of many experiments in which he crossed peas of one characteristic with those of another, Mendel evolved his now famous laws of heredity. In one experiment he crossed dihybrid yellow round peas (which contained green and wrinkled as recessive genes) and obtained 560 peas of the following types: 319 yellow round, 101 yellow wrinkled, 108 green round, and 32 green wrinkled. From his laws of heredity he predicted the ratio $9:3:3:1$. Using the ratio, calculate the theoretical expected number of each type of pea from this cross, and compare it with the experimental results.

MUSIC

15. Starting with a string tuned to a given note, one can move up and down the scale simply by decreasing or increasing its length (while maintaining the same tension) according to simple whole number ratios (see Fig. 7). For example, $\frac{8}{9}$ of the C string gives the next higher note D, $\frac{2}{3}$ of the C string gives G, and $\frac{1}{2}$ of the C string gives C 1 octave higher. (The reciprocals of these fractions, $\frac{9}{8}$, $\frac{3}{2}$, and 2, respectively, are proportional to the frequencies of these notes.) Find the lengths of 7 strings (each less than 30 inches) that will produce the following seven chords when paired with a 30-inch string:

(*A*)	Octave	1 : 2	(*B*)	Fifth	2 : 3
(*C*)	Fourth	3 : 4	(*D*)	Major third	4 : 5
(*E*)	Minor third	5 : 6	(*F*)	Major sixth	3 : 5
(*G*)	Minor sixth	5 : 8			

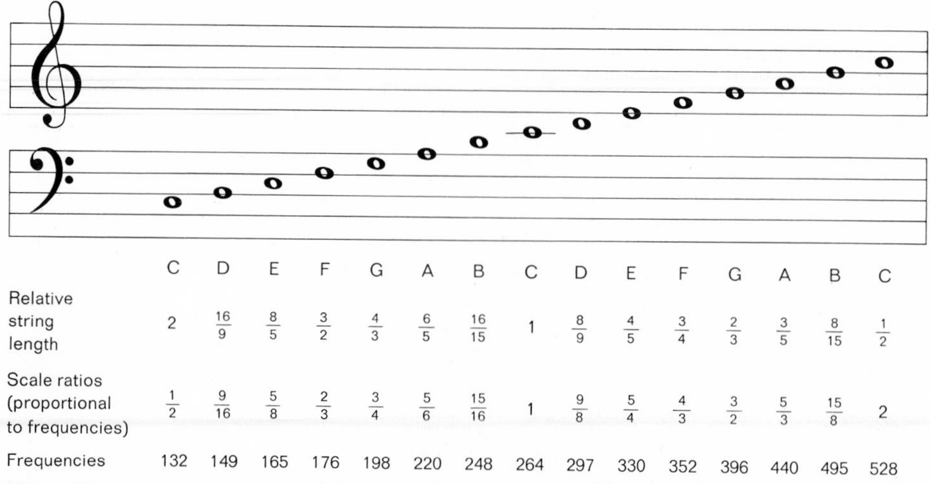

	C	D	E	F	G	A	B	C	D	E	F	G	A	B	C
Relative string length	2	$\frac{16}{9}$	$\frac{8}{5}$	$\frac{3}{2}$	$\frac{4}{3}$	$\frac{6}{5}$	$\frac{16}{15}$	1	$\frac{8}{9}$	$\frac{4}{5}$	$\frac{3}{4}$	$\frac{2}{3}$	$\frac{3}{5}$	$\frac{8}{15}$	$\frac{1}{2}$
Scale ratios (proportional to frequencies)	$\frac{1}{2}$	$\frac{9}{16}$	$\frac{5}{8}$	$\frac{2}{3}$	$\frac{3}{4}$	$\frac{5}{6}$	$\frac{15}{16}$	1	$\frac{9}{8}$	$\frac{5}{4}$	$\frac{4}{3}$	$\frac{3}{2}$	$\frac{5}{3}$	$\frac{15}{8}$	2
Frequencies	132	149	165	176	198	220	248	264	297	330	352	396	440	495	528

Figure 7

PHOTOGRAPHY **16.** A photographic developer is to be kept between 68 and 77°F. What is the range of temperature in Celsius ($F = \frac{9}{5}C + 32$)? Solve using inequality methods.

PHYSICS-ENGINEERING **17.** If the large cross-sectional area in a hydraulic lift (Fig. 8) is approximately 100 square inches and a person wants to lift a weight of 5,000 pounds with a 50-pound force, how large should the small cross-sectional area be?

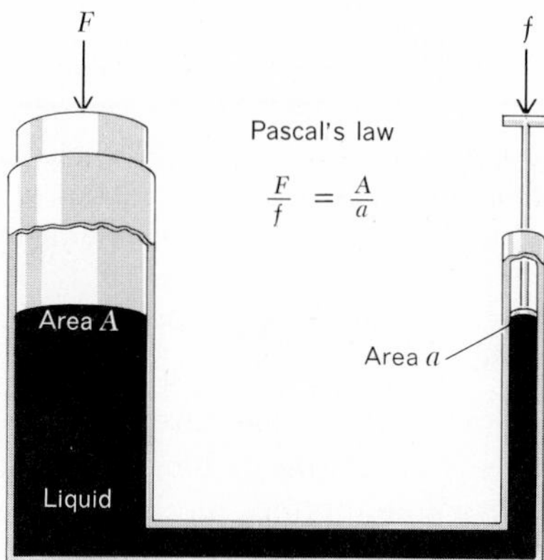

Figure 8

18. It is customary in supersonic studies to specify the velocity of an object relative to the velocity of sound. The ratio between these two velocities is called the Mach number, and it is given by the formula

$$M(\text{Mach number}) = \frac{V(\text{speed of object})}{S(\text{speed of sound})}$$

If a supersonic transport is designed to operate between Mach 1.7 and 2.4, what is the speed range of the transport in miles per hour? (Assume that the speed of sound is 740 miles/hour.) Solve using inequality methods.

PSYCHOLOGY

19. A person's IQ is found by dividing mental age, as indicated by standard tests, by chronological age, and then multiplying this ratio by 100. In terms of a formula,

$$\text{IQ} = \frac{\text{MA} \cdot 100}{\text{CA}}$$

If the IQ range of a group of 12-year-olds is $70 \leq \text{IQ} \leq 120$, what is the mental-age range of this group? Solve using inequality methods.

PUZZLE

20. In crossing the Grand Canyon in Arizona, a group went one-third of the way by foot, 10 miles by boat, and one-sixth of the way by mule. How long was the trip?

21. A pole is located in a pond. One-fifth of the pole is in the sand, 10 feet of it is in the water, and two-thirds of it is in the air. How long is the pole?

★22. Diophantus, an early Greek algebraist (A.D. 280), was the subject for a famous ancient puzzle. See if you can find Diophantus' age at death from the following information: Diophantus was a boy for one-sixth of his life; after one-twelfth more he grew a beard; after one-seventh more he married, and after 5 years of marriage he was granted a son; the son lived one-half as long as his father; and Diophantus died 4 years after his son's death.

6.6 Solving a Formula or Equation for a Particular Variable

One of the immediate applications you will have for algebra in other courses is the changing of formulas or equations to alternate equivalent forms. The following examples are more or less typical.

EXAMPLE 18

Solve the formula $c = wrt/1{,}000$ for t. The formula gives the cost c of using an electrical appliance; w = power in watts, r = rate per kilowatt hour, t = time in hours.

SOLUTION

$$\frac{wrt}{1{,}000} = c \qquad \text{Recall: If } a = b, \text{ then } b = a.$$

$$\frac{1{,}000}{wr} \cdot \frac{wrt}{1{,}000} = \frac{1{,}000}{wr} \cdot c$$

$$t = \frac{1{,}000c}{wr}$$

PROBLEM 18 Solve the formula in Example 8 for w.

ANSWER $w = \dfrac{1{,}000c}{rt}$

EXAMPLE 19 Solve the formula $A = P + Prt$ for r (simple-interest formula).

SOLUTION

$$P + Prt = A$$
$$Prt = A - P$$
$$r = \frac{A - P}{Pt}$$

PROBLEM 19 Solve the formula $A = P + Prt$ for t.

ANSWER $t = \dfrac{A - P}{Pr}$

EXAMPLE 20 Solve the formula $A = P + Prt$ for P.

SOLUTION

$$P + Prt = A$$
$$P(1 + \mathrm{rt}) = A$$
$$P = \frac{A}{1 + rt}$$

PROBLEM 20 Solve $A = xy + xz$ for x.

ANSWER $x = \dfrac{A}{y + z}$

Exercise 35

A

1. Solve $d = rt$ for r DISTANCE-RATE-TIME
2. Solve $d = 1{,}100t$ for t SOUND DISTANCE IN AIR
3. Solve $C = 2\pi r$ for r CIRCUMFERENCE OF A CIRCLE
4. Solve $I = Prt$ for t SIMPLE INTEREST
5. Solve $C = \pi D$ for π CIRCUMFERENCE OF A CIRCLE
6. Solve $e = mc^2$ for m MASS-ENERGY EQUATION
7. Solve $ax + b = 0$ for x FIRST-DEGREE POLYNOMIAL EQUATION
8. Solve $p = 2a + 2b$ for a PERIMETER OF A RECTANGLE

9. Solve $y = 2x - 5$ for x SLOPE-INTERCEPT EQUATION FOR A LINE

10. Solve $y = mx + b$ for m SLOPE-INTERCEPT EQUATION FOR A LINE

B 11. Solve $3x - 4y - 12 = 0$ for y LINEAR EQUATION IN TWO VARIABLES

12. Solve $Ax + By + C = 0$ for y LINEAR EQUATION IN TWO VARIABLES

13. Solve $I = \frac{E}{R}$ for R ELECTRICAL CIRCUITS-OHM'S LAW

14. Solve $m = \frac{b}{a}$ for a OPTICS-MAGNIFICATION

15. Solve $C = \frac{100B}{L}$ for B ANTHROPOLOGY-CEPHALIC INDEX

16. Solve $(\mathrm{IQ}) = \frac{100(\mathrm{MA})}{(\mathrm{CA})}$ for CA PSYCHOLOGY-INTELLIGENCE QUOTIENT

17. Solve $F = G\frac{m_1 m_2}{d^2}$ for G GRAVITATIONAL FORCE BETWEEN TWO MASSES

18. Solve $F = G\frac{m_1 m_2}{d^2}$ for m_1 GRAVITATIONAL FORCE BETWEEN TWO MASSES

19. Solve $F = \frac{9}{5}\mathrm{C} + 32$ for C CELSIUS-FAHRENHEIT

20. Solve $\mathrm{C} = \frac{5}{9}(\mathrm{F} - 32)$ for F CELSIUS-FAHRENHEIT

C 21. Solve $\frac{1}{f} = \frac{1}{a} + \frac{1}{b}$ for f OPTICS-FOCAL LENGTH

22. Solve $\frac{1}{R} = \frac{1}{R_1} + \frac{1}{R_2}$ for R ELECTRICAL CIRCUITS

23. Solve $a_n = a_1 + (n - 1)d$ for n ARITHMETIC PROGRESSION

24. Solve $a_n = a_1 + (n - 1)d$ for d ARITHMETIC PROGRESSION

25. Solve $P_1V_1/T_1 = P_2V_2/T_2$ for T_2 GAS LAW

26. Solve $P_1V_1/T_1 = P_2V_2/T_2$ for V_1 GAS LAW

27. Solve $y = \frac{2x - 3}{3x - 5}$ for x RATIONAL EQUATION

6.7 Complex Fractions

A fractional form with fractions in its numerator, denominator, or both is called a *complex fraction*. It is often necessary to represent a complex fraction as a simple fraction, that is (in all cases we will consider), as the quotient of two polynomials. The process does not involve any new concepts. It is a matter of applying old concepts in the right sequence. We will illustrate two approaches to the problem, each with its own merits depending on the particular problem under consideration. One of the methods makes very effective use of the property $a/b = ak/bk$, with $k \neq 0$.

EXAMPLE 21 Express $\dfrac{1 - \dfrac{1}{c^2}}{1 + \dfrac{1}{c}}$ as a simple fraction.

SOLUTION METHOD 1 Multiply the numerator and denominator by the lcd of all fractions within the numerator and denominator. Thus

$$\frac{1 - \dfrac{1}{c^2}}{1 + \dfrac{1}{c}} = \frac{c^2\left(1 - \dfrac{1}{c^2}\right)}{c^2\left(1 + \dfrac{1}{c}\right)} = \frac{c^2 - 1}{c^2 + c} = \frac{(c+1)(c-1)}{c(c+1)} = \frac{c-1}{c}$$

METHOD 2 Write the numerator and denominator as single fractions. Then treat as a quotient. Thus

$$\frac{1 - \dfrac{1}{c^2}}{1 + \dfrac{1}{c}} = \frac{\dfrac{c^2 - 1}{c^2}}{\dfrac{c+1}{c}} = \frac{(c+1)(c-1)}{c^2} \cdot \frac{c}{(c+1)} = \frac{c-1}{c}$$

PROBLEM 21 Express as a simple fraction in lowest terms. Use two methods.

$$\frac{1 - \dfrac{1}{3x}}{1 - \dfrac{1}{9x^2}}$$

ANSWER $\dfrac{3x}{3x+1}$

EXAMPLE 22 Express $2 - \dfrac{2}{2 - \dfrac{1}{x}}$ as a simple fraction.

SOLUTION

$$2 - \frac{2}{2 - \dfrac{1}{x}} = 2 - \frac{2x}{\left(2 - \dfrac{1}{x}\right)x} = 2 - \frac{2x}{2x - 1} = \frac{2(2x-1) - 2x}{2x - 1} = \frac{2x - 2}{2x - 1}$$

PROBLEM 22 Express $3 - \dfrac{1}{1 + \dfrac{2}{x}}$ as a simple fraction.

ANSWER $\dfrac{2x + 6}{x + 2}$

Exercise 36

Simplify:

A **1.** $\dfrac{1 + \dfrac{3}{x}}{x - \dfrac{9}{x}}$

2. $\dfrac{1 - \dfrac{2}{x}}{x - \dfrac{4}{x}}$

3. $\dfrac{1 - \dfrac{y^2}{x^2}}{1 - \dfrac{y}{x}}$

4. $\dfrac{\dfrac{a^2}{b^2} - 1}{\dfrac{a}{b} - 1}$

5. $\dfrac{\dfrac{1}{x} + \dfrac{1}{y}}{\dfrac{y}{x} - \dfrac{x}{y}}$

6. $\dfrac{b - \dfrac{a^2}{b}}{\dfrac{1}{a} - \dfrac{1}{b}}$

B **7.** $\dfrac{1 + \dfrac{2}{x} - \dfrac{15}{x^2}}{1 + \dfrac{4}{x} - \dfrac{5}{x^2}}$

8. $\dfrac{\dfrac{x}{y} - 2 + \dfrac{y}{x}}{\dfrac{x}{y} - \dfrac{y}{x}}$

9. $\dfrac{\dfrac{a^2}{a - b} - a}{\dfrac{b^2}{a - b} + b}$

10. $\dfrac{n - \dfrac{n^2}{n - m}}{1 + \dfrac{m^2}{n^2 - m^2}}$

11. $\dfrac{\dfrac{m}{m + 2} - \dfrac{m}{m - 2}}{\dfrac{m + 2}{m - 2} - \dfrac{m - 2}{m + 2}}$

12. $\dfrac{\dfrac{y}{x + y} - \dfrac{x}{x - y}}{\dfrac{x}{x + y} + \dfrac{y}{x - y}}$

13. $1 - \dfrac{1}{1 - \dfrac{1}{x}}$

14. $2 - \dfrac{1}{1 - \dfrac{2}{x + 2}}$

C **15.** $1 - \cfrac{x - \cfrac{1}{x}}{1 - \cfrac{1}{x}}$

16. $\cfrac{t - \cfrac{1}{1 + \cfrac{1}{t}}}{t + \cfrac{1}{t - \cfrac{1}{t}}}$

17. $1 + \cfrac{1}{1 + \cfrac{1}{1 + \cfrac{1}{1 + x}}}$

18. $1 - \cfrac{1}{1 - \cfrac{1}{1 - \cfrac{1}{1 - x}}}$

19. A formula for the average rate r for a round trip between two points, where the rate going is r_G and the rate returning is r_R, is given by the complex fraction

$$r = \cfrac{2}{\cfrac{1}{r_G} + \cfrac{1}{r_R}}$$

Express r as a simple fraction.

20. The airspeed indicator on a jet aircraft registers 500 miles/hour. If the plane is traveling with an airstream moving at 100 miles/hour, then the plane's ground-speed would be 600 miles/hour—or would it? According to Einstein, velocities must be added according to the following formula:

$$v = \cfrac{v_1 + v_2}{1 + \cfrac{v_1 v_2}{c^2}}$$

where v is the resultant velocity, c is the speed of light, and v_1 and v_2 are the two velocities to be added. Convert the right side of the equation into a simple fraction.

Exercise 37 Chapter Review

A *Perform the indicated operations and simplify.*

1. $\dfrac{y}{2} + \dfrac{y}{3}$

2. $1 + \dfrac{2}{3x}$

3. $\dfrac{2}{x} - \dfrac{1}{6x} + \dfrac{1}{3}$

4. $\dfrac{-3y}{5xz} \cdot \dfrac{-10z}{15xy}$

5. $\dfrac{4x^2y^3}{3a^2b^2} \div \dfrac{2xy^2}{3ab}$

6. $\dfrac{x + 1}{x + 2} - \dfrac{x + 2}{x + 3}$

7. $(d-2)^2 \div \dfrac{d^2-4}{d-2}$

8. $\dfrac{1-\dfrac{2}{y}}{1+\dfrac{1}{y}}$

Solve:

9. $-\dfrac{3}{5}y = \dfrac{2}{3}$

10. $\dfrac{x}{4} - 3 = \dfrac{x}{5}$

11. $0.4x + 0.3x = 6.3$

12. $\dfrac{x}{4} - 1 \geq \dfrac{x}{3}$

13. $\dfrac{2}{3m} - \dfrac{1}{4m} = \dfrac{1}{12}$

14. $\dfrac{3x}{x-5} - 8 = \dfrac{15}{x-5}$

15. If the ratio of men to women in a small town is 2 : 3 and there are 990 women, how many men are there? Set up a proportion and solve.

16. One-eighth of what number is $-\dfrac{3}{2}$? Write an equation and solve.

17. Solve $W = I^2R$ for R ELECTRICAL CIRCUITS, POWER IN WATTS

18. Solve $A = \dfrac{bh}{2}$ for b AREA OF A TRIANGLE

B *Perform the indicated operations and simplify.*

19. $\dfrac{2}{5b} - \dfrac{4}{3a^3} - \dfrac{1}{6a^2b^2}$

20. $\dfrac{4x^2y}{3ab^2} \div \left(\dfrac{2a^2x^2}{b^2y} \cdot \dfrac{6a}{2y^2}\right)$

21. $1 - \dfrac{m-1}{m+1}$

22. $\dfrac{3s}{3s^2-12s} + \dfrac{1}{2s+4s}$

23. $\dfrac{x}{x^2+4x} + \dfrac{2x}{3x^2-48}$

24. $\dfrac{x^3-x}{x^2-x} \div \dfrac{x^2+2x+1}{x}$

25. $\dfrac{\dfrac{x}{y} - \dfrac{y}{x}}{\dfrac{x}{y} + 1}$

26. $\dfrac{\dfrac{y^2}{x^2-y^2} + 1}{\dfrac{x^2}{x-y} - x}$

Solve:

27. $\dfrac{x}{4} - \dfrac{x-3}{3} = 2$

28. $0.4x - 0.3(x-3) = 5$

29. $\dfrac{x+3}{8} \leq 5 - \dfrac{2-x}{3}$

30. $\dfrac{5}{2x+3} - 5 = \dfrac{-5x}{2x+3}$

31. $\dfrac{3}{x} - \dfrac{2}{x+1} = \dfrac{1}{2x}$

32. $\dfrac{u-3}{2u-2} = \dfrac{1}{6} - \dfrac{1-u}{3u-3}$

33. Solve $S = \dfrac{n(a+L)}{2}$ for L ARITHMETIC PROGRESSION

34. If on a trip your car goes 391 miles on 23 gallons of gas, how many gallons of gas would be required for a 850-mile trip? Set up a proportion and solve.

35. A store has a camera on sale at 20 percent off list price. If the sale price is $64, what is the list price?

36. If an airplane can travel 300 miles against the wind in the same time it travels 400 miles with the wind and the speed of the wind is 25 miles/hour, what is the cruising speed of the airplane in still air?

37. A chemist has 1,200 grams of 60% pure acid. How much should be drained off and replaced with pure acid to obtain a solution that is 75% pure?

C *Perform the indicated operations and simplify.*

38. $-\dfrac{1}{a-b} = \dfrac{?}{b-a}$

39. $\dfrac{4}{s^2-4} + \dfrac{1}{2-s}$

40. $\dfrac{y^2-y-6}{(y+2)^2} \cdot \dfrac{2+y}{3-y}$

41. $\dfrac{y}{x^2} \div \left(\dfrac{x^2+3x}{2x^2+5x-3} \div \dfrac{x^3y-x^2y}{2x^2-3x+1}\right)$

42. $\left(x - \dfrac{1}{1-\dfrac{1}{x}}\right) \div \left(\dfrac{x}{x+1} - \dfrac{x}{1-x}\right)$

43. $\dfrac{1-\dfrac{1}{1+\dfrac{x}{y}}}{1-\dfrac{1}{1-\dfrac{x}{y}}}$

Solve:

44. $\dfrac{x-3}{12} - \dfrac{(x+2)}{9} = \dfrac{1-x}{6} - 1$

45. $\dfrac{3x}{5} - \dfrac{1}{2}(x-3) \leq \dfrac{1}{3}(x+2)$

46. $\dfrac{7}{2-x} = \dfrac{10-4x}{x^2+3x-10}$

47. Given $I = \frac{E}{R}$ and $W = IE$, write a formula for W in terms of E and R.

48. Solve $\frac{1}{f} = \frac{1}{f_1} + \frac{1}{f_2}$ for f_1 OPTICS

49. The three minor chords are composed of notes whose frequencies are in the ratio 10 : 12 : 15. If the first note of a minor chord is A with a frequency of 220 hertz, what are the frequencies of the other two notes?

50. It is known that a carton contains 100 packages and that some of the packages weigh $\frac{1}{2}$ pound each and the rest weigh $\frac{1}{3}$ pound each. To save time counting each type of package in the carton, you can weigh the whole contents of the box (45 pounds) and determine the number of each kind of package by use of algebra. How many are there of each kind?

51. A chemist has 3 kilograms (3,000 grams) of 20% hydrochloric acid. She wishes to increase its strength to 25% by draining off some and replacing it with an 80% solution. How many grams must be drained and replaced with the 80% solution?

Chapter 7
EXPONENTS

7.1 Positive-Integer Exponents

Recall that if n is a positive integer and a is any real number, then a^n represents the product of n factors of a. Symbolically,

$$a^n = aa \cdots a \qquad n \text{ factors of } a$$

As a consequence of this definition, we obtained the first property of exponents for m and n positive integers.

EXAMPLE 1 $$a^3a^4 = \underbrace{(a \cdot a \cdot a)}_{3 \text{ factors}}\underbrace{(a \cdot a \cdot a \cdot a)}_{4 \text{ factors}} = \underbrace{(a \cdot a \cdot a \cdot a \cdot a \cdot a \cdot a)}_{3+4 \text{ factors}} = a^{3+4} = a^7$$

PROPERTY 1 $a^m a^n = a^{m+n}$

PROBLEM 1 $x^7x^9 = ?$

ANSWER x^{16}

By now you are probably using this property with very little effort. When more complicated expressions involving exponents are encountered, other exponent properties combined with the first provide an efficient tool for simplifying and manipulating these expressions. In this section we will introduce and discuss four additional exponent properties.

In each of the following expressions m and n are natural numbers, and a and b are real numbers, excluding division by 0, of course.

EXAMPLE 2

$$(a^3)^4 = a^3 \cdot a^3 \cdot a^3 \cdot a^3 = \overbrace{(a \cdot a \cdot a)(a \cdot a \cdot a)(a \cdot a \cdot a)(a \cdot a \cdot a)}^{\text{4 groups of 3 factors each}}$$

$$= \overbrace{(a \cdot a \cdot a \cdot a \cdot a \cdot a \cdot a \cdot a \cdot a \cdot a \cdot a \cdot a)}^{4 \cdot 3 \text{ factors}} = a^{4 \cdot 3} = a^{12}$$

PROPERTY 2 $(a^n)^m = a^{mn}$

PROBLEM 2 $(x^2)^5 = ?$

ANSWER x^{10}

EXAMPLE 3

$$(ab)^4 = \overbrace{(ab)(ab)(ab)(ab)}^{\text{4 factors of } (ab)} = \overbrace{(a \cdot a \cdot a \cdot a)}^{\text{4 factors}}\overbrace{(b \cdot b \cdot b \cdot b)}^{\text{4 factors}} = a^4b^4$$

PROPERTY 3 $(ab)^m = a^mb^m$

PROBLEM 3 $(xy)^7 = ?$

ANSWER x^7y^7

EXAMPLE 4

$$\left(\frac{a}{b}\right)^5 = \overbrace{\left(\frac{a}{b} \cdot \frac{a}{b} \cdot \frac{a}{b} \cdot \frac{a}{b} \cdot \frac{a}{b}\right)}^{\text{5 factors of } a/b} = \frac{a \cdot a \cdot a \cdot a \cdot a}{b \cdot b \cdot b \cdot b \cdot b} = \frac{a^5}{b^5}$$

PROPERTY 4 $\left(\dfrac{a}{b}\right)^m = \dfrac{a^m}{b^m}$

PROBLEM 4 $\left(\dfrac{x}{y}\right)^3 = ?$

ANSWER $\dfrac{x^3}{y^3}$

EXAMPLE 5

(A) $$\frac{a^7}{a^3} = \frac{a \cdot a \cdot a \cdot a \cdot a \cdot a \cdot a}{a \cdot a \cdot a} = \frac{(a \cdot a \cdot a)(a \cdot a \cdot a \cdot a)}{(a \cdot a \cdot a)} = a^{7-3} = a^4$$

(B) $$\frac{a^3}{a^3} = \frac{a \cdot a \cdot a}{a \cdot a \cdot a} = 1$$

(C) $$\frac{a^4}{a^7} = \frac{a \cdot a \cdot a \cdot a}{a \cdot a \cdot a \cdot a \cdot a \cdot a \cdot a} = \frac{(a \cdot a \cdot a \cdot a)}{(a \cdot a \cdot a \cdot a)(a \cdot a \cdot a)} = \frac{1}{a^{7-4}} = \frac{1}{a^3}$$

PROPERTY 5

$$\frac{a^m}{a^n} = \begin{cases} a^{m-n} & \text{if } m > n \\ 1 & \text{if } m = n \\ \dfrac{1}{a^{n-m}} & \text{if } n > m \end{cases}$$

PROBLEM 5 (A) $x^8/x^3 = ?$ (B) $x^8/x^8 = ?$ (C) $x^3/x^8 = ?$

ANSWER (A) x^5 (B) 1 (C) $1/x^5$

The properties of exponents are theorems, and as such they require proofs. We have only given plausible arguments for each property; formal proofs of these properties require a property of the natural numbers, called the inductive property, which is beyond the scope of this course.

It is very important to observe and remember that the properties of exponents apply to products and quotients, and not to sums and differences. Many mistakes are made in algebra by people applying a property of exponents to the wrong algebraic form. The exponent properties are summarized below for m and n positive integers.

PROPERTIES OF EXPONENTS

1 $a^m a^n = a^{m+n}$

2 $(a^n)^m = a^{mn}$

3 $(ab)^m = a^m b^m$

4 $\left(\dfrac{a}{b}\right)^m = \dfrac{a^m}{b^m}$

5 $\dfrac{a^m}{a^n} = \begin{cases} a^{m-n} & \text{if } m > n \\ 1 & \text{if } m = n \\ \dfrac{1}{a^{n-m}} & \text{if } n > m \end{cases}$

As before, the "dotted boxes" in the examples given below are used to indicate steps that are usually carried out mentally. In addition, we note that

$$-3^2 \neq (-3)^2$$

since -3^2 is to be interpreted to mean $-(3 \cdot 3) = -9$ while $(-3)^2 = (-3)(-3) = 9$.

EXAMPLE 6 (A) $x^{12}x^{13} \boxed{= x^{12+13}} = x^{25}$

(B) $(t^7)^5 = t^{5\cdot 7} = t^{35}$

(C) $(xy)^5 = x^5y^5$

(D) $\left(\frac{u}{v}\right)^3 = \frac{u^3}{v^3}$

(E) $\frac{x^{12}}{x^4} = x^{12-4} = x^8$

(F) $\frac{t^4}{t^9} = \frac{1}{t^{9-4}} = \frac{1}{t^5}$

PROBLEM 6 Simplify:

(A) x^8x^6 (B) $(u^4)^5$ (C) $(xy)^9$

(D) $\left(\frac{x}{y}\right)^4$ (E) x^{10}/x^3 (F) x^3/x^{10}

ANSWER (A) x^{14} (B) u^{20} (C) x^9y^9 (D) $\frac{x^4}{y^4}$ (E) x^7

(F) $1/x^7$

EXAMPLE 7 (A) $(x^2y^3)^4 = (x^2)^4(y^3)^4 = x^8y^{12}$

(B) $\left(\frac{u^3}{v^4}\right)^3 = \frac{(u^3)^3}{(v^4)^3} = \frac{u^9}{v^{12}}$

(C) $\frac{2x^9y^{11}}{4x^{12}y^7} = \frac{2}{4}\cdot\frac{x^9}{x^{12}}\cdot\frac{y^{11}}{y^7} = \frac{1}{2}\cdot\frac{1}{x^3}\cdot\frac{y^4}{1} = \frac{y^4}{2x^3}$

PROBLEM 7 Simplify:

(A) $(u^3v^4)^2$ (B) $\left(\frac{u^3}{v^4}\right)^3$ (C) $\frac{9x^7y^2}{3x^5y^3}$

ANSWER (A) u^6v^8 (B) $\frac{u^9}{v^{12}}$ (C) $\frac{3x^2}{y}$

Knowing the rules of the game of chess doesn't make one good at playing chess; similarly, memorizing the properties of exponents doesn't necessarily make one good at using these properties. To acquire skill in their use, one must use these properties in a fairly large variety of problems. The following exercises should help you acquire this skill.

Exercise 38

A *Replace the question marks with appropriate symbols.*

1. $y^2y^7 = y^?$ **2.** $x^7x^5 = x^?$ **3.** $y^8 = y^3y$

4. $x^{10} = x^?x^6$ **5.** $(u^4)^3 = u^?$ **6.** $(v^2)^3 = ?$

7. $x^{10} = (x^?)^5$ **8.** $y^{12} = (y^6)^?$ **9.** $(uv)^7 = ?$

10. $(xy)^5 = x^5y^?$ **11.** $p^4q^4 = (pq)^?$ **12.** $m^3n^3 = (mn)^?$

13. $\left(\frac{a}{b}\right)^8 = ?$ **14.** $\left(\frac{x}{y}\right)^4 = \frac{x^?}{y^4}$ **15.** $\frac{m^3}{n^3} = \left(\frac{m}{n}\right)^?$

16. $\frac{x^7}{x^7} = \left(\frac{x}{y}\right)^?$ **17.** $\frac{n^{14}}{n^8} = n^?$ **18.** $\frac{x^7}{x^3} = x^?$

19. $m^6 = \frac{m^8}{m^?}$ **20.** $x^3 = \frac{x^?}{x^4}$ **21.** $\frac{x^4}{x^{11}} = \frac{1}{x^?}$

22. $\frac{a^5}{a^9} = \frac{1}{a^?}$ **23.** $\frac{1}{x^8} = \frac{x^4}{x^?}$ **24.** $\frac{1}{u^2} = \frac{u^?}{u^9}$

Simplify, using appropriate properties of exponents.

25. $(5x^2)(2x^9)$ **26.** $(2x^3)(3x^7)$ **27.** $\frac{9x^6}{3x^4}$

28. $\frac{4x^8}{2x^6}$ **29.** $\frac{6m^5}{8m^7}$ **30.** $\frac{4u^3}{2u^7}$

31. $(xy)^{10}$ **32.** $(cd)^{12}$ **33.** $\left(\frac{m}{n}\right)^5$

34. $\left(\frac{x}{y}\right)^6$

B **35.** $(4y^3)(3y)(y^6)$ **36.** $(2x^2)(3x^3)(x^4)$ **37.** $(5 \times 10^8)(7 \times 10^9)$

38. $(2 \times 10^3)(3 + 10^{12})$ **39.** $(10^7)^2$ **40.** $(10^4)^5$

41. $(x^3)^2$ **42.** $(y^4)^5$ **43.** $(m^2n^5)^3$

44. $(x^2y^3)^4$ **45.** $\left(\frac{c^2}{d^5}\right)^3$ **46.** $\left(\frac{a^3}{b^2}\right)^4$

47. $\frac{9u^8v^6}{3u^4v^8}$ **48.** $\frac{2x^3y^8}{6x^7y^2}$ **49.** $(2s^2t^4)^4$

50. $(3a^3b^2)^3$ **51.** $6(xy^3)^5$ **52.** $2(x^2y)^4$

53. $\left(\frac{mn^3}{p^2q}\right)^4$ **54.** $\left(\frac{x^2y}{2w^2}\right)^3$

C **55.** $\dfrac{(4u^3v)^3}{(2uv^2)^6}$ **56.** $\dfrac{(2xy^3)^2}{(4x^2y)^3}$ **57.** $\dfrac{(9x^3)^2}{(-3x)^2}$

58. $\dfrac{(-2x^2)^3}{(2^2x)^4}$ **59.** $\dfrac{-x^2}{(-x)^2}$ **60.** $\dfrac{-2^2}{(-2)^2}$

61. $\dfrac{(-x^2)^2}{(-x^3)^3}$ **62.** $\dfrac{-2^4}{(-2a^2)^4}$

7.2 Integer Exponents in General

How should symbols such as

8^0 and 7^{-3}

be defined? In this section we will extend the meaning of exponent to include 0 and negative integers. Thus, typical scientific expressions such as

The diameter of a red corpuscle is approximately 8×10^{-5} centimeters.

The amount of water found in the air as vapor is about 9×10^{-6} times that found in seas.

The focal length of a thin lens is given by $f^{-1} = a^{-1} + b^{-1}$.

will then make sense.

In extending the concept of exponent beyond the natural numbers, we will require that any new exponent symbol be defined in such a way that all five properties of exponents for natural numbers continue to hold. Thus, we will need only one set of properties for all types of exponents rather than a new set for each new exponent.

We will start by defining the 0 exponent. If all the exponent properties must hold even if some of the exponents are 0, then, in particular, the first exponent property must hold for zero exponents, that is,

$$a^0 \cdot a^2 = a^{0+2} = a^2$$

This suggests that a^0 should be defined as 1 for all nonzero real numbers a, since 1 is the only real number that gives a^2 when multiplied by a^2. If we let $a = 0$ and follow the same reasoning, we find that

$$0^0 \cdot 0^2 = 0^{0+2} = 0^2 = 0$$

and 0^0 could be any real number since $0^2 = 0$; hence 0^0 is not uniquely determined. For this reason 0^0 is not defined.

DEFINITION OF THE ZERO EXPONENT

For all real numbers $a \neq 0$.

$$a^0 = 1$$

0^0 is not defined

(A) $5^0 = 1$

(B) $325^0 = 1$

(C) $\left(\frac{1}{3}\right)^0 = 1$

(D) $t^0 = 1 \quad t \neq 0$

(E) $(x^2y^3)^0 = 1 \quad x \neq 0, y \neq 0$

PROBLEM 8 Simplify:

(A) 12^0 (B) 999^0 (C) $\left(\frac{2}{7}\right)^0$

(D) $x^0 \quad x \neq 0$ (E) $(m^3n^3)^0 \quad m \neq 0, n \neq 0$

ANSWER All are equal to 1.

To get an idea of how a negative-integer exponent should be defined, we can proceed as above. If the first law of exponents is to hold, then a^{-2} $(a \neq 0)$ must be defined so that

$$a^{-2} \cdot a^2 = a^{-2+2} = a^0 = 1$$

Thus,

$$a^{-2}a^2 = 1$$

and after dividing both sides by a^2, we obtain

$$a^{-2} = \frac{1}{a^2}$$

This kind of reasoning leads us to the following general definition.

DEFINITION OF NEGATIVE INTEGER EXPONENT

If n is a positive integer and a is a nonzero real number, then

$$a^{-n} = \frac{1}{a^n}$$

Of course, it follows, using equality properties, that

$$a^n = \frac{1}{a^{-n}}$$

EXAMPLE 9

(A) $a^{-7} = \frac{1}{a^7}$

(B) $\frac{1}{x^{-8}} = x^8$

(C) $10^{-3} = \frac{1}{10^3}$ or $\frac{1}{1{,}000}$ or 0.001

(D) $\frac{x^{-3}}{y^{-5}} = \frac{x^{-3}}{1} \cdot \frac{1}{y^{-5}} = \frac{1}{x^3} \cdot \frac{y^5}{1} = \frac{y^5}{x^3}$

PROBLEM 9 Write using positive exponents:

(A) x^{-5} (B) $\frac{1}{y^{-4}}$

(C) 10^{-2} (D) $\frac{m^{-2}}{n^{-3}}$

ANSWER (A) $\frac{1}{x^5}$ (B) y^4 (C) $\frac{1}{10^2}$ or $\frac{1}{100}$ or 0.01 (D) $\frac{n^3}{m^2}$

With the definition of negative exponent and zero exponent at hand, we can now replace the fifth property of exponents with a simpler form that does not have any restrictions on the relative size of the exponents. Thus

$$\frac{a^m}{a^n} = a^{m-n} = \frac{1}{a^{n-m}} \qquad a \neq 0$$

EXAMPLE 10

(A) $\frac{2^5}{2^8} = 2^{5-8} = 2^{-3}$ or $\frac{2^5}{2^8} = \frac{1}{2^{8-5}} = \frac{1}{2^3}$

(B) $\frac{10^{-3}}{10^6} = 10^{-3-6} = 10^{-9}$ or $\frac{10^{-3}}{10^6} = \frac{1}{10^{6-(-3)}} = \frac{1}{10^{6+3}} = \frac{1}{10^9}$

PROBLEM 10 (A) Combine denominator with numerator:

$$\frac{x^{-2}}{x^3}$$

(*B*) Combine numerator with denominator in part *A*.

ANSWER (*A*) x^{-5} (*B*) $\dfrac{1}{x^5}$

Table 1 provides a summary of all of our work on exponents to this point.

TABLE 1 INTEGER EXPONENTS AND THEIR PROPERTIES (SUMMARY)

DEFINITION OF a^p, p AN INTEGER AND a A REAL NUMBER	PROPERTIES OF EXPONENTS, n AND m INTEGERS, a AND b REAL NUMBERS
1 *If p is a positive integer, then* $a^p = a \cdot a \cdots a$ *p factors of a*	1 $a^m a^n = a^{m+n}$ 2 $(a^n)^m = a^{mn}$
EXAMPLE: $3^5 = 3 \cdot 3 \cdot 3 \cdot 3 \cdot 3$	3 $(ab)^m = a^m b^m$
2 *If* $p = 0$ *and* $a \neq 0$, *then* $a^p = 1$; 0^0 *is not defined*	4 $\left(\dfrac{a}{b}\right)^m = \dfrac{a^m}{b^m}$
EXAMPLE: $8^0 = 1$	5 $\dfrac{a^m}{a^n} = a^{m-n} = \dfrac{1}{a^{n-m}}$
3 *If p is a negative integer, then* $a^p = \dfrac{1}{a^{-p}}$ $\quad a \neq 0$	
EXAMPLE: $3^{-4} = \dfrac{1}{3^{-(-4)}} = \dfrac{1}{3^4}$	

EXAMPLE 11 Simplify and express answers using positive exponents only.

(*A*) $a^5 a^{-2} = a^{5-2} = a^3$

(*B*) $(a^{-3}b^2)^{-2} = (a^{-3})^{-2}(b^2)^{-2} = a^6 b^{-4} = \dfrac{a^6}{b^4}$

(*C*) $\left(\dfrac{a^{-5}}{a^{-2}}\right)^{-1} = \dfrac{(a^{-5})^{-1}}{(a^{-2})^{-1}} = \dfrac{a^5}{a^2} = a^3$

(*D*) $\dfrac{4x^{-3}y^{-5}}{6x^{-4}y^3} = \dfrac{2x^{-3-(-4)}}{3y^{3-(-5)}} = \dfrac{2x^{-3+4}}{3y^{3+5}} = \dfrac{2x}{3y^8}$

or, changing to positive exponents first,

$$\frac{4x^{-3}y^{-5}}{6x^{-4}y^3} = \frac{2x^4}{3x^3y^3y^5} = \frac{2x}{3y^8}$$

(E) $$\frac{10^{-4}\cdot 10^{2}}{10^{-3}\cdot 10^{5}} = \frac{10^{-4+2}}{10^{-3+5}} = \frac{10^{-2}}{10^{2}} = \frac{1}{10^{4}} = \frac{1}{10{,}000} = 0.0001$$

(F) $$\left(\frac{m^{-3}m^{3}}{n^{-2}}\right)^{-2} = \left(\frac{m^{-3+3}}{n^{-2}}\right)^{-2} = \left(\frac{m^{0}}{n^{-2}}\right)^{-2} = \left(\frac{1}{n^{-2}}\right)^{-2} = \frac{1^{-2}}{(n^{-2})^{-2}} = \frac{1}{n^{4}}$$

PROBLEM 11 Simplify and express answers using positive exponents only:

(A) $x^{-2}x^{6}$ (B) $(x^{3}y^{-2})^{-2}$ (C) $\left(\dfrac{x^{-6}}{x^{-2}}\right)^{-1}$

(D) $\dfrac{8m^{-2}n^{-4}}{6m^{-5}n^{2}}$ (E) $\dfrac{10^{-3}\cdot 10^{5}}{10^{-2}\cdot 10^{6}}$

ANSWER (A) x^{4} (B) y^{4}/x^{6} (C) x^{4} (D) $4m^{3}/3n^{6}$

(E) $1/10^{2}$ or 0.01

EXAMPLE 12 Simplify and express answers using positive exponents only.

(A) $$\frac{3^{-2}+2^{-1}}{11} = \frac{\dfrac{1}{3^{2}}+\dfrac{1}{2}}{11} = \frac{\dfrac{2}{18}+\dfrac{9}{18}}{11} = \frac{11}{18} \div 11 = \frac{11}{18}\cdot\frac{1}{11} = \frac{1}{18}$$

(B) $$(a^{-1}-b^{-1})^{2} = \left(\frac{1}{a}-\frac{1}{b}\right)^{2} = \left(\frac{b-a}{ab}\right)^{2} = \frac{b^{2}-2ab+a^{2}}{a^{2}b^{2}}$$

PROBLEM 12 Simplify and express answers using positive exponents only:

(A) $\dfrac{2^{-2}+3^{-1}}{5}$ (B) $(x^{-1}+y^{-1})^{2}$

ANSWER (A) $\dfrac{7}{60}$ (B) $\dfrac{(x+y)^{2}}{x^{2}y^{2}}$ or $\dfrac{x^{2}+2xy+y^{2}}{x^{2}y^{2}}$

Exercise 39

A *Simplify and write answers using positive exponents only.*

1. 23^{0} **2.** 10^{0} **3.** y^{0}

4. x^{0} **5.** 3^{-3} **6.** 2^{-2}

7. m^{-7} **8.** x^{-4} **9.** $\dfrac{1}{4^{-3}}$

10. $\frac{1}{3^{-2}}$ **11.** $\frac{1}{y^{-5}}$ **12.** $\frac{1}{x^{-3}}$

13. $10^7 \cdot 10^{-5}$ **14.** $10^{-4} \cdot 10^6$ **15.** $y^{-3}y^4$

16. x^6x^{-2} **17.** u^5u^{-5} **18.** $m^{-3}m^3$

19. $\frac{10^3}{10^{-7}}$ **20.** $\frac{10^8}{10^{-3}}$ **21.** $\frac{x^9}{x^{-2}}$

22. $\frac{a^8}{a^{-4}}$ **23.** $\frac{z^{-2}}{z^3}$ **24.** $\frac{b^{-3}}{b^5}$

25. $\frac{10^{-1}}{10^6}$ **26.** $\frac{10^{-4}}{10^2}$ **27.** $(10^{-4})^{-3}$

28. $(2^{-3})^{-2}$ **29.** $(y^{-2})^{-4}$ **30.** $(x^{-5})^{-2}$

31. $(u^{-5}v^{-3})^{-2}$ **32.** $(x^{-3}y^{-2})^{-1}$ **33.** $(x^2y^{-3})^2$

34. $(x^{-2}y^3)^2$ **35.** $(x^{-2}y^3)^{-1}$ **36.** $(x^2y^{-3})^{-1}$

B **37.** $(m^2)^0$ **38.** $1{,}231^0$ **39.** $\frac{10^{-3}}{10^{-5}}$

40. $\frac{10^{-2}}{10^{-4}}$ **41.** $\frac{y^{-2}}{y^{-3}}$ **42.** $\frac{x^{-3}}{x^{-2}}$

43. $\frac{10^{-13} \cdot 10^{-4}}{10^{-21} \cdot 10^3}$ **44.** $\frac{10^{23} \cdot 10^{-11}}{10^{-3} \cdot 10^{-2}}$ **45.** $\frac{18 \times 10^{12}}{6 \times 10^{-4}}$

46. $\frac{8 \times 10^{-3}}{2 \times 10^{-5}}$ **47.** $\left(\frac{y}{y^{-2}}\right)^3$ **48.** $\left(\frac{x^2}{x^{-1}}\right)^2$

49. $\frac{1}{(3mn)^{-2}}$ **50.** $(2cd^2)^{-3}$ **51.** $(2mn^{-3})^3$

52. $(3x^3y^{-2})^2$ **53.** $(m^4n^{-5})^{-3}$ **54.** $(x^{-3}y^2)^{-2}$

55. $(2^23^{-3})^{-1}$ **56.** $(2^{-3}3^2)^{-2}$ **57.** $(10^{12} \cdot 10^{-12})^{-1}$

58. $(10^2 \cdot 3^0)^{-2}$ **59.** $\frac{8x^{-3}y^{-1}}{6x^2y^{-4}}$ **60.** $\frac{9m^{-4}n^3}{12m^{-1}n^{-1}}$

61. $\frac{2a^6b^{-2}}{16a^{-3}b^2}$ **62.** $\frac{4x^{-2}y^{-3}}{2x^{-3}y^{-1}}$ **63.** $\left(\frac{x^{-1}}{x^{-8}}\right)^{-1}$

64. $\left(\frac{n^{-3}}{n^{-2}}\right)^{-2}$ **65.** $\left(\frac{m^{-2}n^3}{m^4n^{-1}}\right)^2$ **66.** $\left(\frac{x^4y^{-1}}{x^{-2}y^3}\right)^2$

67. $\left(\frac{6nm^{-2}}{3m^{-1}n^2}\right)^{-3}$ **68.** $\left(\frac{2x^{-3}y^2}{4xy^{-1}}\right)^{-2}$

C **69.** $(a^2 - b^2)^{-1}$ **70.** $(x + 2)^{-2}$ **71.** $\dfrac{x^{-1} + y^{-1}}{x + y}$

72. $\dfrac{2^{-1} + 3^{-1}}{25}$ **73.** $\dfrac{c - d}{c^{-1} - d^{-1}}$ **74.** $\dfrac{12}{2^{-2} + 3^{-1}}$

75. $(x^{-1} + y^{-1})^{-1}$ **76.** $(2^{-2} + 3^{-2})^{-1}$ **77.** $(x^{-1} - y^{-1})^2$

78. $(10^{-2} + 10^{-3})^{-1}$

7.3 Scientific Notation and Applications

Work in science and engineering often involves the use of very, very large numbers:

> The estimated free oxygen of the earth weighs approximately 1,500,000,000,000,000,000,000 grams.

Also involved is the use of very, very small numbers:

> The probable mass of a hydrogen atom is 0.000 000 000 000 000 000 000 001 7 grams.

Writing and working with numbers of this type in standard decimal notation is generally awkward. It is often convenient to represent numbers of this type in *scientific notation*; that is, as the product of a number between 1 and 10 and a power of 10.

EXAMPLE 13 **DECIMAL FRACTIONS AND SCIENTIFIC NOTATION**

$$5 = 5 \times 10^0 \qquad 0.7 = 7 \times 10^{-1}$$

$$35 = 3.5 \times 10 \qquad 0.083 = 8.3 \times 10^{-2}$$

$$430 = 4.3 \times 10^2 \qquad 0.0043 = 4.3 \times 10^{-3}$$

$$5{,}870 = 5.87 \times 10^3 \qquad 0.000687 = 6.87 \times 10^{-4}$$

$$8{,}910{,}000 = 8.91 \times 10^6 \qquad 0.00000036 = 3.6 \times 10^{-7}$$

Can you discover a simple mechanical rule that relates the number of decimal places the decimal is moved with the power of 10 that is used?

PROBLEM 13 Write as a product of a number between 1 and 10 and a power of 10:

(*A*) 450 (*B*) 27,000

(*C*) 0.05 (*D*) 0.0000063

ANSWER (*A*) 4.5×10^2 (*B*) 2.7×10^4 (*C*) 5×10^{-2}

(*D*) 6.3×10^{-6}

Figure 1 shows the relative size of a number of familiar objects on a power-of-ten scale. Note that 10^{10} is not just double 10^5 (Why?).

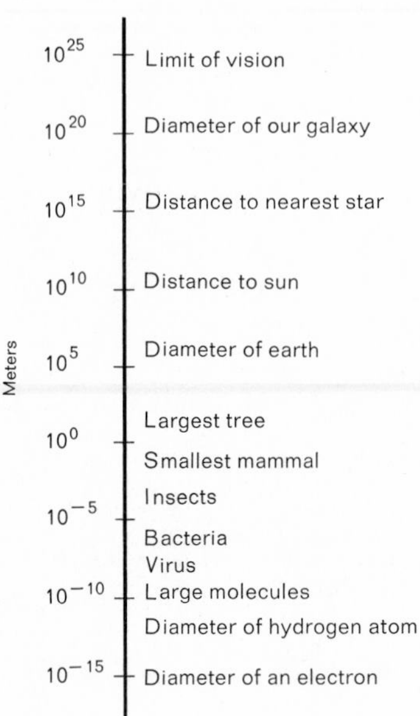

Figure 1

EXAMPLE 14 **EVALUATION OF A COMPLICATED ARITHMETIC PROBLEM**

$$\frac{(0.26)(720)}{(48{,}000{,}000)(0.0013)} = \frac{(2.6\times 10^{-1})(7.2\times 10^{2})}{(4.8\times 10^{7})(1.3\times 10^{-3})}$$

$$= \frac{\overset{2}{(\cancel{2.6})}\overset{6}{(\cancel{7.2})}}{\underset{4}{(\cancel{4.8})}\underset{1}{(\cancel{1.3})}} \cdot \frac{(10^{-1})(10^{2})}{(10^{7})(10^{-3})}$$

$$= 3 \times 10^{-3} \text{ or } 0.003$$

PROBLEM 14 Convert to scientific notation and evaluate: $\dfrac{(42{,}000)(0.009)}{(600)(0.000021)}$

ANSWER 3×10^4 or 30,000

We are able to look back into time by looking out into space. Since light travels at a fast but finite rate, we see heavenly bodies not as they exist now, but as they existed sometime in the past. If the distance between the sun and the earth is approximately 9.3×10^7 miles and if light travels at the rate of approximately 1.86×10^5 miles/second, we see the sun as it was how many minutes ago?

$$d = rt \qquad t = \frac{9.3 \times 10^7}{1.86 \times 10^5} = 5 \times 10^2 = 500 \text{ seconds} \quad \text{or}$$
$$t = d/r$$

$$\frac{500}{60} = 8.3 \text{ minutes}$$

Hence, we always see the sun as it was 8.3 minutes ago.

Exercise 40

A *Write in scientific notation:*

1. 70 **2.** 50 **3.** 800
4. 600 **5.** 80,000 **6.** 600,000
7. 0.008 **8.** 0.06 **9.** 0.00000008
10. 0.00006 **11.** 52 **12.** 35
13. 0.63 **14.** 0.72 **15.** 340
16. 270 **17.** 0.085 **18.** 0.032
19. 6,300 **20.** 5,200 **21.** 0.0000068
22. 0.00072

Write as a decimal fraction:

23. 8×10^2 **24.** 5×10^2 **25.** 4×10^{-2}
26. 8×10^{-2} **27.** 3×10^5 **28.** 6×10^6
29. 9×10^{-4} **30.** 2×10^{-5} **31.** 5.6×10^4
32. 7.1×10^3 **33.** 9.7×10^{-3} **34.** 8.6×10^{-4}
35. 4.3×10^5 **36.** 8.8×10^6 **37.** 3.8×10^{-7}
38. 6.1×10^{-6}

B *Write in scientific notation:*

39. 5,460,000,000 **40.** 42,700,000
41. 0.0000000729 **42.** 0.0000723

43. The energy of a laser beam can go as high as 10,000,000,000,000 watts.

44. The distance that light travels in 1 year is called a light-year. It is approximately 5,870,000,000,000 miles.

45. The nucleus of an atom has a diameter of a little more than 1/100,000 that of the whole atom.

46. The mass of one water molecule is 0.00000000000000000000003 grams.

Write as a decimal fraction.

47. 8.35×10^{10}

48. 3.46×10^{9}

49. 6.14×10^{-12}

50. 6.23×10^{-7}

51. The diameter of the sun is approximately 8.65×10^{5} miles.

52. The distance from the earth to the sun is approximately 9.3×10^{7} miles.

53. The probable mass of a hydrogen atom is 1.7×10^{-24} grams.

54. The diameter of a red corpuscle is approximately 7.5×10^{-5} centimeters.

Simplify and express answer in scientific notation.

55. $(3 \times 10^{-6})(3 \times 10^{10})$

56. $(4 \times 10^{5})(2 \times 10^{-3})$

57. $(2 \times 10^{3})(3 \times 10^{-7})$

58. $(4 \times 10^{-8})(2 \times 10^{5})$

59. $\dfrac{6 \times 10^{12}}{2 \times 10^{7}}$

60. $\dfrac{9 \times 10^{8}}{3 \times 10^{5}}$

61. $\dfrac{15 \times 10^{-2}}{3 \times 10^{-6}}$

62. $\dfrac{12 \times 10^{3}}{4 \times 10^{-4}}$

Convert each numeral to scientific notation and simplify. Express answer in scientific notation and as a decimal fraction.

63. $\dfrac{(90{,}000)(0.000002)}{0.006}$

64. $\dfrac{(0.0006)(4{,}000)}{0.00012}$

65. $\dfrac{(60{,}000)(0.000003)}{(0.0004)(1{,}500{,}000)}$

66. $\dfrac{(0.000039)(140)}{(130{,}000)(0.00021)}$

C **67.** If the mass of the earth is 6×10^{27} grams and each gram is 1.1×10^{-6} tons, find the mass of the earth in tons.

68. In 1929 Vernadsky, a biologist, estimated that all of the free oxygen of the earth is 1.5×10^{21} grams and that it is produced by life alone. If one grain is approximately 2.2×10^{-3} pounds, what is the amount of free oxygen in pounds?

69. Some of the designers of high-speed computers are currently thinking of single-addition times of 10^{-7} seconds (100 nanoseconds). How many additions would such a computer be able to perform in 1 second? In 1 minute?

70. If electricity travels in a computer circuit at the speed of light (1.86×10^{6}

miles/second), how far will it travel in the time it takes the computer in the preceding problem to complete a single addition? (Size of circuits is becoming a critical problem in computer design.) Give the answer in miles and in feet.

7.4 Rational Exponents

REAL ROOTS OF REAL NUMBERS

What do we mean by a root of a real number? Perhaps you recall that a *square root* of a number b is a number a such that $a^2 = b$; a *cube root* of a number b is a number a such that $a^3 = b$. Thus a square root of 4 is 2 or -2, since $2^2 = 4$ and $(-2)^2 = 4$. A cube root of -8 is -2, since $(-2)^3 = -8$. In general, for n a natural number,

$$a \text{ is an } n\text{th root of } b \text{ if } a^n = b$$

How many real fourth roots of 7 exist? Of -7? How many real fifth roots of 12 exist? Of -12? In general, what can we say about the number of real nth roots of a real number b? The following important theorem (which we state without proof) answers these questions completely.

THEOREM **NUMBER OF REAL ROOTS OF A REAL NUMBER**

(*A*) There are two real nth roots of b if n is even and b is positive; one is the negative of the other.

(*B*) There are no real nth roots of b if n is even and b is negative.

(*C*) There is exactly one real nth root of b if n is odd; the root is positive if b is positive and negative if b is negative.

Thus, 7 has two real fourth roots (which are irrational and each is the negative of the other); -7 has no real fourth roots (since no real number raised to a fourth power can be negative); 12 has exactly one real fifth root, which is positive and irrational; and -12 has exactly one real fifth root, which is negative and irrational.

NOTE: When a real nth root of b exists, we say that the nth root is irrational if there is no rational number a such that $a^n = b$.

RATIONAL EXPONENTS

If all exponent laws are to continue to hold even if some of the exponents are rational numbers, then how should symbols such as $7^{1/2}$ or $5^{1/3}$ be defined? Applying one of the laws of exponents, we note that

$$(7^{1/2})^2 = 7^{2/2} = 7 \qquad \text{and} \qquad (5^{1/3})^3 = 5^{3/3} = 5$$

Hence $7^{1/2}$ must name a square root of 7, since $7^{1/2}$ satisfies the equation $a^2 = 7$; and $5^{1/3}$ must name a cube root of 5, since $5^{1/3}$ satisfies the equation $a^3 = 5$. In general, for n a natural number and b not negative when n is even,

$$(b^{1/n})^n = b^{n/n} = b$$

Therefore, $b^{1/n}$ must name an nth root of b, since $b^{1/n}$ satisfies the equation $a^n = b$. But which nth root? We would like $b^{1/n}$ to be uniquely defined. If n is even and b is positive, then, according to the theorem above, there are two real nth roots of b, one the negative of the other. We will use $b^{1/n}$ to name only the positive one. The following definition takes care of all cases.

DEFINITION $b^{1/n}$ $\quad$ n a natural number and b a real number

n	b POSITIVE	b NEGATIVE	$b = 0$
Even	$b^{1/n}$ is the positive nth root of b	$b^{1/n}$ is not a real number	$0^{1/n} = 0$
Odd	$b^{1/n}$ is *the* nth real root of b (positive)	$b^{1/n}$ is *the* nth real root of b (negative)	$0^{1/n} = 0$

EXAMPLE 15

(A) $4^{1/2} = 2$

(B) $-4^{1/2} = -2$

(C) $(-4)^{1/2}$ is not a real number

(D) $8^{1/3} = 2$

(E) $(-8)^{1/3} = -2$

(F) $0^{1/5} = 0$

PROBLEM 15 Find each of the following.

(A) $9^{1/2}$ $\quad$ (B) $-9^{1/2}$ $\quad$ (C) $(-9)^{1/2}$

(D) $27^{1/3}$ $\quad$ (E) $(-27)^{1/3}$ $\quad$ (F) $0^{1/4}$

ANSWER (A) 3 $\quad$ (B) -3 $\quad$ (C) not a real number $\quad$ (D) 3

(E) -3 $\quad$ (F) 0

How should a symbol such as $5^{2/3}$ be defined? If the properties of exponents are to continue to hold for all rational exponents, then $5^{2/3} = (5^{1/3})^2$; that is, $5^{2/3}$ must represent the square of the cube root of 5. We define

$$b^{m/n} = (b^{1/n})^m$$

where m and n are natural numbers and b is any real number, except b cannot be negative when n is even. Using the same restrictions on b, we define

$$b^{-m/n} = \frac{1}{b^{m/n}} \qquad b \neq 0$$

We now know what $b^{m/n}$ means for all rational numbers m/n and real numbers b. It can be shown, though we will not, that all five properties of exponents continue to hold for rational exponents as long as we avoid even roots of negative numbers. With the latter restriction in effect, the following useful relationship is an immediate consequence of the exponent properties:

$$b^{m/n} = (b^{1/n})^m = (b^m)^{1/n}$$

EXAMPLE 16

(A) $8^{2/3} = (8^{1/3})^2 = 2^2 = 4$ or $8^{2/3} = (8^2)^{1/3} = 64^{1/3} = 4$

(B) $(-8)^{5/3} = [(-8)^{1/3}]^5 = (-2)^5 = -32$

(C) $(3x^{1/3})(2x^{1/2}) = 6x^{1/3+1/2} = 6x^{5/6}$

(D) $(2x^{1/3}y^{-2/3})^3 = 8xy^{-2}$ or $8x/y^2$

(E) $\left(\dfrac{4x^{1/3}}{x^{1/2}}\right)^{1/2} = \dfrac{4^{1/2}x^{1/6}}{x^{1/4}} = \dfrac{2}{x^{1/4-1/6}} = \dfrac{2}{x^{1/12}}$ or $2x^{-1/12}$

(F) $(2a^{1/2} + b^{1/2})(a^{1/2} + 3b^{1/2}) = 2a + 7a^{1/2}b^{1/2} + 3b$

PROBLEM 16

Simplify, and express the answers using positive exponents only.

(A) $9^{3/2}$ (B) $(-27)^{4/3}$

(C) $(5y^{3/4})(2y^{1/3})$ (D) $(2x^{-3/4}y^{1/4})^4$

(E) $\left(\dfrac{8x^{1/2}}{x^{2/3}}\right)^{1/3}$ (F) $(x^{1/2} - 2y^{1/2})(3x^{1/2} + y^{1/2})$

ANSWER (A) 27 (B) 81 (C) $10y^{13/12}$ (D) $16y/x^3$

(E) $2/x^{1/18}$ (F) $3x - 5x^{1/2}y^{1/2} - 2y$

The properties of exponents can be used as long as we are dealing with symbols that name real numbers. Can you resolve the following contradiction?

$$-1 = (-1)^{2/2} = [(-1)^2]^{1/2} = 1^{1/2} = 1$$

The second member of the equality chain, $(-1)^{2/2}$, involves the even root of a negative number, which is not real. Thus we see that the properties of exponents do not necessarily hold when we are dealing with nonreal quantities.

Exercise 41

All variables represent positive real numbers.

A *Most of the following are integers. Find them.*

1. $25^{1/2}$ **2.** $36^{1/2}$ **3.** $(-25)^{1/2}$

4. $(-36)^{1/2}$ **5.** $8^{1/3}$ **6.** $27^{1/3}$

7. $(-8)^{1/3}$ **8.** $(-27)^{1/3}$ **9.** $-8^{1/3}$

10. $-27^{1/3}$ **11.** $16^{3/2}$ **12.** $25^{3/2}$

13. $8^{2/3}$ **14.** $27^{2/3}$

Simplify, and express the answer using positive exponents only.

15. $x^{1/4}x^{3/4}$ **16.** $y^{1/5}y^{2/5}$ **17.** $\dfrac{x^{2/5}}{x^{3/5}}$

18. $\dfrac{a^{2/3}}{a^{1/3}}$ **19.** $(x^4)^{1/2}$ **20.** $(y^{1/2})^4$

21. $(a^3b^9)^{1/3}$ **22.** $(x^4y^2)^{1/2}$ **23.** $\left(\dfrac{x^9}{y^{12}}\right)^{1/3}$

24. $\left(\dfrac{m^{12}}{n^{16}}\right)^{1/4}$ **25.** $(x^{1/3}y^{1/2})^6$ **26.** $\left(\dfrac{u^{1/2}}{v^{1/3}}\right)^{12}$

B *Most of the following are rational numbers. Find them.*

27. $\left(\frac{4}{25}\right)^{1/2}$ **28.** $\left(\frac{9}{4}\right)^{1/2}$ **29.** $\left(\frac{4}{25}\right)^{3/2}$

30. $\left(\frac{9}{4}\right)^{3/2}$ **31.** $\left(\frac{1}{8}\right)^{2/3}$ **32.** $\left(\frac{1}{27}\right)^{2/3}$

33. $36^{-1/2}$ **34.** $25^{-1/2}$ **35.** $25^{-3/2}$

36. $16^{-3/2}$ **37.** $5^{3/2} \cdot 5^{1/2}$ **38.** $7^{2/3} \cdot 7^{4/3}$

39. $(3^6)^{-1/3}$ **40.** $(4^{-8})^{3/16}$

Simplify, and express the answer using positive exponents only.

41. $x^{1/4}x^{-3/4}$ **42.** $\dfrac{d^{2/3}}{d^{-1/3}}$ **43.** $n^{3/4}n^{-2/3}$

44. $m^{1/2}m^{-1/3}$ **45.** $(x^{-2/3})^{-6}$ **46.** $(y^{-8})^{1/16}$

47. $(4u^{-2}v^4)^{1/2}$ **48.** $(8x^3y^{-6})^{1/3}$ **49.** $(x^4y^6)^{-1/2}$

50. $(4x^{1/2}y^{3/2})^2$ **51.** $\left(\dfrac{x^{-2/3}}{y^{-1/2}}\right)^{-6}$ **52.** $\left(\dfrac{m^{-3}}{n^2}\right)^{-1/6}$

53. $\left(\dfrac{25x^5y^{-1}}{16x^{-3}y^{-5}}\right)^{1/2}$ **54.** $\left(\dfrac{8a^{-4}b^3}{27a^2b^{-3}}\right)^{1/3}$ **55.** $\left(\dfrac{8y^{1/3}y^{-1/4}}{y^{-1/12}}\right)^2$

56. $\left(\dfrac{9x^{1/3}x^{1/2}}{x^{-1/6}}\right)^{1/2}$

Multiply, and express answer using positive exponents only.

57. $3m^{3/4}(4m^{1/4} - 2m^8)$ **58.** $2x^{1/3}(3x^{2/3} - x^6)$

59. $(2x^{1/2} + y^{1/2})(x^{1/2} + y^{1/2})$ **60.** $(x^{1/2} + y^{1/2})(x^{1/2} - y^{1/2})$

61. $(x^{1/2} + y^{1/2})^2$ **62.** $(x^{1/2} - y^{1/2})^2$

C *Simplify, and express answer using positive exponents only.*

63. $(-16)^{-3/2}$ **64.** $-16^{-3/2}$

65. $(a^{-1/2} + 3b^{-1/2})(2a^{-1/2} - b^{-1/2})$ **66.** $(x^{-1/2} - y^{-1/2})^2$

67. $(a^{n/2}b^{n/3})^{1/n}$, $n > 0$ **68.** $(a^{3/n}b^{3/m})^{1/3}$, $n > 0$, $m > 0$

69. $\left(\dfrac{x^{m+2}}{x^m}\right)^{1/2}$, $m > 0$ **70.** $\left(\dfrac{a^m}{a^{m-2}}\right)^{1/2}$, $m > 0$

71. $(x^{m/4}x^{m/4})^{-2}$, $m > 0$ **72.** $(y^{m^2+1}y^{2m})^{1/(m+1)}$, $m > 0$

73. (*A*) Find a value of x such that $(x^2)^{1/2} \neq x$.

(*B*) Find a real number x and a natural number n such that $(x^n)^{1/n} \neq x$.

Exercise 42 Chapter Review

All variables represent positive real numbers.

A *Evaluate:*

1. $\left(\dfrac{1}{3}\right)^0$ **2.** 3^{-2} **3.** $\dfrac{1}{2^{-3}}$

4. $4^{-1/2}$ **5.** $(-9)^{3/2}$ **6.** $(-8)^{2/3}$

7. Write in scientific notation: (*A*) 4,280,000,000; (*B*) 0.0000318

8. Write as a decimal fraction: (*A*) 7.29×10^5; (*B*) 6.03×10^{-4}

Simplify, and write answers using positive exponents only.

9. $(3x^3y^2)(2xy^5)$ **10.** $\dfrac{9u^8v^6}{3u^4v^8}$ **11.** $6(xy^3)^5$

12. $\left(\frac{c^2}{d^5}\right)^3$ 13. $\left(\frac{2x^2}{3y^3}\right)^2$ 14. $(x^{-3})^{-4}$

15. $\frac{y^{-3}}{y^{-5}}$ 16. $(x^2y^{-3})^{-1}$ 17. $(x^9)^{1/3}$

18. $(x^4)^{-1/2}$ 19. $x^{1/3}x^{-2/3}$ 20. $\frac{u^{5/3}}{u^{2/3}}$

B 21. Convert each number to scientific notation, simplify, and write answer in scientific notation and as a decimal fraction:

$$\frac{0.000052}{130(0.0002)}$$

Simplify, and write answers using positive exponents only.

22. $\frac{3m^4n^{-7}}{6m^2n^{-2}}$ 23. $(x^{-3}y^2)^{-2}$

24. $\frac{1}{(2x^2y^{-3})^{-2}}$ 25. $\left(-\frac{a^2b}{c}\right)^2\left(\frac{c}{b^2}\right)^3\left(\frac{1}{a^3}\right)^2$

26. $\left(\frac{8u^{-1}}{2^2u^2v^0}\right)^{-2}\left(\frac{u^{-5}}{u^{-3}}\right)^3$ 27. $\left(\frac{9m^3n^{-3}}{3m^{-2}n^2}\right)^{-2}$

28. $(x-y)^{-2}$ 29. $(9a^4b^{-2})^{1/2}$

30. $\left(\frac{27x^2y^{-3}}{8x^{-4}y^3}\right)^{1/3}$ 31. $\frac{m^{-1/4}}{m^{3/4}}$

32. $(2x^{1/2})(3x^{-1/3})$ 33. $\frac{3x^{-1/4}}{6x^{-1/3}}$

34. $\frac{5^0}{3^2}+\frac{3^{-2}}{2^{-2}}$ 35. $(x^{1/2}+y^{1/2})^2$

36. If a is a square root of b, then does $a^2=b$ or does $b^2=a$?

C *Simplify, and write answers using positive exponents only.*

37. $(x^{-1}+y^{-1})^{-1}$ 38. $\left(\frac{a^{-2}}{b^{-1}}+\frac{b^{-2}}{a^{-1}}\right)^{-1}$

39. $(x^{1-2m}x^{m^2})^{1/(m-1)}$, $m>1$ 40. $\left(\frac{x^{m^2}}{x^{2m-1}}\right)^{1/(m-1)}$, $m>1$

41. The volume of mercury increases linearly with temperature over a fairly wide temperature range (this is why mercury is often used in thermometers). If 1 cubic centimeter of mercury at 0°C is heated to a temperature of T degrees centigrade, its volume is given by the formula

$$V = 1 + (1.8 \times 10^{-4})T$$

Express the volume of this sample at (2×10^2)°C as a decimal fraction.

Chapter 8

RADICALS; COMPLEX NUMBERS

8.1 Radical Forms and Rational Exponents

In the last chapter we introduced the symbol $b^{1/n}$ to represent an nth root of b, and found that the symbol could be combined with other exponent forms using the properties of exponents. Another symbol is also used to represent an nth root, and it involves the use of a radical sign. Both symbols are widely used and you should become familiar with both and their respective properties.

For n a natural number greater than 1 and b any real number, we define

$$\sqrt[n]{b} = b^{1/n}$$

The symbol "$\sqrt{}$" is called a *radical*, n is called the *index*, and b is called the *radicand*. If $n = 2$, we write

$$\sqrt{b} \qquad \text{and not} \qquad \sqrt[2]{b}$$

and refer to $\sqrt{b}$ as "the positive square root of b." Thus, it follows that

$$\sqrt{3} = 3^{1/2}$$

$$\sqrt[8]{5} = 5^{1/8}$$

$$\sqrt[3]{x^2} = (x^2)^{1/3} = x^{2/3}$$

$$(\sqrt[3]{x})^2 = (x^{1/3})^2 = x^{2/3}$$

There are occasions when it is more convenient to work with radicals than with rational exponents, and vice versa. It is often an advantage to be able to shift back and forth between the two forms. The following relationships, suggested by the above examples, are useful in this regard.

$$b^{m/n} = (b^m)^{1/n} = \sqrt[n]{b^m}$$

$$b^{m/n} = (b^{1/n})^m = (\sqrt[n]{b})^m$$

where b is not negative when n is even. The following examples and problems should make the process of changing from one form to the other clear. All variables represent positive real numbers.

EXAMPLE 1 From rational exponent form to radical form.

(A) $5^{1/2} = \sqrt{5}$ positive square root of 5

(B) $x^{1/7} = \sqrt[7]{x}$

(C) $m^{2/3} = \sqrt[3]{m^2}$ or $(\sqrt[3]{m})^2$ first form is usually preferred

(D) $(3u^2v^3)^{3/5} = \sqrt[5]{(3u^2v^3)^3}$ or $(\sqrt[5]{3u^2v^3})^3$ first form is usually preferred

(E) $y^{-2/3} = \dfrac{1}{y^{2/3}} = \dfrac{1}{\sqrt[3]{y^2}}$ or $\dfrac{1}{(\sqrt[3]{y})^2}$ index cannot be negative

(F) $(x^2 + y^2)^{1/2} = \sqrt{x^2 + y^2}$ $\neq x + y$. Why?

PROBLEM 1 Convert to radical form.

(A) $7^{1/2}$ (B) $u^{1/5}$ (C) $x^{3/5}$

(D) $(2x^3y^2)^{2/3}$ (E) $x^{-3/4}$ (F) $(x^3 + y^3)^{1/3}$

ANSWER (A) $\sqrt{7}$ (B) $\sqrt[5]{u}$ (C) $\sqrt[5]{x^3}$ or $(\sqrt[5]{x})^3$

(D) $\sqrt[3]{(2x^3y^2)^2}$ or $(\sqrt[3]{2x^3y^2})^2$

(E) $\dfrac{1}{\sqrt[4]{x^3}}$ or $\dfrac{1}{(\sqrt[4]{x})^3}$ (F) $\sqrt[3]{x^3 + y^3}$

EXAMPLE 2 Convert from radical form to rational exponent form.

(A) $\sqrt{13} = 13^{1/2}$

(B) $\sqrt[5]{x} = x^{1/5}$

(C) $\sqrt[4]{w^3} = w^{3/4}$

(D) $\sqrt[5]{(3x^2y^2)^4} = (3x^2y^2)^{4/5}$

(*E*) $\frac{1}{\sqrt[3]{x^2}} = \frac{1}{x^{2/3}} = x^{-2/3}$

(*F*) $\sqrt[4]{x^4 + y^4} = (x^4 + y^4)^{1/4} \quad \neq x + y$. Why?

PROBLEM 2 Convert to rational exponent form.

(*A*) $\sqrt{17}$ (*B*) $\sqrt[7]{m}$ (*C*) $\sqrt[5]{x^2}$

(*D*) $(\sqrt[7]{5m^3n^4})^3$ (*E*) $\frac{1}{\sqrt[6]{u^5}}$ (*F*) $\sqrt[5]{x^5 - y^5}$

ANSWER (*A*) $17^{1/2}$ (*B*) $m^{1/7}$ (*C*) $x^{2/5}$ (*D*) $(5m^3n^4)^{3/7}$

(*E*) $u^{-5/6}$ (*F*) $(x^5 - y^5)^{1/5}$

Exercise 43

All variables are restricted to avoid even roots of negative numbers.

A *Change to radical form. (Do not simplify.)*

1. $11^{1/2}$ **2.** $7^{1/2}$ **3.** $5^{1/3}$

4. $6^{1/4}$ **5.** $u^{3/5}$ **6.** $x^{3/4}$

7. $4y^{3/7}$ **8.** $5m^{2/3}$ **9.** $(4y)^{3/7}$

10. $(5m)^{2/3}$ **11.** $(4ab^3)^{2/5}$ **12.** $(7x^2y)^{2/3}$

13. $(a + b)^{1/2}$ **14.** $(a^2 + b^2)^{1/2}$

Change to rational exponent form. (Do not simplify.)

15. $\sqrt{6}$ **16.** $\sqrt{3}$ **17.** $\sqrt[4]{m}$

18. $\sqrt[7]{m}$ **19.** $\sqrt[5]{y^3}$ **20.** $\sqrt[3]{a^2}$

21. $\sqrt[4]{(xy)^3}$ **22.** $\sqrt[5]{(7m^3n^3)^4}$ **23.** $\sqrt{x^2 - y^2}$

24. $\sqrt{1 + y^2}$

B *Change to radical form. (Do not simplify.)*

25. $-5y^{2/5}$ **26.** $-3x^{1/2}$ **27.** $(1 + m^2n^2)^{3/7}$

28. $(x^2y^2 - w^3)^{4/5}$ **29.** $w^{-2/3}$ **30.** $y^{-3/5}$

31. $(3m^2n^3)^{-3/5}$ **32.** $(2xy)^{-2/3}$ **33.** $a^{1/2} + b^{1/2}$

34. $x^{-1/2} + y^{-1/2}$ **35.** $(a^3 + b^3)^{2/3}$ **36.** $(x^{1/2} + y^{-1/2})^{1/3}$

Change to rational exponent form. (Do not simplify.)

37. $\sqrt[3]{(a + b)^2}$ **38.** $\sqrt[5]{(x - y)^2}$ **39.** $-3x\sqrt[4]{a^3b}$

40. $-5\sqrt[3]{2x^2y^2}$ **41.** $\sqrt[9]{-2x^3y^7}$ **42.** $\sqrt[5]{-4m^2n^3}$

43. $\dfrac{3}{\sqrt[3]{y}}$ **44.** $\dfrac{2x}{\sqrt{y}}$ **45.** $\dfrac{-2x}{\sqrt{x^2+y^2}}$

46. $\dfrac{2}{\sqrt{x}}+\dfrac{3}{\sqrt{y}}$ **47.** $\sqrt[3]{m^2}-\sqrt{n}$ **48.** $\dfrac{-5u^2}{\sqrt{u}+\sqrt[5]{v^3}}$

C **49.** Show that $(x^2+y^2)^{1/2} \neq x+y$.

50. Show that $\sqrt{a^2+b^2} \neq a+b$.

51. (A) Find a value of x such that $\sqrt{x^2} \neq x$.

(B) Find a positive integer n and a real number x such that $\sqrt[n]{x^n} \neq x$.

52. Which of the following statements is true for all real x (n is a positive integer)?

(A) $\sqrt{x^2}=|x|$; (B) $\sqrt[2n]{x^{2n}}=|x|$

8.2 Changing and Simplifying Radical Expressions

Changing and simplifying radical expressions is aided by the introduction of several properties of radicals that follow directly from the exponent properties considered earlier. To start, consider the following examples:

1 $\sqrt[5]{2^5}=(2^5)^{1/5}=2^{5/5}=2^1=2$

2 $\sqrt{4\cdot 9}=\sqrt{36}=6$ and $\sqrt{4}\sqrt{9}=2\cdot 3=6$

3 $\sqrt{\dfrac{36}{4}}=\sqrt{9}=3$ and $\dfrac{\sqrt{36}}{\sqrt{4}}=\dfrac{6}{2}=3$

4 $\sqrt[6]{2^4}=(2^4)^{1/6}=2^{4/6}=2^{2/3}=(2^2)^{1/3}=\sqrt[3]{2^2}$

These examples suggest the following general properties of radicals.

PROPERTIES OF RADICALS

n, m, and k are natural numbers ≥ 2, x and y are positive real numbers

1 $\sqrt[n]{x^n}=x$

2 $\sqrt[n]{xy}=\sqrt[n]{x}\sqrt[n]{y}$

3 $\sqrt[n]{\dfrac{x}{y}}=\dfrac{\sqrt[n]{x}}{\sqrt[n]{y}}$

4 $\sqrt[kn]{x^{km}}=\sqrt[n]{x^m}$

These properties are easily proved, as follows:

1 $\sqrt[n]{x^n} = (x^n)^{1/n} = x^{n/n} = x$

2 $\sqrt[n]{xy} = (xy)^{1/n} = x^{1/n}y^{1/n} = \sqrt[n]{x}\,\sqrt[n]{y}$

3 $\sqrt[n]{\dfrac{x}{y}} = \left(\dfrac{x}{y}\right)^{1/n} = \dfrac{x^{1/n}}{y^{1/n}} = \dfrac{\sqrt[n]{x}}{\sqrt[n]{y}}$

4 $\sqrt[kn]{x^{km}} = (x^{km})^{1/km} = x^{km/kn} = x^{m/n} = \sqrt[n]{x^m}$

The following example and problem illustrate how these properties are used. All variables represent positive real numbers.

EXAMPLE 3

(*A*) $\sqrt[5]{(3x^2y)^5} = 3x^2y$ property 1

(*B*) $\sqrt{10}\,\sqrt{5} = \sqrt{50} = \sqrt{25 \cdot 2} = \sqrt{25}\,\sqrt{2} = 5\sqrt{2}$ property 2

(*C*) $\sqrt[3]{\dfrac{x}{27}} = \dfrac{\sqrt[3]{x}}{\sqrt[3]{27}} = \dfrac{\sqrt[3]{x}}{3}$ or $\dfrac{1}{3}\sqrt[3]{x}$ property 3

(*D*) $\sqrt[6]{x^4} = \sqrt[2\cdot 3]{x^{2\cdot 2}} = \sqrt[3]{x^2}$ property 4

PROBLEM 3

Simplify as in Example 3.

(*A*) $\sqrt[7]{(u^2+v^2)^7}$ (*B*) $\sqrt{6}\,\sqrt{2}$

(*C*) $\sqrt[3]{x^2/8}$ (*D*) $\sqrt[8]{y^6}$

ANSWER (*A*) $u^2 + v^2$ (*B*) $2\sqrt{3}$ (*C*) $\sqrt[3]{x^2}/2$ or $\frac{1}{2}\sqrt[3]{x^2}$ (*D*) $\sqrt[4]{y^3}$

The laws of radicals provide us with the means of changing algebraic expressions containing radicals to a variety of equivalent forms. One form often useful is the simplest radical form. An algebraic expression that contains radicals is said to be in the *simplest radical form* if all four of the following conditions are satisfied:

SIMPLEST RADICAL FORM

1 A radicand (the expression within the radical sign) contains no polynomial factor to a power greater than or equal to the index of the radical. ($\sqrt{x^3}$ violates this condition)

2 The power of the radicand and the index of the radical have no common factor other than 1. ($\sqrt[6]{x^4}$ violates this condition)

3 No radical appears in a denominator. ($3/\sqrt{5}$ violates this condition)

4 No fraction appears within a radical. ($\sqrt{\frac{2}{3}}$ violates this condition)

It should be understood that forms other than the simplest radical form may be more useful on occasion. The choice depends on the situation.

EXAMPLE 4 Change to simplest radical form:

(A) $\sqrt{72} = \sqrt{6^2 \cdot 2} = \sqrt{6^2}\sqrt{2} = 6\sqrt{2}$ $\quad \sqrt{72}$ violates condition 1

(B) $\sqrt{8x^3} = \sqrt{(4x^2)(2x)} = \sqrt{4x^2}\sqrt{2x} = 2x\sqrt{2x}$ $\quad \sqrt{8x^3}$ violates condition 1

(C) $\sqrt[9]{x^6} = \sqrt[3]{x^2}$ $\quad \sqrt[9]{x^6}$ violates condition 2

(D) $\dfrac{3x}{\sqrt{3}} = \dfrac{3x}{\sqrt{3}} \cdot \dfrac{\sqrt{3}}{\sqrt{3}} = \dfrac{3x\sqrt{3}}{3} = x\sqrt{3}$ $\quad \dfrac{3x}{\sqrt{3}}$ violates condition 3

(E) $\sqrt{\dfrac{x}{2}} = \sqrt{\dfrac{x}{2} \cdot \dfrac{2}{2}} = \sqrt{\dfrac{2x}{4}} = \dfrac{\sqrt{2x}}{\sqrt{4}} = \dfrac{\sqrt{2x}}{2}$ or $\dfrac{1}{2}\sqrt{2x}$ $\quad \sqrt{\dfrac{x}{2}}$ violates condition 4

NOTE: The process of removing radicals from a denominator is called *rationalizing the denominator.*

PROBLEM 4 Change to simplest radical form:

(A) $\sqrt{32}$ (B) $\sqrt{18y^3}$ (C) $\sqrt[12]{y^8}$

(D) $\dfrac{2x}{\sqrt{2}}$ (E) $\sqrt{\dfrac{y}{3}}$

ANSWER (A) $4\sqrt{2}$ (B) $3y\sqrt{2y}$ (C) $\sqrt[3]{y^2}$ (D) $x\sqrt{2}$

(E) $\dfrac{\sqrt{3y}}{3}$ or $\dfrac{1}{3}\sqrt{3y}$

EXAMPLE 5 Change to simplest radical form.

(A) $\sqrt[3]{54} = \sqrt[3]{3^3 \cdot 2} = \sqrt[3]{3^3} \cdot \sqrt[3]{2} = 3\sqrt[3]{2}$

(B) $\sqrt{12x^3y^5z^2} = \sqrt{(2^2x^2y^4z^2)(3xy)} = \sqrt{2^2x^2y^4z^2}\,\sqrt{3xy} = 2xy^2z\sqrt{3xy}$

(C) $\sqrt[6]{16x^4y^2} = \sqrt[6]{(2^2x^2y)^2} = \sqrt[3]{4x^2y}$

(D) $\dfrac{3}{\sqrt{5}} = \dfrac{3\sqrt{5}}{\sqrt{5}\sqrt{5}} = \dfrac{3\sqrt{5}}{5}$ or $\dfrac{3}{5}\sqrt{5}$

(E) $\dfrac{6x^2}{\sqrt[3]{3x}} = \dfrac{6x^2\sqrt[3]{3^2x^2}}{\sqrt[3]{3x}\sqrt[3]{3^2x^2}} = \dfrac{6x^2\sqrt[3]{9x^2}}{\sqrt[3]{3^3x^3}} = \dfrac{6x^2\sqrt[3]{9x^2}}{3x} = 2x\sqrt[3]{9x^2}$

(F) $\sqrt[3]{\dfrac{2a^2}{3b}} = \sqrt[3]{\dfrac{(2a^2)(3^2b^2)}{(3b)(3^2b^2)}} = \sqrt[3]{\dfrac{18a^2b^2}{3^3b^3}} = \dfrac{\sqrt[3]{18a^2b^2}}{\sqrt[3]{3^3b^3}} = \dfrac{\sqrt[3]{18a^2b^2}}{3b}$

PROBLEM 5 Change to simplest radical form.

(A) $\sqrt[3]{16}$ (B) $\sqrt{18x^5y^2z^3}$ (C) $\sqrt[9]{8x^6y^3}$

(D) $\dfrac{6}{\sqrt{2x}}$ (E) $\dfrac{10x^3}{\sqrt[3]{2x^2}}$ (F) $\sqrt[3]{\dfrac{3y^2}{2x^4}}$

ANSWER (*A*) $2\sqrt[3]{2}$ (*B*) $3x^2yz\sqrt{2xz}$ (*C*) $\sqrt[3]{2x^2y}$ (*D*) $3\sqrt{2x}/x$
(*E*) $5x^2\sqrt[3]{4x}$ (*F*) $\sqrt[3]{12x^2y^2}/2x^2$

Exercise 44

Simplify, and write in simplest radical form. All variables represent positive real numbers.

A

1. $\sqrt{y^2}$
2. $\sqrt{x^2}$
3. $\sqrt{4u^2}$
4. $\sqrt{9m^2}$
5. $\sqrt{49x^4y^2}$
6. $\sqrt{25x^2y^4}$
7. $\sqrt{18}$
8. $\sqrt{8}$
9. $\sqrt{m^3}$
10. $\sqrt{x^3}$
11. $\sqrt{8x^3}$
12. $\sqrt{18y^3}$
13. $\sqrt{\dfrac{1}{9}}$
14. $\sqrt{\dfrac{1}{4}}$
15. $\dfrac{1}{\sqrt{y^2}}$
16. $\dfrac{1}{\sqrt{x^2}}$
17. $\dfrac{1}{\sqrt{5}}$
18. $\dfrac{1}{\sqrt{3}}$
19. $\sqrt{\dfrac{1}{5}}$
20. $\sqrt{\dfrac{1}{3}}$
21. $\dfrac{1}{\sqrt{y}}$
22. $\dfrac{1}{\sqrt{x}}$
23. $\sqrt{\dfrac{1}{y}}$
24. $\sqrt{\dfrac{1}{x}}$
25. $\sqrt{9x^3y^5}$
26. $\sqrt{4x^5y^3}$
27. $\sqrt{18x^8y^5}$
28. $\sqrt{8x^7y^6}$
29. $\dfrac{1}{\sqrt{2x}}$
30. $\dfrac{1}{\sqrt{3y}}$
31. $\dfrac{6x^2}{\sqrt{3x}}$
32. $\dfrac{4xy}{\sqrt{2y}}$
33. $\dfrac{3a}{\sqrt{2ab}}$
34. $\dfrac{2x^2y}{\sqrt{3xy}}$
35. $\sqrt{\dfrac{6x}{7y}}$
36. $\sqrt{\dfrac{3m}{2n}}$

B

37. $\sqrt{\dfrac{9m^5}{2n}}$
38. $\sqrt{\dfrac{4a^3}{3b}}$
39. $\sqrt[4]{16x^8y^4}$
40. $\sqrt[5]{32m^5n^{15}}$
41. $\sqrt[3]{2^4x^4y^7}$
42. $\sqrt[4]{2^4a^5b^8}$
43. $\sqrt[4]{x^2}$
44. $\sqrt[10]{x^6}$
45. $\sqrt{2}\sqrt{8}$
46. $\sqrt[3]{3}\sqrt[3]{9}$
47. $\sqrt{18m^3n^4}\sqrt{2m^3n^2}$
48. $\sqrt[3]{9x^2y}\sqrt[3]{3xy^2}$
49. $\dfrac{6}{\sqrt[3]{3}}$
50. $\dfrac{2}{\sqrt[3]{2}}$
51. $\dfrac{\sqrt{4a^3}}{\sqrt{3b}}$
52. $\dfrac{\sqrt{9m^5}}{\sqrt{2n}}$
53. $\sqrt{a^2+b^2}$
54. $\sqrt[3]{x^3+y^3}$

55. $\sqrt[3]{\dfrac{8x^3}{27y^6}}$ **56.** $\sqrt[4]{\dfrac{a^8b^4}{16c^{12}}}$ **57.** $-m\sqrt[5]{3^6m^7n^{11}}$

58. $-2x\sqrt[3]{8x^8y^{13}}$ **59.** $\sqrt[6]{x^4(x-y)^2}$ **60.** $\sqrt[8]{2^6(x+y)^6}$

61. $\sqrt[3]{2x^2y^3}\sqrt[3]{3x^5y}$ **62.** $\sqrt[4]{6u^3v^4}\sqrt[4]{4u^5v}$ **63.** $\dfrac{4x^3y^2}{\sqrt[3]{2xy^2}}$

64. $\dfrac{8u^3v^5}{\sqrt[3]{4u^2v^2}}$ **65.** $-2x\sqrt[3]{\dfrac{3y^2}{4x}}$ **66.** $6c\sqrt[3]{\dfrac{2ab}{9c^2}}$

C **67.** $\dfrac{x-y}{\sqrt[3]{x-y}}$ **68.** $\dfrac{1}{\sqrt[3]{(x-y)^2}}$ **69.** $\sqrt[4]{\dfrac{3y^3}{4x}}$

70. $\sqrt[5]{\dfrac{4n^2}{16m^3}}$ **71.** $-\sqrt{x^4+2x^2}$ **72.** $\sqrt[4]{m^4+4m^6}$

73. $\sqrt[4]{16x^4}\sqrt[3]{16x^{24}y^4}$ **74.** $\sqrt[3]{8\sqrt{16x^6y^4}}$ **75.** $\sqrt[3]{3m^2n^2}\sqrt[4]{3m^3n^2}$

76. $\sqrt{2x^5y^3}\sqrt[3]{16x^7y^7}$ **77.** $\sqrt[3]{x^{3n}(x+y)^{3n+6}}$ **78.** $\sqrt[n]{x^{2n}y^{n^2+n}}$

8.3 Sums and Differences Involving Radicals

Algebraic expressions involving radicals can often be simplified by adding or subtracting terms that contain exactly the same radical expressions. We proceed in essentially the same way as when we combine like terms in polynomials. You will recall that the distributive property of real numbers played a central role in this process.

EXAMPLE 6 Combine as many terms as possible.

(A) $5\sqrt{3}+4\sqrt{3} \boxed{=(5+4)\sqrt{3}} = 9\sqrt{3}$

(B) $2\sqrt[3]{xy^2}-7\sqrt[3]{xy^2} \boxed{=(2-7)\sqrt[3]{xy^2}} = -5\sqrt[3]{xy^2}$

(C) $$3\sqrt{xy}-2\sqrt[3]{xy}+4\sqrt{xy}-7\sqrt[3]{xy} = 3\sqrt{xy}+4\sqrt{xy}-2\sqrt[3]{xy}-7\sqrt[3]{xy}$$
$$= 7\sqrt{xy}-9\sqrt[3]{xy}$$

PROBLEM 6 Combine as many terms as possible.

(A) $6\sqrt{2}+2\sqrt{2}$ (B) $3\sqrt[5]{2x^2y^3}-8\sqrt[5]{2x^2y^3}$

(C) $5\sqrt[3]{mn^2}-3\sqrt{mn}-2\sqrt[3]{mn^2}+7\sqrt{mn}$

ANSWER (A) $8\sqrt{2}$ (B) $-5\sqrt[5]{2x^2y^3}$ (C) $3\sqrt[3]{mn^2}+4\sqrt{mn}$

Thus we see that if two terms contain exactly the same radical—having the same index and the same radicand—they can be combined into a single term. Occasionally, terms containing radicals can be combined after they have been expressed in simplest radical form.

EXAMPLE 7 Express terms in simplest radical form and combine where possible.

(*A*) $4\sqrt{8} - 2\sqrt{18} = 4\sqrt{4 \cdot 2} - 2\sqrt{9 \cdot 2}$

$= 8\sqrt{2} - 6\sqrt{2}$

$= 2\sqrt{2}$

(*B*) $2\sqrt{12} - \sqrt{\frac{1}{3}} = 2 \cdot \sqrt{4} \cdot \sqrt{3} - \frac{1 \cdot \sqrt{3}}{\sqrt{3} \cdot \sqrt{3}}$

$= 4\sqrt{3} - \frac{\sqrt{3}}{3}$

$= \left(4 - \frac{1}{3}\right)\sqrt{3}$

$= \frac{11}{3}\sqrt{3}$ or $\frac{11\sqrt{3}}{3}$

(*C*) $\sqrt[3]{81} - \sqrt[3]{\frac{1}{9}} = \sqrt[3]{3^3 \cdot 3} - \sqrt[3]{\frac{3}{3^3}} = 3\sqrt[3]{3} - \frac{1}{3}\sqrt[3]{3}$

$= \left(3 - \frac{1}{3}\right)\sqrt[3]{3} = \frac{8}{3}\sqrt[3]{3}$

PROBLEM 7 Express terms in simplest radical form and combine where possible.

(*A*) $\sqrt{12} - \sqrt{48}$ (*B*) $3\sqrt{8} - \sqrt{\frac{1}{2}}$ (*C*) $\sqrt[3]{\frac{1}{4}} - \sqrt[3]{16}$

ANSWER (*A*) $-2\sqrt{3}$ (*B*) $\frac{11\sqrt{2}}{2}$ (*C*) $-\frac{3}{2}\sqrt[3]{2}$.

Exercise 45

Express in simplest radical form and combine where possible.

A

1. $7\sqrt{3} + 2\sqrt{3}$
2. $5\sqrt{2} + 3\sqrt{2}$
3. $2\sqrt{a} - 7\sqrt{a}$
4. $\sqrt{y} - 4\sqrt{y}$
5. $\sqrt{n} - 4\sqrt{n} - 2\sqrt{n}$
6. $2\sqrt{x} - \sqrt{x} + 3\sqrt{x}$
7. $\sqrt{5} - 2\sqrt{3} + 3\sqrt{5}$
8. $3\sqrt{2} - 2\sqrt{3} - \sqrt{2}$
9. $\sqrt{m} - \sqrt{n} - 2\sqrt{n}$
10. $2\sqrt{x} - \sqrt{y} + 3\sqrt{y}$
11. $\sqrt{18} + \sqrt{2}$
12. $\sqrt{8} - \sqrt{2}$
13. $\sqrt{8} - 2\sqrt{32}$
14. $\sqrt{27} - 3\sqrt{12}$

B **15.** $\sqrt{8mn} + 2\sqrt{18mn}$

16. $\sqrt{4x} - \sqrt{9x}$

17. $\sqrt{8} - \sqrt{20} + 4\sqrt{2}$

18. $\sqrt{24} - \sqrt{12} + 3\sqrt{3}$

19. $\sqrt[5]{a} - 4\sqrt[5]{a} + 2\sqrt[5]{a}$

20. $3\sqrt[3]{u} - 2\sqrt[3]{u} - 2\sqrt[3]{u}$

21. $2\sqrt[3]{x} + 3\sqrt[3]{x} - \sqrt{x}$

22. $5\sqrt[5]{y} - 2\sqrt[5]{y} + 3\sqrt[4]{y}$

23. $\sqrt{\frac{1}{8}} + \sqrt{8}$

24. $\sqrt{\frac{2}{3}} - \sqrt{\frac{3}{2}}$

25. $\sqrt{\dfrac{3uv}{2}} - \sqrt{24uv}$

26. $\sqrt{\dfrac{xy}{2}} + \sqrt{8xy}$

C **27.** $\dfrac{\sqrt{3}}{3} + 2\sqrt{\dfrac{1}{3}} + \sqrt{12}$

28. $\sqrt{\dfrac{1}{2}} + \dfrac{\sqrt{2}}{2} + \sqrt{8}$

29. $\sqrt[3]{\frac{1}{3}} + \sqrt[3]{3^5}$

30. $\sqrt[4]{32} - \sqrt[4]{\frac{1}{8}}$

8.4 Products and Quotients Involving Radicals

We will now consider several types of special products and quotients that involve radicals. The distributive property of real numbers plays a central role in our approach to these problems. In the discussion that follows all variables represent positive real numbers.

SPECIAL PRODUCTS

EXAMPLE 8 Multiply and simplify:

(A) $\sqrt{2}(\sqrt{10} - 3) = \sqrt{2}\sqrt{10} - \sqrt{2}\cdot 3 = \sqrt{20} - 3\sqrt{2} = 2\sqrt{5} - 3\sqrt{2}$

(B)
$$\begin{aligned}(\sqrt{2} - 3)(\sqrt{2} + 5) &= \sqrt{2}\sqrt{2} - 3\sqrt{2} + 5\sqrt{2} - 15\\ &= 2 + 2\sqrt{2} - 15\\ &= 2\sqrt{2} - 13\end{aligned}$$

(C)
$$\begin{aligned}(\sqrt{x} - 3)(\sqrt{x} + 5) &= \sqrt{x}\sqrt{x} - 3\sqrt{x} + 5\sqrt{x} - 15\\ &= x + 2\sqrt{x} - 15\end{aligned}$$

(D)
$$\begin{aligned}(\sqrt[3]{m} + \sqrt[3]{n^2})(\sqrt[3]{m^2} - \sqrt[3]{n}) &= \sqrt[3]{m^3} - \sqrt[3]{mn} + \sqrt[3]{m^2n^2} - \sqrt[3]{n^3}\\ &= m - \sqrt[3]{mn} + \sqrt[3]{m^2n^2} - n\end{aligned}$$

PROBLEM 8 Multiply and simplify:

(A) $\sqrt{3}(\sqrt{6} - 4)$

(B) $(\sqrt{3} - 2)(\sqrt{3} + 4)$

(C) $(\sqrt{y} - 2)(\sqrt{y} + 4)$

(D) $(\sqrt[3]{x^2} - \sqrt[3]{y^2})(\sqrt[3]{x} + \sqrt[3]{y})$

ANSWER (*A*) $3\sqrt{2} - 4\sqrt{3}$ (*B*) $2\sqrt{3} - 5$ (*C*) $y + 2\sqrt{y} - 8$

(*D*) $x + \sqrt[3]{x^2y} - \sqrt[3]{xy^2} - y$

EXAMPLE 9 Show that $(2 - \sqrt{3})$ is a solution of the equation $x^2 - 4x + 1 = 0$.

SOLUTION

$$x^2 - 4x + 1 = 0$$
$$(2 - \sqrt{3})^2 - 4(2 - \sqrt{3}) + 1 \stackrel{?}{=} 0$$
$$4 - 4\sqrt{3} + 3 - 8 + 4\sqrt{3} + 1 \stackrel{?}{=} 0$$
$$0 \stackrel{\checkmark}{=} 0$$

PROBLEM 9 Show that $(2 + \sqrt{3})$ is a solution of $x^2 - 4x + 1 = 0$.

ANSWER $(2 + \sqrt{3})^2 - 4(2 + \sqrt{3}) + 1 = 4 + 4\sqrt{3} + 3 - 8 - 4\sqrt{3} + 1 = 0$

SPECIAL QUOTIENTS—RATIONALIZING DENOMINATORS

Recall that to express $\sqrt{2}/\sqrt{3}$ in simplest radical form, we multiplied the numerator and denominator by $\sqrt{3}$ to clear the denominator of the radical:

$$\frac{\sqrt{2}}{\sqrt{3}} = \frac{\sqrt{2} \cdot \sqrt{3}}{\sqrt{3} \cdot \sqrt{3}} = \frac{\sqrt{6}}{3}$$

The denominator is thus converted to a rational number. Also recall that the process of converting irrational denominators to rational forms is called *rationalizing the denominator.*

How can we rationalize the binomial denominator in

$$\frac{1}{\sqrt{3} - \sqrt{2}}$$

Multiplying the numerator and denominator by $\sqrt{3}$ or $\sqrt{2}$ does not help. Try it! Recalling the special product

$$(a - b)(a + b) = a^2 - b^2$$

suggests that we multiply the numerator and denominator by the denominator, only with the middle sign changed. Thus,

$$\frac{1}{\sqrt{3} - \sqrt{2}} = \frac{1(\sqrt{3} + \sqrt{2})}{(\sqrt{3} - \sqrt{2})(\sqrt{3} + \sqrt{2})} = \frac{\sqrt{3} + \sqrt{2}}{3 - 2} = \sqrt{3} + \sqrt{2}$$

EXAMPLE 10 Rationalize denominators and simplify:

(*A*)
$$\frac{\sqrt{2}}{\sqrt{6} - 2} = \frac{\sqrt{2}(\sqrt{6} + 2)}{(\sqrt{6} - 2)(\sqrt{6} + 2)} = \frac{\sqrt{12} + 2\sqrt{2}}{6 - 4}$$
$$= \frac{2\sqrt{3} + 2\sqrt{2}}{2} = \frac{2(\sqrt{3} + \sqrt{2})}{2} = \sqrt{3} + \sqrt{2}$$

(*B*) $$\frac{\sqrt{x}-\sqrt{y}}{\sqrt{x}+\sqrt{y}} = \frac{(\sqrt{x}-\sqrt{y})(\sqrt{x}-\sqrt{y})}{(\sqrt{x}+\sqrt{y})(\sqrt{x}-\sqrt{y})} = \frac{x-2\sqrt{xy}+y}{x-y}$$

PROBLEM 10 Rationalize denominators and simplify:

(*A*) $\dfrac{\sqrt{2}}{\sqrt{2}+3}$ (*B*) $\dfrac{\sqrt{x}+\sqrt{y}}{\sqrt{x}-\sqrt{y}}$

ANSWER (*A*) $\dfrac{2-3\sqrt{2}}{-7}$ (*B*) $\dfrac{x+2\sqrt{xy}+y}{x-y}$

Exercise 46

Multiply and simplify where possible:

A

1. $\sqrt{7}(\sqrt{7}-2)$ **2.** $\sqrt{5}(\sqrt{5}-2)$

3. $\sqrt{2}(3-\sqrt{2})$ **4.** $\sqrt{3}(2-\sqrt{3})$

5. $\sqrt{y}(\sqrt{y}-8)$ **6.** $\sqrt{x}(\sqrt{x}-3)$

7. $\sqrt{n}(4-\sqrt{n})$ **8.** $\sqrt{m}(3-\sqrt{m})$

9. $\sqrt{3}(\sqrt{3}+\sqrt{6})$ **10.** $\sqrt{5}(\sqrt{10}+\sqrt{5})$

11. $(2-\sqrt{3})(3+\sqrt{3})$ **12.** $(\sqrt{2}-1)(\sqrt{2}+3)$

13. $(\sqrt{5}+2)^2$ **14.** $(\sqrt{3}-3)^2$

15. $(\sqrt{m}-3)(\sqrt{m}-4)$ **16.** $(\sqrt{x}+2)(\sqrt{x}-3)$

Rationalize denominators and simplify:

17. $\dfrac{1}{\sqrt{5}+2}$ **18.** $\dfrac{1}{\sqrt{11}-3}$

19. $\dfrac{2}{\sqrt{5}+1}$ **20.** $\dfrac{4}{\sqrt{6}-2}$

21. $\dfrac{\sqrt{2}}{\sqrt{10}-2}$ **22.** $\dfrac{\sqrt{2}}{\sqrt{6}+2}$

23. $\dfrac{\sqrt{y}}{\sqrt{y}+3}$ **24.** $\dfrac{\sqrt{x}}{\sqrt{x}-2}$

B *Multiply and simplify where possible:*

25. $(4\sqrt{3}-1)(3\sqrt{3}-2)$ **26.** $(2\sqrt{7}-\sqrt{3})(2\sqrt{7}+\sqrt{3})$

27. $(\sqrt{x}-\sqrt{y})(\sqrt{x}+\sqrt{y})$ **28.** $(2\sqrt{x}+3)(2\sqrt{x}-3)$

29. $(5\sqrt{m}+2)(2\sqrt{m}-3)$ **30.** $(3\sqrt{u}-2)(2\sqrt{u}+4)$

31. $(\sqrt[3]{4} + \sqrt[3]{9})(\sqrt[3]{2} + \sqrt[3]{3})$

32. $\sqrt[3]{4}(\sqrt[3]{2} - \sqrt[3]{16})$

33. $(\sqrt[3]{x} - \sqrt[3]{y^2})(\sqrt[3]{x^2} + 2\sqrt[3]{y})$

34. $(\sqrt[5]{u^2} - \sqrt[5]{v^3})(\sqrt[5]{u^3} + \sqrt[5]{v^2})$

35. Show that $2 - \sqrt{3}$ is a solution to $x^2 - 4x + 1 = 0$.

36. Show that $2 + \sqrt{3}$ is a solution to $x^2 - 4x + 1 = 0$.

Rationalize denominators and simplify:

37. $\dfrac{\sqrt{3} + 2}{\sqrt{3} - 2}$

38. $\dfrac{\sqrt{2} - 1}{\sqrt{2} + 1}$

39. $\dfrac{\sqrt{2} + \sqrt{3}}{\sqrt{3} - \sqrt{2}}$

40. $\dfrac{3 - \sqrt{a}}{\sqrt{a} - 2}$

41. $\dfrac{2 + \sqrt{x}}{\sqrt{x} - 3}$

42. $\dfrac{\sqrt{5} - \sqrt{2}}{\sqrt{5} + \sqrt{2}}$

43. $\dfrac{3\sqrt{x}}{2\sqrt{x} - 3}$

44. $\dfrac{5\sqrt{a}}{3 - 2\sqrt{a}}$

C *Rationalize denominators and simplify:*

45. $\dfrac{2\sqrt{5} - 3\sqrt{2}}{5\sqrt{5} + 2\sqrt{2}}$

46. $\dfrac{3\sqrt{2} + 2\sqrt{3}}{2\sqrt{2} - 3\sqrt{3}}$

47. $\dfrac{1}{\sqrt[3]{x} + \sqrt[3]{y}}$

48. $\dfrac{1}{\sqrt[3]{x} - \sqrt[3]{y}}$

49. $\dfrac{1}{\sqrt{x} + \sqrt{y} - \sqrt{z}}$

50. $\dfrac{1}{\sqrt{x} - \sqrt{y} + \sqrt{z}}$

HINT: In Problem 49 start by multiplying numerator and denominator by $(\sqrt{x} + \sqrt{y}) + \sqrt{z}$

8.5 Complex Numbers

INTRODUCTION

The Pythagoreans (500–275 B.C.) proved that the elementary polynomial equation

$$x^2 = 2 \tag{1}$$

has no rational number solutions. (Recall that a rational number is any number that can be expressed as the ratio of two integers, excluding division by 0.) A new kind of number had to be invented if (1) were to have a solution. What kind of number? The irrational numbers. The irrational numbers did not appear overnight. Actually, their evolution took place over a couple of thousand years, and it was not until the last century that they were finally placed on a firm foundation. The rational and irrational numbers together constitute the real number system (see inside front cover).

Do the real numbers satisfy all of our number needs? No! Consider the following simple equation

$$x^2 = -1 \tag{2}$$

What real number squared is -1? None, since the square of any real number cannot be negative. If (2) is to have a solution, then a new kind of number system must be invented—a number system that has the possibility of producing negative real numbers when some numbers are squared. This new number system will be called the complex numbers. The complex numbers evolved over a long period of time, dating back to the Italian mathematician, Cardono (1545).

Early resistance to these new numbers was suggested in the words used to name them: "complex" and "imaginary." In spite of this early resistance, complex numbers have come into widespread use in both pure and applied mathematics. They are used extensively, for example, in electrical engineering, physics, chemistry, statistics, and aeronautical engineering. Our first use of complex numbers will be in connection with solutions of second-degree equations in the next chapter.

THE COMPLEX NUMBER SYSTEM

A *complex number* is any number of the form

$$\boxed{a + bi}$$

where a and b are real numbers; i is called the *imaginary unit*. Thus

$$5 + 2i \qquad \tfrac{1}{4} + 2i \qquad \sqrt{2} - \tfrac{1}{3}i$$

$$0 + 5i \qquad 6 + 0i \qquad 0 + 0i$$

are all complex numbers. Particular kinds of complex numbers are given special names:

$a + 0i = a$	real number
$0 + bi = bi$	pure imaginary number
$0 + 0i = 0$	zero
$1i = i$	imaginary unit
$a - bi$	conjugate of $a + bi$

Just as every integer is a rational number, every real number is a complex number. That is, the set of real numbers is a subset of the set of complex numbers (see inside front cover).

For complex numbers to be useful, we must know how to add, subtract, multiply, and divide them. We start by defining equality, addition, and multiplication.

EQUALITY	$a + bi = c + di$ if only if $a = c$ and $b = d$
ADDITION	$(a + bi) + (c + di) = (a + c) + (b + d)i$
MULTIPLICATION	$(a + bi)(c + di) = (ac - bd) + (ad + bc)\,i$

These definitions, particularly the one for multiplication, probably seem strange to you. But if we want the possibility of a number squared being negative, and the basic field properties for the real numbers to continue to hold (see inside back cover), then it turns out that we have no choice in the matter.

Fortunately, you do not need to memorize these definitions. One can show that under these definitions, the complex numbers have the same field properties as the real numbers; hence we can manipulate complex numbers as if they were algebraic forms in the real numbers system. Before we use this fact, let us find out what happens to i when it is squared. Using the definition of multiplication above and the fact that $i = 0 + 1i$, we obtain

$$\begin{aligned} i^2 = (0 + 1i)(0 + 1i) &= (0 \cdot 0 - 1 \cdot 1) + (0 \cdot 1 + 1 \cdot 0)i \\ &= -1 + 0i \\ &= -1 \end{aligned}$$

Thus,

$$\boxed{i^2 = -1}$$

a very important result. We now have a number whose square is negative! We also write

$$i = \sqrt{-1} \quad \text{and} \quad -i = -\sqrt{-1}$$

Now to the process of adding, subtracting, multiplying, and dividing complex numbers. Assuming the field properties listed on the inside back cover hold for complex numbers (or can be shown), we will carry out the four basic operations in a rather mechanical way. That is, it follows that we can treat complex numbers as we treat ordinary binomial forms in the algebra of real numbers, with the exception that whenever

i^2 occurs, we replace it with -1. The following example illustrates the procedure.

EXAMPLE 11 Write each of the following in the form $a + bi$:

(*A*) $(3 + 2i) + (2 - i)$ (*B*) $(3 + 2i) - (2 - i)$

(*C*) $(3 + 2i)(2 - i)$ (*D*) $(3 + 2i)/(2 - i)$

SOLUTION (*A*)
$$\begin{aligned}(3 + 2i) + (2 - i) &= 3 + 2i + 2 - i \\ &= 5 + i\end{aligned}$$

(*B*)
$$\begin{aligned}(3 + 2i) - (2 - i) &= 3 + 2i - 2 + i \\ &= 1 + 3i\end{aligned}$$

(*C*)
$$\begin{aligned}(3 + 2i)(2 - i) &= 6 + i - 2i^2 \\ &= 6 + i - 2(-1) \\ &= 6 + i + 2 \\ &= 8 + i\end{aligned}$$

(*D*) In order to eliminate i from the denominator, we multiply the numerator and denominator by the conjugate of $2 - i$, that is, by $2 + i$:

$$\frac{3 + 2i}{2 - i} \cdot \frac{2 + i}{2 + \mathrm{i}} = \frac{6 + 7i + 2i^2}{4 - i^2} = \frac{6 + 7i + 2(-1)}{4 - (-1)}$$

$$= \frac{4 + 7i}{5} = \frac{4}{5} + \frac{7}{5}i$$

Recall that subtraction and division are defined, in general, as follows:

$$A - B = C \quad \text{if and only if} \quad A = B + C$$

$$A \div B = C \quad \text{if and only if} \quad A = BC,\ B \neq 0$$

The results obtained by the procedures illustrated in Example 11*B* and 11*D* above are consistent with these definitions, as can easily be checked. And with a little extra work, these procedures can be shown to hold in general (see the **C** problems in Exercise 47).

PROBLEM 11 Carry out the following operations and write each answer in the $a + bi$ form.

(*A*) $(3 + 2i) + (6 - 4i)$ (*B*) $(3 - 5i) - (1 - 3i)$

(*C*) $(2 - 4i)(3 + 2i)$ (*D*) $\dfrac{2 + 4i}{3 + 2i}$

ANSWER (*A*) $9 - 2i$ (*B*) $2 - 2i$ (*C*) $14 - 8i$ (*D*) $\frac{14}{13} + \frac{8}{13}i$

COMPLEX NUMBERS AND RADICALS

Recall that we say y is a square root of x if $y^2 = x$. It can be shown that if x is a positive real number, then x has two real square roots, one the negative of the other; if x is negative, then x has two complex square roots, one also the negative of the other. In particular, if we let $x = -a$, $a > 0$, then[†]

$$\boxed{\sqrt{-a} = i\sqrt{a} \qquad a > 0}$$

To check this, we square $i\sqrt{a}$ and obtain $-a$:

$$(i\sqrt{a})^2 = i^2(\sqrt{a})^2 = (-1)a = -a$$

EXAMPLE 12 Write in the form $a + bi$.

(A) $\sqrt{-4}$

SOLUTION $\sqrt{-4} = i\sqrt{4} = 2i$ or $0 + 2i$

(B) $4 + \sqrt{-4}$

SOLUTION $4 + \sqrt{-4} = 4 + i\sqrt{4} = 4 + 2i$

(C) $\dfrac{-3 - \sqrt{-7}}{2}$

SOLUTION $\dfrac{-3 - \sqrt{-7}}{2} = \dfrac{-3 - i\sqrt{7}}{2} = -\dfrac{3}{2} - \dfrac{\sqrt{7}}{2}i$

PROBLEM 12 Write in the form $a + bi$:

(A) $\sqrt{-16}$ (B) $5 + \sqrt{-16}$ (C) $\dfrac{-5 - \sqrt{-2}}{2}$

ANSWER (A) $4i$ or $0 + 4i$ (B) $5 + 4i$ (C) $-\dfrac{5}{2} - \dfrac{\sqrt{2}}{2}i$

Exercise 47

A *Perform the indicated operations and write each answer in the form $a + bi$.*

1. $(5+2i) + (3+i)$

2. $(6+i) + (2+3i)$

3. $(-8+5i) + (3-2i)$

4. $(2-3i) + (5-2i)$

5. $(8+5i) - (3+2i)$

6. $(9+7i) - (2+5i)$

[†]Note that if in $a + bi$, $b = \sqrt{k}$, then we often write $a + i\sqrt{k}$ instead of $a + \sqrt{k}i$ so that i will not accidentally end up under the radical sign.

7. $(4+7i)-(-2-6i)$

8. $(9-3i)-(12-5i)$

9. $(3-7i)+5i$

10. $12+(5-2i)$

11. $(5i)(3i)$

12. $(2i)(4i)$

13. $-2i(5-3i)$

14. $-3i(2-4i)$

15. $(2-3i)(3+3i)$

16. $(3-5i)(-2-3i)$

17. $(7-6i)(2-3i)$

18. $(2-i)(3+2i)$

19. $(7+4i)(7-4i)$

20. $(5-3i)(5+3i)$

21. $\dfrac{1}{2+i}$

22. $\dfrac{1}{3-i}$

23. $\dfrac{3+i}{2-3i}$

24. $\dfrac{2-i}{3+2i}$

25. $\dfrac{13+i}{2-i}$

26. $\dfrac{15-3i}{2-3i}$

B *Convert square roots of negative numbers to complex form; perform the indicated operations; and express answers in the form $a+bi$.*

27. $(5-\sqrt{-9})+(2-\sqrt{-4})$

28. $(-8+\sqrt{-25})+(3-\sqrt{-4})$

29. $(9-\sqrt{-9})-(12-\sqrt{-25})$

30. $(4+\sqrt{-49})-(-2-\sqrt{-36})$

31. $(-2+\sqrt{-49})(3-\sqrt{-4})$

32. $(5+\sqrt{-9})(2-\sqrt{-1})$

33. $\dfrac{5-\sqrt{-4}}{3}$

34. $\dfrac{6-\sqrt{-64}}{2}$

35. $\dfrac{1}{2-\sqrt{-9}}$

36. $\dfrac{1}{3-\sqrt{-16}}$

37. $\dfrac{2}{5i}$

38. $\dfrac{1}{3i}$

39. $\dfrac{1+3i}{2i}$

40. $\dfrac{2-i}{3i}$

41. $(2-i)^2+3(2-i)-5$

42. $(2-3i)^2-2(2-3i)+9$

43. Evaluate x^2-2x+2 for $x=1-i$.

44. Evaluate x^2-2x for $x=1+i$.

45. Simplify: i^2, i^3, i^4, i^5, i^6, i^7, and i^8

46. Simplify: i^{12}, i^{13}, i^{14}, i^{15}, and i^{16}

47. For what real values of x and y will $(3x)+(y-2)i=(5-2x)+(3y-8)i$?

48. For what real values of x and y will $(2x-1)+(3y+2)i=5-4i$?

C *Perform the indicated operations and write each answer in the form $a + bi$.*

49. $(a+bi)+(c+di)$

50. $(a+bi)-(c+di)$

51. $(a+bi)(a-bi)$

52. $(u-vi)(u+vi)$

53. $(a+bi)(c+di)$

54. $\dfrac{a+bi}{c+di}$

55. $\left(-\dfrac{1}{2}-\dfrac{\sqrt{3}}{2}i\right)^3$

56. $\left(-\dfrac{1}{2}+\dfrac{\sqrt{3}}{2}i\right)^3$

Solve each equation:

57. $y^2=-36$

58. $x^2=-25$

59. $(x-9)^2=-9$

60. $(x-3)^2=-4$

61. For what values of x will $\sqrt{x-10}$ be real?

62. When will $\dfrac{-b \pm \sqrt{b^2-4ac}}{2a}$ represent a complex number, assuming a, b, and c are all real numbers $(a \neq 0)$?

Exercise 48 Chapter Review

A **1.** Change to radical form: (*A*) $(3m)^{1/2}$ (*B*) $3m^{1/2}$

Change to rational exponent form: (*C*) $\sqrt{2x}$ (*D*) $\sqrt{a+b}$

Simplify, and write in simplest radical form. All variables represent positive real numbers.

2. $\sqrt{4x^2y^4}$

3. $\sqrt{\dfrac{25}{y^2}}$

4. $\sqrt{36x^4y^7}$

5. $\dfrac{1}{\sqrt{2y}}$

6. $\dfrac{6ab}{\sqrt{3a}}$

7. $\sqrt{2x^2y^5}\sqrt{18x^3y^2}$

8. $\sqrt{\dfrac{y}{2x}}$

9. $4\sqrt{x}-7\sqrt{x}$

10. $\sqrt{7}+2\sqrt{3}-4\sqrt{3}$

11. $\sqrt{5}(\sqrt{5}+2)$

12. $(\sqrt{3}-1)(\sqrt{3}+2)$

13. $\dfrac{\sqrt{5}}{3-\sqrt{5}}$

Perform the indicated operations and write the answer in the form $a + bi$.

14. $(-3 + 2i) + (6 - 8i)$

15. $(3 - 3i)(2 + 3i)$

16. $\dfrac{13 - i}{5 - 3i}$

B **17.** Change to radical form: (*A*) $(2mn)^{2/3}$; (*B*) $3x^{2/5}$

Change to rational exponent form: (*C*) $\sqrt[7]{x^5}$; (*D*) $-3\sqrt[3]{(xy)^2}$

Simplify, and write in simplest radical form. All variables represent positive real numbers.

18. $\sqrt[3]{(2x^2y)^3}$

19. $3x\sqrt[3]{x^5y^4}$

20. $\dfrac{\sqrt{8m^3n^4}}{\sqrt{12m^2}}$

21. $\sqrt[8]{y^6}$

22. $-2x\sqrt[5]{3^6x^7y^{11}}$

23. $\dfrac{2x^2}{\sqrt[3]{4x}}$

24. $\sqrt[5]{\dfrac{3y^2}{8x^2}}$

25. $(2\sqrt{x} - 5\sqrt{y})(\sqrt{x} + \sqrt{y})$

26. $\dfrac{\sqrt{x} - 2}{\sqrt{x} + 2}$

27. $\dfrac{3\sqrt{x}}{2\sqrt{x} - \sqrt{y}}$

28. $\sqrt{\frac{2}{3}} + \sqrt{\frac{3}{2}}$

Perform the indicated operations and write the answer in the form $a + bi$.

29. $(2 - 2\sqrt{-4}) - (3 - \sqrt{-9})$

30. $\dfrac{2 - \sqrt{-1}}{3 + \sqrt{-4}}$

31. $(3 + i)^2 - 2(3 + i) + 3$

32. i^{27}

C *Simplify, and write in simplest radical form. All variables represent positive real numbers.*

33. $\sqrt[9]{8x^6y^{12}}$

34. $\sqrt[3]{3} - \dfrac{6}{\sqrt[3]{9}} + 3\sqrt[3]{\dfrac{1}{9}}$

35. $\sqrt[3]{8\sqrt{64\sqrt{xy}}}$

36. $\sqrt{9m^4 + 9m^2n^2}$

37. $\sqrt{2xy\sqrt[3]{4x^2y^2}}$

38. $\sqrt[(n+1)]{x^{n^2}x^{2n+1}}$

39. $\dfrac{1}{\sqrt[3]{x^2} - \sqrt[3]{y^2}}$

40. Describe the set $\{x \mid \sqrt{x^2} = |x|, x \text{ a real number}\}$.

41. Evaluate $(a + bi)\left(\dfrac{a}{a^2+b^2} - \dfrac{b}{a^2+b^2}i\right)$, $a \neq 0$ and $b \neq 0$, thus showing that each nonzero complex number $a + bi$ has an inverse relative to multiplication.

Chapter 9

SECOND-DEGREE EQUATIONS AND INEQUALITIES

9.1 Quadratic Equations

The equation

$$\tfrac{1}{2}x - \tfrac{1}{3}(x + 3) = 2 - x$$

though complicated-looking, is actually a first-degree equation in one variable, since it can be transformed into the equivalent equation

$$7x - 18 = 0$$

which is a special case of

$$ax + b = 0 \qquad a \neq 0$$

We have solved many equations of this type and found that they always have a single solution. From a mathematical point of view we have essentially taken care of the problem of solving first-degree equations in one variable.

In this chapter we will consider the next class of polynomial equations called second-degree equations or quadratic equations. A *quadratic equation* in one variable is any equation that can be written in the form

QUADRATIC EQUATION

$$ax^2 + bx + c = 0 \qquad a \neq 0 \tag{1}$$

where x is a variable and a, b, and c are constants. We will refer to this form as the *standard form* for the quadratic equation. The equations

$$2x^2 - 3x + 5 = 0$$

$$15 = 180t - 16t^2$$

are both quadratic equations since they are either in the standard form or can be transformed into this form.

Applications that give rise to quadratic equations are many and varied. A brief glance at Sec. 9.4 will give you some indication of the variety.

9.2 Solution of $ax^2 + c = 0$ by Square Root and of $ax^2 + bx + c = 0$ by Factoring

SOLUTION BY SQUARE ROOT

The easiest type of quadratic equation to solve is the special form where the first-degree term is missing; that is, when (1) is of the form

$$ax^2 + c = 0 \qquad a \neq 0$$

The method of solution makes direct use of the definition of square root. The process is illustrated in the following example.

EXAMPLE 1 Solve by the square root method:

(*A*) $x^2 - 8 = 0$

SOLUTION $x^2 - 8 = 0$

$x^2 = 8$ What number squared is 8?

$x = \pm\sqrt{8}$ or $\pm 2\sqrt{2}$ $\quad$ $\pm 2\sqrt{2}$ is a short way of writing $-2\sqrt{2}$ and $+2\sqrt{2}$

(*B*) $2x^2 - 3 = 0$

SOLUTION $2x^2 - 3 = 0$

$2x^2 = 3$

$x^2 = \frac{3}{2}$ What number squared is $\frac{3}{2}$?

$$x = \pm\sqrt{\frac{3}{2}} \quad \text{or} \quad \pm\frac{\sqrt{6}}{2}$$

(*C*) $3x^2 + 27 = 0$

SOLUTION $3x^2 + 27 = 0$

$$3x^2 = -27$$

$$x^2 = -9 \qquad \text{What number squared is } -9?$$

$$x = \pm\sqrt{-9} \quad \text{or} \quad \pm 3i$$

(*D*) $(x + \frac{1}{2})^2 = \frac{5}{4}$

SOLUTION $(x + \frac{1}{2})^2 = \frac{5}{4}$ What number squared is $\frac{5}{4}$?

$$x + \frac{1}{2} = \pm\sqrt{\frac{5}{4}}$$

$$x = -\frac{1}{2} \pm \frac{\sqrt{5}}{2}$$

$$= \frac{-1 \pm \sqrt{5}}{2}$$

PROBLEM 1 Solve by the square root method.

(*A*) $x^2 - 12 = 0$

(*B*) $3x^2 - 5 = 0$

(*C*) $2x^2 + 8 = 0$

(*D*) $(x + \frac{1}{3})^2 = \frac{2}{9}$

ANSWER (*A*) $\pm 2\sqrt{3}$ (*B*) $\pm\sqrt{\frac{5}{3}}$ or $\pm\frac{\sqrt{15}}{3}$ (*C*) $\pm 2i$

(*D*) $(-1 \pm \sqrt{2})/3$

SOLUTION BY FACTORING

If the coefficients a, b, and c in the quadratic equation

$$ax^2 + bx + c = 0$$

are such that $ax^2 + bx + c$ can be written as the product of two first-degree factors with integer coefficients, then the quadratic equation can be quickly and easily solved. The method of solution by factoring rests on the following property of the real numbers:

If a and b are real numbers, then

$$ab = 0 \qquad \textit{if and only if} \qquad a = 0 \textit{ or } b = 0 \textit{ (or both)}$$

This property is easily proved: If $a = 0$, we are through. If $a \neq 0$, we multiply both sides of $ab = 0$ by $1/a$ to obtain $b = 0$.

EXAMPLE 2 Solve by factoring, if possible.

(*A*) $x^2 + 2x - 15 = 0$

SOLUTION

$$x^2 + 2x - 15 = 0$$

$$(x - 3)(x + 5) = 0$$

$(x - 3)(x + 5) = 0$ if and only if $(x - 3) = 0$ or $(x + 5) = 0$

$$x - 3 = 0 \quad \text{or} \quad x + 5 = 0$$

$$x = 3 \quad \text{or} \quad x = -5$$

(*B*) $2x^2 = 3x$

SOLUTION

$$2x^2 = 3x$$

If both sides are divided by x, we lose one solution.

$$2x^2 - 3x = 0$$

$$x(2x - 3) = 0$$

$x(2x - 3) = 0$ if and only if $x = 0$ or $2x - 3 = 0$

$$x = 0 \quad \text{or} \quad 2x - 3 = 0$$

$$x = 0 \quad \text{or} \quad x = \tfrac{3}{2}$$

(*C*) $2x^2 - 8x + 3 = 0$

SOLUTION The polynomial cannot be factored in the integers; hence another method must be used.

PROBLEM 2 Solve by factoring, if possible.

(*A*) $x^2 - 2x - 8 = 0$ (*B*) $3t^2 = 2t$ (*C*) $x^2 - 3x - 3 = 0$

ANSWER (*A*) $x = 4, -1$ (*B*) $t = 0, \frac{2}{3}$

(*C*) Cannot be factored in the integers

Exercise 49

A *Solve by square root method:*

1. $x^2 - 16 = 0$

2. $x^2 - 25 = 0$

3. $x^2 + 16 = 0$

4. $x^2 + 25 = 0$

5. $y^2 - 45 = 0$

6. $m^2 - 12 = 0$

7. $4x^2 - 9 = 0$

8. $9y^2 - 16 = 0$

9. $16y^2 = 9$

10. $9x^2 = 4$

Solve by factoring:

11. $u^2 + 5u = 0$

12. $v^2 - 3v = 0$

13. $3A^2 = -12A$
14. $4u^2 = 8u$
15. $x^2 - 4x - 12 = 0$
16. $y^2 - 6y + 5 = 0$
17. $x^2 + 4x - 5 = 0$
18. $x^2 - 4x - 12 = 0$
19. $3Q^2 - 10Q - 8 = 0$
20. $2d^2 + 15d - 8 = 0$

B *Solve by square root method:*

21. $y^2 = 2$
22. $x^2 = 3$
23. $16a^2 + 9 = 0$
24. $4x^2 + 25 = 0$
25. $9x^2 - 7 = 0$
26. $4t^2 - 3 = 0$
27. $(m - 3)^2 = 25$
28. $(n + 5)^2 = 9$
29. $(t + 1)^2 = -9$
30. $(d - 3)^2 = -4$
31. $\left(x - \frac{1}{3}\right)^2 = \frac{4}{9}$
32. $\left(x - \frac{1}{2}\right)^2 = \frac{9}{4}$

Solve by factoring. (Write equations in standard form first.)

33. $u^2 = 2u + 3$
34. $m^2 + 2m = 15$
35. $3x^2 = x + 2$
36. $2x^2 = 3 - 5x$
37. $y^2 = 5y - 2$
38. $3 = t^2 + 7t$
39. $2x(x - 1) = 3(x + 1)$
40. $3x(x - 2) = 2(x - 2)$
41. $\frac{t}{2} = \frac{2}{t}$
42. $y = \frac{9}{y}$
43. $\frac{m}{4}(m + 1) = 3$
44. $\frac{A^2}{2} = A + 4$
45. $2y = \frac{2}{y} + 3$
46. $L = \frac{15}{L - 2}$
47. $2 + \frac{2}{x^2} = \frac{5}{x}$
48. $1 - \frac{3}{x} = \frac{10}{x^2}$

49. The width of a rectangle is 8 inches less than its length. If its area is 33 square inches, find its dimensions.

50. Find the base and height of a triangle with area 2 square feet if its base is 3 feet longer than its height $\left(A = \frac{1}{2}bh\right)$.

C *Solve by square root method:*

51. $\left(y + \frac{5}{2}\right)^2 = \frac{5}{2}$
52. $\left(x - \frac{3}{2}\right)^2 = \frac{3}{2}$

Solve for the indicated letters in terms of the other letters. Use positive square roots only.

53. $a^2 + b^2 = c^2$ (Solve for a.)
54. $s = \frac{1}{2}gt^2$ (Solve for t.)

55. In a given city on a given day, the demand equation for gasoline is $d = 600/p$ and the supply equation is $s = p - 50$, where d and s denote the number of gallons demanded and supplied (in thousands), respectively, at a price of p cents per gallon. Find the price at which supply is equal to demand.

56. To find the critical velocity at the top of the loop necessary to keep a steel ball on the track (see Fig. 1), the centripetal force mv^2/r is equated to the force due to gravity, mg. The mass m cancels out of the equation, and we are left with $v^2 = gr$. For a loop of radius 0.25 feet, find the critical velocity (in feet per second) at the top of the loop that is required to keep the ball on the track. Use $g = 32$ and compute your answer to two decimal places using a square root table.

Figure 1

9.3 The Quadratic Formula

The factoring and square root methods discussed in the last section are fast and easy to use when they apply. Many quadratic equations, however, cannot be solved by either method. For example, the simple-looking polynomial in the equation

$$x^2 + 6x - 2 = 0$$

cannot be factored in the integers. The equation requires a new method if it can be solved at all.

There is a method called "solution by completing the square" that will work for all quadratic equations. After briefly describing this method, we will then use it to develop the famous quadratic formula, a formula that will enable us to solve any quadratic equation quite mechanically.

SOLUTION BY COMPLETING THE SQUARE

The method of completing the square is based on the process of transforming the standard quadratic equation

$$ax^2 + bx + c = 0$$

into the form

$$(x + A)^2 = B$$

where A and B are constants. This last equation can easily be solved by the square root method discussed in the last section. Thus,

$$(x + A)^2 = B \qquad \text{What number squared is } B?$$
$$x + A = \pm\sqrt{B}$$
$$x = -A \pm \sqrt{B}$$

Before considering how the first part is accomplished, let's pause for a moment and consider a related problem: What number must be added to $x^2 + 6x$ so that the result is the square of a linear expression? There is an easy mechanical rule for finding this number based on the squares of binomials:

$$(x + m)^2 = x^2 + 2mx + m^2$$
$$(x - m)^2 = x^2 - 2mx + m^2$$

In either case, we see that the third term on the right is the square of one-half of the coefficient of x in the second term on the right. This observation leads directly to the rule:

> To *complete the square* of a quadratic of the form
>
> $$x^2 + bx$$
>
> add the square of one-half of the coefficient of x, that is
>
> $$\left(\frac{b}{2}\right)^2$$

Now let us see how completing the square can lead to a solution of a quadratic equation.

EXAMPLE 3 Solve $2x^2 - 4x - 3 = 0$ by the method of completing the square.

SOLUTION

$2x^2 - 4x - 3 = 0$ — Divide through by 2 to make coefficient of x^2 equal 1.

$x^2 - 2x - \frac{3}{2} = 0$ — Add $\frac{3}{2}$ to both sides.

$x^2 - 2x = \frac{3}{2}$ — Complete the square of the left side by adding $(-2/2)^2 = 1$ to both sides.

$x^2 - 2x + 1 = \frac{3}{2} + 1$ — Factor left side.

$(x - 1)^2 = \frac{5}{2}$ — Solve by square root method

$$x - 1 = \pm\sqrt{\tfrac{5}{2}}$$

$$x = 1 \pm \frac{\sqrt{10}}{2} \quad \text{or} \quad \frac{2 \pm \sqrt{10}}{2}$$

PROBLEM 3 Solve $2x^2 + 8x + 3 = 0$ by method of completing the square.

ANSWER $x = -2 \pm \sqrt{\dfrac{5}{2}}$ or $\dfrac{-4 \pm \sqrt{10}}{2}$

QUADRATIC FORMULA

The method of completing the square can be used to solve any quadratic equation, but the process is often tedious. If you had a very large number of quadratic equations to solve by completing the square, before you finished you would probably ask yourself if the process could not be made more efficient. Why not take the general equation

$$ax^2 + bx + c = 0 \qquad a \neq 0$$

and solve it once and for all for x in terms of the coefficients a, b, and c by the method of completing the square, and thus, obtain a formula that could be memorized and used whenever a, b, and c are known?

We start by making the leading coefficient 1. How? Multiply both sides of the equation by $1/a$. Thus

$$x^2 + \frac{b}{a}x + \frac{c}{a} = 0$$

Adding $-c/a$ to both members and then completing the square of the left member, we have

$$x^2 + \frac{b}{a}x + \frac{b^2}{4a^2} = \frac{b^2}{4a^2} - \frac{c}{a}$$

We now factor the left member and solve by the square-root method.

$$\left(x + \frac{b}{2a}\right)^2 = \frac{b^2 - 4ac}{4a^2}$$

$$x + \frac{b}{2a} = \pm\sqrt{\frac{b^2 - 4ac}{4a^2}}$$

$$x = -\frac{b}{2a} \pm \frac{\sqrt{b^2 - 4ac}}{2a}$$

Thus,

QUADRATIC FORMULA

$$x = \frac{-b \pm \sqrt{b^2 - 4ac}}{2a} \qquad a \neq 0$$

The last equation is called the *quadratic formula*. It should be memorized and used to solve quadratic equations when simpler methods fail. Note that $b^2 - 4ac$, called the *discriminant*, gives us the following useful information about roots.

$b^2 - 4ac$	$ax^2 + bx + c = 0$
positive	two real solutions
zero	one real solution
negative	two complex solutions

EXAMPLE 4 Solve $2x^2 - 4x - 3 = 0$ by use of the quadratic formula.

SOLUTION $2x^2 - 4x - 3 = 0$

$$x = \frac{-b \pm \sqrt{b^2 - 4ac}}{2a} \qquad \begin{aligned} a &= 2 \\ b &= -4 \\ c &= -3 \end{aligned}$$

Write down the quadratic formula, and identify a, b, and c.

$$= \frac{-(-4) \pm \sqrt{(-4)^2 - 4(2)(-3)}}{2(2)}$$

Substitute into formula and simplify.

$$= \frac{4 \pm \sqrt{40}}{4} = \frac{4 \pm 2\sqrt{10}}{4}$$

$$= \frac{2 \pm \sqrt{10}}{2}$$

PROBLEM 4 Solve $x^2 - 2x - 1 = 0$ using the quadratic formula.

ANSWER $x = 1 \pm \sqrt{2}$

EXAMPLE 5 Solve $x^2 + 11 = 6x$ using the quadratic formula.

SOLUTION $x^2 + 11 = 6x$

$x^2 - 6x + 11 = 0$ Write in standard form.

$$x = \frac{-b \pm \sqrt{b^2 - 4ac}}{2a} \qquad \begin{aligned} a &= 1 \\ b &= -6 \\ c &= 11 \end{aligned}$$

$$= \frac{-(-6) \pm \sqrt{(-6)^2 - 4(1)(11)}}{2(1)}$$

$$= \frac{6 \pm \sqrt{-8}}{2}$$

$$= \frac{6 \pm 2i\sqrt{2}}{2} = 3 \pm i\sqrt{2}$$

PROBLEM 5 Solve $2x^2 + 3 = 4x$ using the quadratic formula.

ANSWER $x = 1 \pm \frac{i\sqrt{2}}{2}$ or $\frac{2 \pm i\sqrt{2}}{2}$

WHICH METHOD?

In normal practice the quadratic formula is used whenever the square root method or the factoring method do not produce results. These latter methods are generally faster when they apply, and should be used.

Note that any equation of the form

$$ax^2 + c = 0$$

can always be solved by the square root method. And any equation of the form

$$ax^2 + bx = 0$$

can always be solved by factoring since $ax^2 + bx = x(ax + b)$.

It is important to realize, however, that the quadratic formula can always be used and will produce the same results as any of the other methods.

Exercise 50

A *Specify the constants a, b, and c for each quadratic equation when written in the standard form $ax^2 + bx + c = 0$.*

1. $2x^2 - 5x + 3 = 0$

2. $3x^2 - 2x + 1 = 0$

3. $m = 1 - 3m^2$

4. $2u^2 = 1 - 3u$

5. $3y^2 - 5 = 0$

6. $2x^2 - 5x = 0$

Solve by use of the quadratic formula:

7. $x^2 + 8x + 3 = 0$

8. $x^2 + 4x + 2 = 0$

9. $y^2 - 10y - 3 = 0$

10. $y^2 - 6y - 3 = 0$

B **11.** $u^2 = 1 - 3u$

12. $t^2 = 1 - t$

13. $y^2 + 3 = 2y$

14. $x^2 + 8 = 4x$

15. $2m^2 + 3 = 6m$

16. $2x^2 + 1 = 4x$

17. $p = 1 - 3p^2$

18. $3q + 2q^2 = 1$

Solve each of the following equations by any method, excluding completing the square.

19. $(x - 5)^2 = 7$

20. $(y + 4)^2 = 11$

21. $x^2 + 2x = 2$

22. $x^2 - 1 = 3x$

23. $2u^2 + 3u = 0$

24. $2n^2 = 4n$

25. $x^2 - 2x + 9 = 2x - 4$

26. $x^2 + 15 = 2 - 6x$

27. $\dfrac{2}{u} = \dfrac{3}{u^2} + 1$

28. $1 + \dfrac{8}{x^2} = \dfrac{4}{x}$

29. $\dfrac{1.2}{y-1} + \dfrac{1.2}{y} = 1$

30. $\dfrac{24}{10+m} + 1 = \dfrac{24}{10-m}$

Solve for the indicated letter in terms of the other letters.

31. $d = \frac{1}{2}gt^2$ for t (positive)

32. $a^2 + b^2 = c^2$ for a (positive)

33. $A = P(1 + r)^2$ for r (positive)

34. $P = EI - RI^2$ for I (positive)

C *Solve by completing the square:*

35. $y^2 - 10y - 3 = 0$

36. $x^2 - 6x - 3 = 0$

37. $2d^2 - 4d + 1 = 0$

38. $2y^2 - 6y + 3 = 0$

39. $x^2 + mx + n = 0$

40. $ax^2 + bx + c = 0,\ a \neq 0$

41. Show that if r_1 and r_2 are the two roots of $ax^2 + bx + c = 0$, then $r_1r_2 = c/a$.

42. For r_1 and r_2 in the preceding problem, show that $r_1 + r_2 = -b/a$.

9.4 Applications

We will now consider a number of applications from several fields. Since quadratic equations often have two solutions, it is important to check both of the solutions in the original problem to see if one or the other must be rejected.

EXAMPLE 6 The sum of a number and its reciprocal is $\frac{5}{2}$. Find the number.

SOLUTION Let $x =$ the number, then

$$x + \frac{1}{x} = \frac{5}{2}$$

$$2x^2 + 2 = 5x$$

$$2x^2 - 5x + 2 = 0$$

$$(2x - 1)(x - 2) = 0$$

$$x = \tfrac{1}{2} \text{ or } 2$$

PROBLEM 6

If the reciprocal of a number is subtracted from the original number, the difference is $\frac{8}{3}$. Find the number.

ANSWER 3

EXAMPLE 7

A tank can be filled in 4 hours by two pipes when both are used. How many hours are required for each pipe to fill the tank alone if the smaller pipe requires 3 hours more than the larger one?

SOLUTION Quantity of work = rate × time

Let

$$x = \text{time for the larger pipe to fill tank alone}$$

$$x + 3 = \text{time for the smaller pipe to fill tank alone}$$

Then

$$\frac{1}{x} = \text{rate for larger pipe} \qquad \frac{1}{x} \text{ of the tank per hour}$$

$$\frac{1}{x+3} = \text{rate for smaller pipe} \qquad \frac{1}{x+3} \text{ of the tank per hour}$$

Thus

$$\frac{4}{x} + \frac{4}{x+3} = 1 \qquad \text{whole tank}$$

$$4x + 12 + 4x = x^2 + 3x$$

$$x^2 - 5x - 12 = 0$$

$$x = \frac{5 \pm \sqrt{73}}{2}$$ Why should we discard the negative answer?

$$x = \frac{5 + \sqrt{73}}{2} \approx 6.78 \text{ hours} \qquad \text{larger pipe}$$

$$x + 3 \approx 9.78 \text{ hours} \qquad \text{smaller pipe}$$

PROBLEM 7

Two pipes can fill a tank in 3 hours when used together. Alone, one can fill the tank 2 hours faster than the other. How long will it take each pipe to fill the tank alone? Compute the answers to two decimal places, using the square root table at the end of the book.

ANSWER 5.16 hours and 7.16 hours

Exercise 51

*These problems are not grouped from easy (**A**) to difficult or theoretical (**C**). They are grouped somewhat according to type. The most difficult problems are double-starred (⋆⋆), those of moderate difficulty single-starred (⋆), and the easier ones are not marked.*

NUMBER PROBLEMS

1. Find two consecutive positive even integers whose product is 168.

2. Find two positive numbers having a sum of 21 and a product of 104.

3. Find all numbers with the property that when the number is added to itself the sum is the same as when the number is multiplied by itself.

4. The sum of a number and its reciprocal is $\frac{10}{3}$. Find the number.

GEOMETRY

The following theorem may be used where needed:

PYTHAGOREAN THEOREM *A triangle is a right triangle if and only if the square of the longest side is equal to the sum of the squares of the two shorter sides.*

$$c^2 = a^2 + b^2$$

⋆ **5.** Approximately how far would the horizon be from an airplane 2 miles high? Assume the radius of the earth is 4,000 miles and use the square root table to estimate the answer to the nearest mile (Fig. 2).

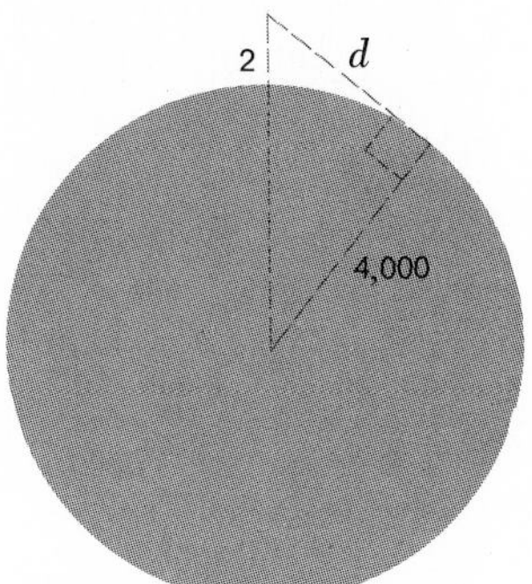

Figure 2

6. Find the base and height of a triangle with area 2 square feet if its base is 3 feet longer than its height. ($A = \frac{1}{2}bh$)

⋆ **7.** If the length and width of a 4- by 2-inch rectangle are each increased by the same amount, the area of the new rectangle will be twice the old. What are the dimensions to two decimal places of the new rectangle?

8. The width of a rectangle is 2 inches less than its length. Find its dimensions to two decimal places if its area is 12 square inches.

★ **9.** A flag has a white cross of uniform width on a red background. Find the width of the cross so that it takes up exactly one-half the total area of a 4- by 3-foot flag.

PHYSICS-ENGINEERING

10. The pressure p in pounds per square foot of wind blowing at v miles/hour is $p = 0.003v^2$. If a pressure gauge on a bridge registers a wind pressure of 14.7 pounds per square foot, what is the velocity of the wind?

11. One method of measuring the velocity of water in a stream or river is to use an L-shaped tube as indicated in Fig. 3. Torricelli's law in physics tells us that the height (in feet) that the water is pushed up into the tube above the surface is related to the water's velocity (in feet per second) by the formula $v^2 = 2gh$, where g is approximately 32 feet per second per second. (NOTE: The device can also be used as a simple speedometer for a boat.) How fast is a stream flowing if $h = 0.5$ foot? Find the answer to two decimal points.

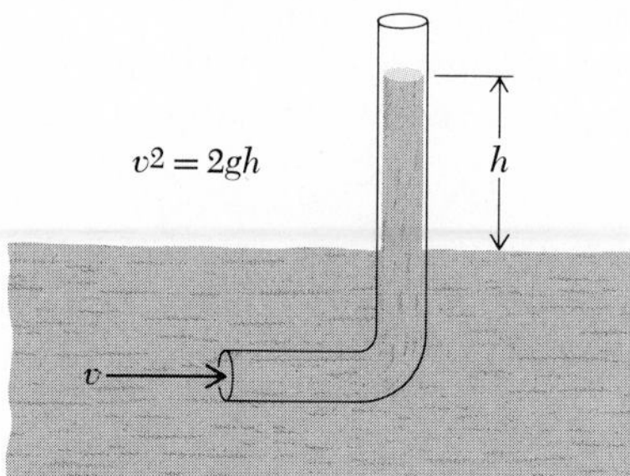

Figure 3

12. At 20 miles/hour a car collides with a stationary object with the same force it would have if it had been dropped $13\frac{1}{2}$ feet, that is, if it had been pushed off the roof of an average one-story house. In general, a car moving at r miles/hour hits a stationary object with a force of impact that is equivalent to that force with which it would hit the ground when falling from a certain height h given by the formula $h = 0.0336r^2$. Approximately how fast would a car have to be moving if it crashed as hard as if it had been pushed off the top of a 12-story building 121 feet high?

★**13.** For a car traveling at a speed of v miles/hour, the least number of feet d under the best possible conditions that is necessary to stop a car (including a reaction time) is given by the empirical formula $d = 0.044v^2 + 1.1v$. Estimate the speed of a car requiring 165 feet to stop after danger is realized.

★**14.** If an arrow is shot vertically in the air (from the ground) with an initial velocity of 176 feet/second its distance y above the ground t seconds after it is released (neglecting air resistance) is given by $y = 176t - 16t^2$.
(A) Find the time when y is 0, and interpret physically.
(B) Find the times when the arrow is 16 feet off the ground. Compute answers to two decimal places.

★**15.** A barrel 2 feet in diameter and 4 feet in height has a 1-inch-diameter drainpipe in the bottom. It can be shown that the height h of the surface of the water above the bottom of the barrel at time t minutes after the drain has been opened is given by the formula $h = [\sqrt{h_0} - \frac{5}{12}t]^2$, where h_0 is the water level above the drain at time $t = 0$. If the barrel is full and the drain opened, how long will it take to empty one-half of the contents? HINT: The problem is very easily solved if the right side of the equation is not squared.

RATE-TIME

★★16. One pipe can fill a tank in 5 hours less than another; together they fill the tank in 5 hours. How long would it take each alone to fill the tank? Compute the answer to two decimal places.

★★17. A new printing press can do a job in 1 hour less than an older press. Together they can do the same job in 1.2 hours. How long would it take each other alone to do the job?

18. Two boats travel at right angles to each other after leaving the same dock at the same time. 1 hour later they are 13 miles apart. If one travels 7 miles/hour faster than the other, what is the rate of each? HINT: Use the pythagorean theorem near the beginning of this exercise.

★ 19. A speedboat takes 1 hour longer to go 24 miles up a river than to return. If the boat cruises at 10 miles/hour in still water, what is the rate of the current?

ECONOMICS-BUSINESS

20. If P dollars is invested at r percent compounded annually, at the end of 2 years it will grow to $A = P(1 + r)^2$. At what interest rate will \$100 grow to \$144 in 2 years? NOTE: If $A = 144$ and $P = 100$, find r.

★21. In a certain city the demand equation for popular records is $d = 3{,}000/p$, where d would be the quantity of records demanded on a given day if the selling price were p dollars per record. (Notice as the price goes up, the number of records the people are willing to buy goes down, and vice versa.) On the other hand, the supply equation is $s = 1{,}000p - 500$, where s is the quantity of records a supplier is willing to supply at p dollars per record. (Notice as the price goes up, the number of records a supplier is willing to sell goes up, and vice versa.) At what price will supply equal demand; that is, at what price will $d = s$? In economic theory the price at which supply equals demand is called the *equilibrium point*, the point at which the price ceases to change.

9.5 Equations Reducible to Quadratic Form

EQUATIONS INVOLVING RADICALS

Consider the equation

$$x - 1 = \sqrt{x + 11}$$

What can we do to solve this equation? Perhaps doing something to the equation to eliminate the radical will help. What? Let us square both members to see what happens—certainly if $a = b$, then $a^2 = b^2$ (why?).

Thus,

$$(x-1)^2 = (\sqrt{x+11})^2$$

$$x^2 - 2x + 1 = x + 11$$

$$x^2 - 3x - 10 = 0$$

$$(x+2)(x-5) = 0$$

$$x = -2, 5$$

Check

$x = -2 \qquad -2 - 1 \stackrel{?}{=} \sqrt{-2+11}$

$-3 \stackrel{?}{=} \sqrt{9}$ Recall that "$\sqrt{9}$" names the positive square root of 9.

$-3 \neq 3$

Hence, $x = -2$ is not a solution.

$x = 5 \qquad 5 - 1 \stackrel{?}{=} \sqrt{5+11}$

$4 \stackrel{?}{=} \sqrt{16}$

$4 \stackrel{\checkmark}{=} 4$

Hence, $x = 5$ is a solution.

Therefore, 5 is a solution and -2 is not. The process of squaring introduced an "extraneous" solution. In general one can prove the following important theorem.

THEOREM 1 If both members of an equation are raised to a natural number power, then the solution set of the original equation is a subset of the solution set of the new equation.

Thus, any new equation obtained by raising both members of an equation to the same natural number power may have solutions (called *extraneous solutions*) that are not solutions of the original equation. On the other hand, any solution of the original equation must be among those of the new equation. We need only check all of the solutions at the end of the process to eliminate the so-called extraneous ones.

EXAMPLE 8 Solve: $x + \sqrt{x-4} = 4$

SOLUTION

$$x + \sqrt{x-4} = 4$$

$$\sqrt{x-4} = 4 - x \qquad \text{Isolate radical on one side.}$$

$$x - 4 = 16 - 8x + x^2 \qquad \text{Square both members.}$$

$$x^2 - 9x + 20 = 0$$

$$(x-5)(x-4)=0$$

$$x=4,\ 5$$

Checking, we find 4 is a solution and 5 is extraneous.

PROBLEM 8 Solve: $x=5+\sqrt{x-3}$

ANSWER $x=7$

OTHER FORMS REDUCIBLE TO QUADRATIC

EXAMPLE 9 If asked to solve the equation

$$x^4-x^2-12=0$$

you might at first have trouble. But if you recognize that the equation is quadratic in x^2, you can solve for x^2 first, then solve for x. You might find it convenient to make the substitution $u=x^2$, and then solve the equation

$$u^2-u-12=0$$

$$(u-4)(u+3)=0$$

$$u=4,\ -3$$

Replacing u with x^2, we obtain

$$x^2=4 \qquad x^2=-3$$

$$x=\pm 2 \qquad x=\pm i\sqrt{3}$$

PROBLEM 9 Solve $x^6+6x^3-16=0$

ANSWER $x=-2,\ \sqrt[3]{2}$

In general, if an equation that is not quadratic can be transformed into the form

$$au^2+bu+c=0$$

where u is an expression in some other variable, then the equation is said to be in *quadratic form.* Once recognized as a quadratic form, an equation can often be solved using quadratic methods.

EXAMPLE 10 Solve $x^{2/3}-x^{1/3}-6=0$

SOLUTION Let $u=x^{1/3}$, then

$$u^2-u-6=0$$

$$(u-3)(u+2)=0$$

$$u=3,\ -2$$

Replacing u with $x^{1/3}$, we obtain

$$x^{1/3} = 3 \qquad x^{1/3} = -2$$

$$x = 27 \qquad x = -8$$

PROBLEM 10 Solve $x^{2/3} - x^{1/3} - 12 = 0$

ANSWER $64, -27$

Exercise 52

Solve:

A

1. $x - 2 = \sqrt{x}$

2. $\sqrt{x} = x - 6$

3. $m - 13 = \sqrt{m + 7}$

4. $\sqrt{5n + 9} = n - 1$

5. $x^4 - 10x^2 + 9 = 0$

6. $x^4 - 13x^2 + 36 = 0$

B

7. $1 + \sqrt{x + 5} = x$

8. $x - \sqrt{x + 10} = 2$

9. $x^4 - 7x^2 - 18 = 0$

10. $y^4 - 2y^2 - 8 = 0$

11. $x^6 - 7x^3 - 8 = 0$ (real solutions)

12. $x^6 + 3x^3 - 10 = 0$ (real solutions)

13. $x^{2/3} - 3x^{1/3} - 10 = 0$

14. $2x^{2/3} + 3x^{1/3} - 2 = 0$

15. $y^{1/2} - 3y^{1/4} + 2 = 0$

16. $y^{1/2} - 5y^{1/4} + 6 = 0$

17. $6x^{-2} - 5x^{-1} - 6 = 0$

18. $3n^{-2} - 11n^{-1} - 20 = 0$

19. $(x^2 + 2x)^2 - (x^2 + 2x) = 6$

20. $(m^2 - m)^2 - 4(m^2 - m) = 12$

C

21. $\sqrt[4]{x - 3} = 2$

22. $\sqrt[3]{x + 5} = 3$

23. $\sqrt{3w - 2} - \sqrt{w} = 2$

24. $\sqrt{3x + 4} = 2 + \sqrt{x}$

25. $\sqrt{2x - 1} - \sqrt{x - 4} = 2$

26. $\sqrt{3y - 2} = 3 - \sqrt{3y + 1}$

27. $(x - 3)^4 + 3(x - 3)^2 = 4$

28. $(m - 5)^4 + 36 = 13(m - 5)^2$

29. $4x^{-4} - 17x^{-2} + 4 = 0$

30. $9y^{-4} - 10y^{-2} + 1 = 0$

9.6 Systems Involving Second-Degree Equations in Two Variables

In this section we will investigate several special types of systems that involve at least one second-degree equation in two variables. The methods used to solve these systems is best illustrated through examples.

EXAMPLE 11 Solve the system:

$$4x^2 + y^2 = 25$$
$$2x + y = 7$$

SOLUTION The substitution principle is effective. Solve the linear equation for one variable in terms of the other, and then substitute into the nonlinear equation to obtain a quadratic equation in one variable. Thus,

$$y = 7 - 2x$$

is substituted into the second-degree equation to obtain

$$4x^2 + (7 - 2x)^2 = 25$$

Now we solve for x. Squaring $7 - 2x$ and collecting all terms on the left side, we obtain

$$8x^2 - 28x + 24 = 0$$
$$2x^2 - 7x + 6 = 0$$
$$(2x - 3)(x - 2) = 0$$
$$x = \tfrac{3}{2}, 2$$

These values are substituted back into the linear equation to find the corresponding values for y. (Note that if we substitute these values back into the second-degree equations, we may obtain "extraneous" roots; try it and see why. Recall that a solution of a system must satisfy both equations.) For $x = \frac{3}{2}$,

$$2\left(\tfrac{3}{2}\right) + y = 7$$
$$y = 4$$

For $x = 2$,

$$2(2) + y = 7$$
$$y = 3$$

Thus $\left(\frac{3}{2}, 4\right)$ and $(2, 3)$ are solutions to the system, as can easily be checked.

PROBLEM 11 Solve the system:

$$2x^2 - y^2 = 1$$
$$3x + y = 2$$

ANSWER $(1, -1), \left(\frac{5}{7}, -\frac{1}{7}\right)$

EXAMPLE 12 Solve the system:

$$x^2 - y^2 = 5$$
$$x^2 + 2y^2 = 17$$

Proceed as with linear equations—subtract to eliminate x.

SOLUTION

$$\begin{aligned} x^2 - y^2 &= 5 \\ x^2 + 2y^2 &= 17 \\ \hline -3y^2 &= -12 \\ y^2 &= 4 \\ y &= \pm 2 \end{aligned}$$

For $y = 2$,

$$x^2 - (2)^2 = 5$$
$$x = \pm 3$$

For $y = -2$,

$$x^2 - (-2)^2 = 5$$
$$x^2 = 9$$
$$x = \pm 3$$

Thus $(3, -2)$, $(3, 2)$, $(-3, -2)$, and $(-3, 2)$ are the four solutions to the system. The reader should check these solutions.

PROBLEM 12 Solve the system:

$$2x^2 - 3y^2 = 5$$
$$3x^2 + 4y^2 = 16$$

ANSWER $(2, 1)$, $(2, -1)$, $(-2, 1)$, $(-2, -1)$

EXAMPLE 13 Solve the system:

$$x^2 + 3xy + y^2 = 20 \qquad (1)$$
$$xy - y^2 = 0 \qquad (2)$$

SOLUTION

$$y(x - y) = 0$$

Factor Eq. (2).

$$y = 0 \quad \text{or} \quad x - y = 0$$
$$y = x$$

Substitute each of these in turn into Eq. (1), and proceed as before.

For $y = 0$,

$$x^2 + 3x(0) + (0)^2 = 20$$
$$x = \pm 2\sqrt{5}$$

For $y = x$,

$$x^2 + 3xx + x^2 = 20$$

$$x = \pm 2$$ Substitute these values back into $y = x$ to find corresponding values of y.

For $x = 2$, $y = 2$; for $x = -2$, $y = -2$.

Thus $(2\sqrt{5}, 0)$, $(-2\sqrt{5}, 0)$, $(2, 2)$, and $(-2, -2)$ are the four solutions to the system. The reader should check these solutions.

PROBLEM 13 Solve the system:

$$x^2 + xy - y^2 = 4$$

$$2x^2 - xy = 0$$

ANSWER $(0, 2i)$, $(0, -2i)$, $(2i, 4i)$, $(-2i, -4i)$

Example 13 is somewhat specialized. However, it suggests a procedure that, when used alone or in combination with other procedures, is effective for some problems.

Exercise 53

Solve each system:

A

1. $x^2 + y^2 = 25$, $y = -4$

2. $x^2 + y^2 = 169$, $x = -12$

3. $y^2 = 2x$, $x = y - \frac{1}{2}$

4. $8x^2 - y^2 = 16$, $y = 2x$

5. $x^2 + 4y^2 = 32$, $x + 2y = 0$

6. $2x^2 - 3y^2 = 25$, $x + y = 0$

7. $x^2 = 2y$, $3x = y + 5$

8. $y^2 = -x$, $x - 2y = 5$

9. $x^2 - y^2 = 3$, $x^2 + y^2 = 5$

10. $2x^2 + y^2 = 24$, $x^2 - y^2 = -12$

11. $x^2 - 2y^2 = 1$, $x^2 + 4y^2 = 25$

12. $x^2 + y^2 = 10$, $16x^2 + y^2 = 25$

B

13. $xy - 6 = 0$
$x - y = 4$

14. $xy = -4$
$y - x = 2$

15. $x^2 + xy - y^2 = -5$
$y - x = 3$

16. $x^2 - 2xy + y^2 = 1$
$x - 2y = 2$

17. $2x^2 - 3y^2 = 10$
$x^2 + 4y^2 = -17$

18. $2x^2 + 3y^2 = -4$
$4x^2 + 2y^2 = 8$

19. $x^2 + y^2 = 20$
$x^2 = y$

20. $x^2 - y^2 = 2$
$y^2 = x$

21. $x^2 + y^2 = 16$
$y^2 = 4 - x$

22. $x^2 + y^2 = 5$
$x^2 = 4(2 - y)$

23. Find the dimensions of a rectangle with area 32 square feet and perimeter 36 square feet.

24. Find two numbers such that their sum is 1 and their product is 1.

C

25. $2x^2 + y^2 = 18$
$xy = 4$

26. $x^2 - y^2 = 3$
$xy = 2$

27. $x^2 + 2xy + y^2 = 36$
$x^2 - xy = 0$

28. $2x^2 - xy + y^2 = 8$
$(x - y)(x + y) = 0$

29. $x^2 - 2xy + 2y^2 = 16$
$x^2 - y^2 = 0$

30. $x^2 + xy - 3y^2 = 3$
$x^2 + 4xy + 3y^2 = 0$

9.7 Second-Degree Inequalities in One Variable

You have now had quite a bit of experience solving first-degree inequalities in one variable such as

$$2x - 3 \leq 4(x - 4)$$

But how do we solve second-degree inequalities such as

$$x^2 + 2x < 8$$

Using the quadratic formula directly doesn't work. However, if we move all terms to the left and factor, then we will be able to observe something that will lead to a solution. Thus,

$$x^2 + 2x - 8 < 0$$

$$(x + 4)(x - 2) < 0$$

We are looking for values of x that will make the left side less than 0, that is, negative. What will the signs of $(x+4)$ and $(x-2)$ have to be so that their product is negative? They must have opposite signs!

Something interesting happens to a linear form $ax+b$, $a \neq 0$, relative to sign changes as x takes on all real values along the real number line, going from left to right:

THEOREM 2 The value of x at which $ax+b$ is 0 is called a *critical point*. To the left of this point, $ax+b$ has one sign for all values of x and the opposite sign to the right of the point.

The proof of this theorem is not difficult and is left to the "C" exercises. Let us see how it is used to solve the problem above.

$x+4=0 \qquad x+4<0 \qquad x+4>0$

$x=-4 \qquad x<-4 \qquad x>-4$

critical point

$x-2=0 \qquad x-2<0 \qquad x-2>0$

$x=2 \qquad x<2 \qquad x>2$

critical point

We represent these results graphically along a real number line

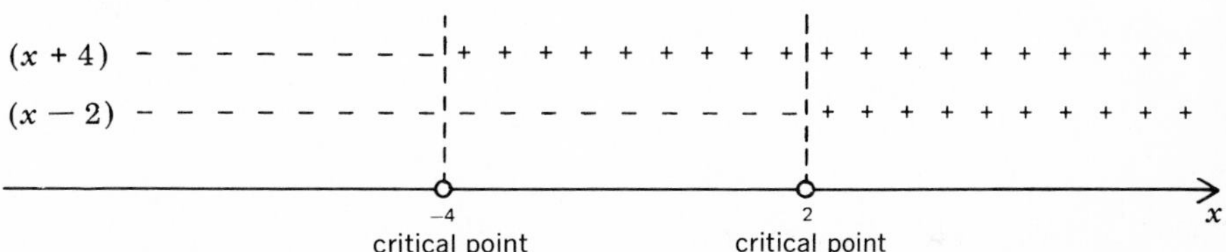

Now it is very easy to see that the factors have opposite signs for x between -4 and 2. Thus, the solution of $x^2+2x<8$ is

$-4<x<2$

EXAMPLE 14 Solve and graph: $x^2 \geq x+6$

SOLUTION

$$x^2 \geq x+6$$

$$x^2-x-6 \geq 0$$

$$(x-3)(x+2) \geq 0$$

Find values of x so that the factors on the left have the same sign. Critical points are -2 and 3.

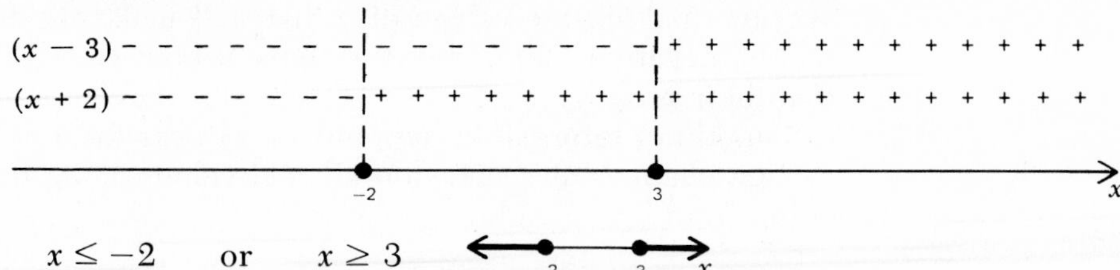

$x \le -2$ or $x \ge 3$

PROBLEM 14 Solve and graph:

(A) $x^2 < x + 12$ (B) $x^2 \ge x + 12$

ANSWER (A) $-3 < x < 4$

(B) $x \le -3$ or $x \ge 4$

The procedures discussed above can also be used on some inequalities that are not quadratic.

EXAMPLE 15 Solve and graph:

$$\frac{x^2 - x + 1}{2 - x} \ge 1$$

SOLUTION

$$\frac{x^2 - x + 1}{2 - x} \ge 1$$

Since we do not know the sign of $2 - x$, we do not multiply both sides by it: instead, we subtract 1 from each side.

$$\frac{x^2 - x + 1}{2 - x} - 1 \ge 0$$

$$\frac{x^2 - x + 1}{2 - x} - \frac{2 - x}{2 - x} \ge 0$$

$$\frac{x^2 - 1}{2 - x} \ge 0$$

$$\frac{(x - 1)(x + 1)}{2 - x} \ge 0$$

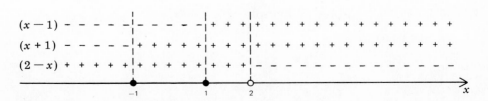

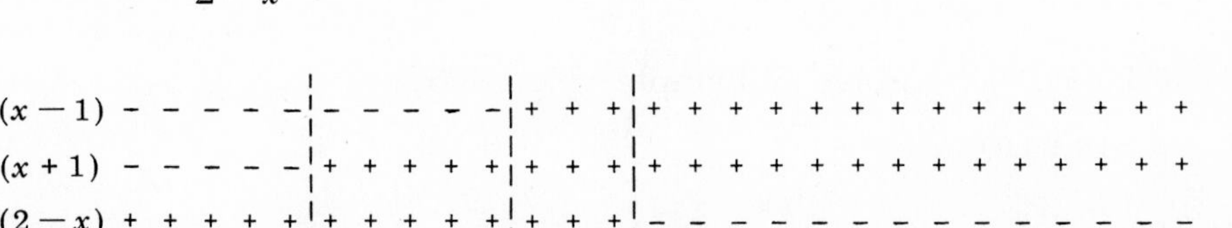

We must have an even number of factors with negative signs. From the graph it is easy to see that the solution for the original inequality is

$x \le -1 \quad$ or $\quad 1 \le x < 2$

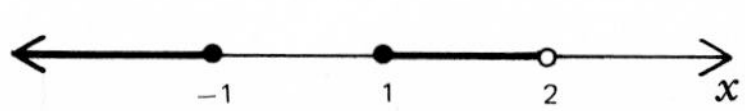

PROBLEM 15 Solve and graph:

$$\frac{3}{2-x} \le \frac{1}{x+4}$$

ANSWER $-4 < x \le -\frac{5}{2}$ or $x > 2$

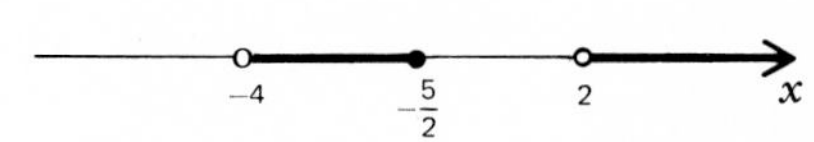

Exercise 54

Solve and graph:

A

1. $(x-3)(x+4) < 0$
2. $(x+2)(x-4) < 0$
3. $(x-3)(x+4) \ge 0$
4. $(x+2)(x-4) > 0$
5. $x^2 + x < 12$
6. $x^2 < 10 - 3x$
7. $x^2 + 21 > 10x$
8. $x^2 + 7x + 10 > 0$

B

9. $x(x+6) \ge 0$
10. $x(x-8) \le 0$
11. $x^2 \ge 9$
12. $x^2 > 4$
13. $\dfrac{x-5}{x+2} \le 0$
14. $\dfrac{x+2}{x-3} < 0$
15. $\dfrac{x-5}{x+2} > 0$
16. $\dfrac{x+2}{x-3} \ge 0$
17. $\dfrac{x-4}{x(x+2)} \le 0$
18. $\dfrac{x(x+5)}{x-3} \ge 0$
19. $\dfrac{1}{x} < 4$
20. $\dfrac{5}{x} > 3$
21. $x^2 + 4 \ge 4x$
22. $6x \le x^2 + 9$
23. $x^2 + 9 < 6x$
24. $x^2 + 4 < 4x$

C

25. $x^2 \ge 3$
26. $x^2 < 2$
27. $\dfrac{2}{x-3} \le -2$
28. $\dfrac{2x}{x+3} \ge 1$
29. $\dfrac{2}{x-3} \le \dfrac{2}{x+2}$
30. $\dfrac{2}{x+1} \ge \dfrac{1}{x-2}$

31. For what values of x will $\sqrt{x^2 - 3x + 2}$ be a real number?

32. For what values of x will $\sqrt{\dfrac{x-3}{x+5}}$ be a real number?

33. If an object is shot straight up from the ground with an initial velocity of 160 feet/second, its distance d in feet above the ground at the end of t seconds (neglecting air resistance) is given by $d = 160t - 16t^2$. Find the duration of time for which $d \geq 256$.

34. Repeat the preceding problem for $d \geq 0$.

35. Prove Theorem 2 for the case $a > 0$.

36. Prove Theorem 2 for the case $a < 0$.

Exercise 55 Chapter Review

A *Find all solutions by factoring or square root methods.*

1. $x^2 - 3x = 0$

2. $x^2 = 25$

3. $x^2 - 5x + 6 = 0$

4. $x^2 - 2x - 15 = 0$

5. $x^2 - 7 = 0$

6. Write $4x = 2 - 3x^2$ in standard form $ax^2 + bx + c = 0$, and identify a, b, and c.

7. Write down the quadratic formula associated with $ax^2 + bx + c = 0$.

8. Use the quadratic formula to solve $x^2 + 3x + 1 = 0$.

9. Find two positive numbers whose product is 27 if one is 6 more than the other.

Solve and graph:

10. $x^2 + x < 20$

11. $x^2 + x \geq 20$

Solve each system:

12. $x^2 + y^2 = 2$
$2x - y = 3$

13. $x^2 - y^2 = 7$
$x^2 + y^2 = 25$

B *Find all solutions by factoring or square root methods:*

14. $10x^2 = 20x$

15. $3x^2 = 36$

16. $3x^2 + 27 = 0$

17. $(x - 2)^2 = 16$

18. $3t^2 - 8t - 3 = 0$

19. $2x = \dfrac{3}{x} - 5$

Solve using the quadratic formula:

20. $3x^2 = 2(x + 1)$

21. $2x(x - 1) = 3$

Solve using any method:

22. $2x^2 - 2x = 40$

23. $\dfrac{8m^2 + 15}{2m} = 13$

24. $m^2 + m - 1 = 0$

25. $u + \dfrac{3}{u} = 2$

26. $\sqrt{5x - 6} - x = 0$

27. $8\sqrt{x} = x + 15$

28. $m^4 + 5m^2 - 36 = 0$

29. $2x^{2/3} - 5x^{1/3} - 12 = 0$

Solve and graph:

30. $x^2 \geq 4x + 21$

31. $\dfrac{1}{x} < 2$

32. $10x > x^2 + 25$

33. $x^2 + 16 \geq 8x$

Solve each system:

34. $3x^2 - y^2 = -6$
$2x^2 + 3y^2 = 29$

35. $x^2 = y$
$y = 2x - 2$

36. The perimeter of a rectangle is 22 inches. If its area is 30 square inches find the length of each side.

C **37.** Solve $x^2 - 6x - 3 = 0$ by completing-the-square method.

Solve using any method:

38. $\left(t - \dfrac{3}{2}\right)^2 = -\dfrac{3}{2}$

39. $3x - 1 = \dfrac{2(x + 1)}{x + 2}$

40. $y^8 - 17y^4 + 16 = 0$

41. $\sqrt{y - 2} - \sqrt{5y + 1} = -3$

42. $\dfrac{3}{x - 4} \leq \dfrac{2}{x - 3}$

43. If the hypotenuse of a right triangle is 15 inches and its area is 54 square inches, what are the lengths of the two sides?

44. Cost equations for manufacturing companies are often quadratic in nature. (At very high or very low outputs the costs are more per unit because of inefficiency of plant operation at these extremes.) If the cost equation for manufacturing transistor radios is $C = x^2 - 10x + 31$, where C is the cost of manufacturing x units per week (both in thousands), find (*A*) the output for a \$15,000 weekly cost; (*B*) the output for a \$6,000 weekly cost.

Chapter 10

RELATIONS AND FUNCTIONS

10.1 Relations and Functions

Seeking relationships between various types of phenomena is undoubtedly one of the most important aspects of science. A psychologist wants to know the relationship between rates of learning and reward systems; an engineer wants to know the relationship between air pressures on a structure and wind velocities; a chemist wants to know the relationship between the rates of chemical reactions and concentrations of certain chemicals; . . . . The list can be continued without end.

Establishing and working with sets of ordered pairs of numbers—whether through tables, graphs, or equations—is so fundamental to both pure and applied science that people have found it necessary to describe this activity in the precise language of mathematics.

RELATIONS

A *relation* is any set of ordered pairs of elements (generally numbers, in this book). The choice of the word relation is quite natural, since any set of ordered pairs of numbers establishes a relationship or correspondence between the set of first components called the *domain* and the set of second components called the *range*. The *graph of a relation* is the set of all points in a cartesian coordinate system that have the elements of the relation as coordinates.

EXAMPLE 1 **A RELATION SPECIFIED BY A SET**

$R = \{(0, 0), (1, 1), (1, -1), (4, 2), (4, -2)\}$

Domain of $R = \{0, 1, 4\}$ — The set of first components

Range of $R = \{-2, -1, 0, 1, 2\}$ — The set of second components

PROBLEM 1 Graph R in Example 1.

ANSWER

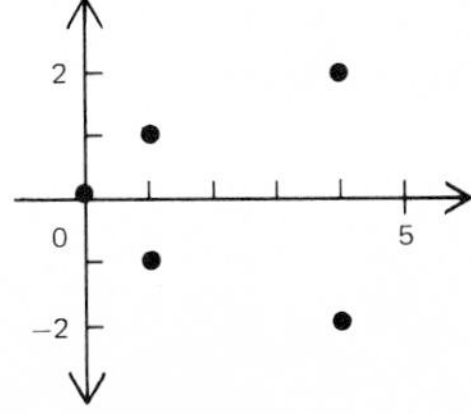

EXAMPLE 2 **A RELATION SPECIFIED BY A TABLE OR A GRAPH**

A laboratory experiment

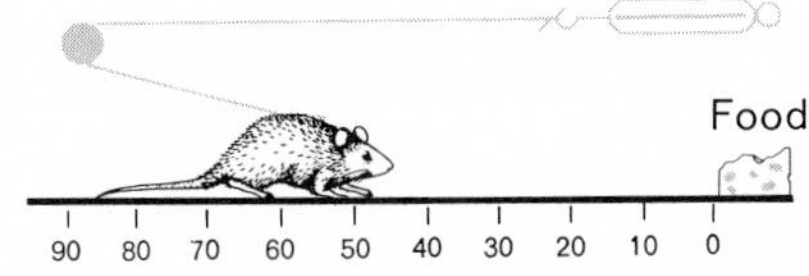

yields a table

d (distance in centimeters)	p (pull in grams)
30	64
50	60
70	56
90	52
110	48
130	44
150	40
170	36

or a graph

The graph (or table) in Example 2 enables us to pair elements from the domain (distance from the food) with elements from the range (strength of pull toward the food), yielding a set of ordered pairs of matched numbers (a relation). (Note that it is the usual practice to associate the domain of a relation with the horizontal axis and the range with the vertical axis.)

PROBLEM 2 Write down the domain and range for the relation in Example 2.

ANSWER Domain = {30, 50, 70, 90, 110, 130, 150, 170}

Range = {36, 40, 44, 48, 52, 56, 60, 64}

EXAMPLE 3 **A RELATION SPECIFIED BY AN EQUATION**

The distance s that an object falls (neglecting air resistance) in t seconds is given by

$$s = 16t^2 \qquad t \geq 0$$

The domain and range of this relation is the set of all nonnegative real numbers.

PROBLEM 3 Graph the relation in Example 3. (Plot points for $t = 0, 1, 2, 3, 4, 5$; then join these points with a smooth curve.)

ANSWER

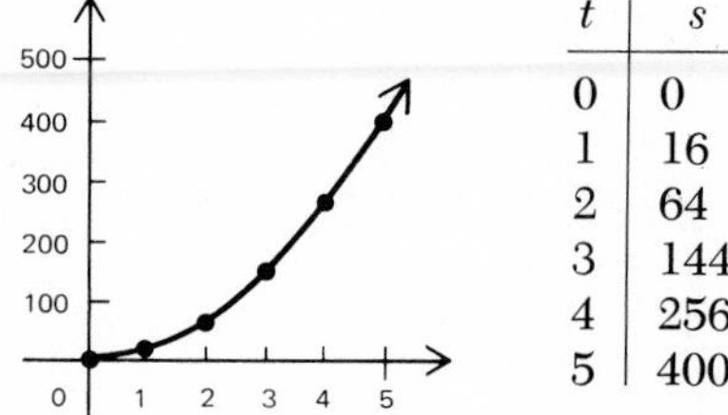

t	s
0	0
1	16
2	64
3	144
4	256
5	400

We emphasize the fact that *any* set of ordered pairs of elements is a relation; whether or not it has physical meaning is entirely beside the point. The concept is purely mathematical in nature, and as such it is completely free to be applied to a variety of practical and theoretical problems.

FUNCTION

A *function* is a relation with the added restriction that no two distinct ordered pairs can have the same first component. All functions are relations, but some relations are not functions.

If in an equation in two variables, say, x and y there corresponds exactly one range value y for each domain value x, then the set of ordered pairs (x, y) that satisfies the equation is a function. Under these conditions, y is frequently referred to as a function of x. If for at least one value of x there corresponds two or more values of y, then the solution set of the equation is a relation, but not a function.

The set of all ordered pairs (x, y) such that

$$x^2 + y^2 = 25$$

is a relation but not a function, since, for example, (4, 3) and (4, −3) both satisfy the equation and they have the same first component 4. In other words, when x is 4, then y can be either −3 or 3. On the other hand, the set of all ordered pairs (x, y) such that

$$y = \frac{x}{4} + 2$$

is a function as well as a relation, since for each value of x there corresponds exactly one value of y, and no more. For example, if $x = 8$, then $y = 4$, and that is the only value that y can have if x is 8.

EXAMPLE 4

If an arrow is shot straight upward from the ground with an initial velocity of 160 feet/second, its distance d in feet above the ground at the end of t seconds (neglecting air resistance) is given by

$$d = 160t - 16t^2 \qquad 0 \le t \le 10$$

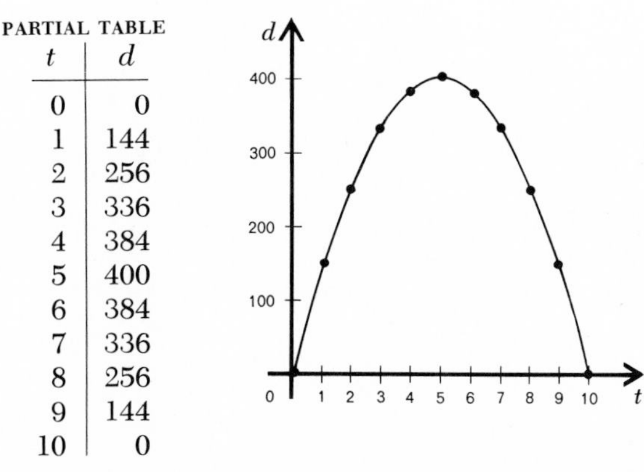

PARTIAL TABLE

t	d
0	0
1	144
2	256
3	336
4	384
5	400
6	384
7	336
8	256
9	144
10	0

DOMAIN SET OF ALL REAL t	RANGE SET OF ALL REAL d
$0 \le t \le 10$	$0 \le d \le 400$

PROBLEM 4

Is the relation in Example 4 a function? Explain.

ANSWER Yes, since each number in the domain (time) is associated with exactly one number in the range (distance above the ground). Note that there are values in the range that are associated with more than one value in the domain; for example, 336 is associated with both 3 and 7. What is the physical meaning of this?

It is very easy to determine whether a relation is a function if one has its graph:

> A relation is a function if each vertical line in the coordinate system passes through at most one point of the graph of the relation.

Example 1 does not meet this test; Examples 2, 3, and 4 do.

Exercise 56

A *Graph each relation, state its domain and range, and indicate which are functions.*

1. $F = \{(1, 1), (2, 1), (3, 2), (3, 3)\}$
2. $f = \{(2, 4), (4, 2), (2, 0), (4, -2)\}$
3. $G = \{(-1, -2), (0, -1), (1, 0), (2, 1), (3, 2), (4, 1)\}$
4. $g = \{(-2, 0), (0, 2), (2, 0)\}$
5. R: $y = 6 - 2x$, $0 \le x \le 4$, x an integer
6. H: $y = \dfrac{x}{2} - 4$, $0 \le x \le 4$, x an integer

B *State the domain and range for each of the following relations and indicate whether the relation is a function.*

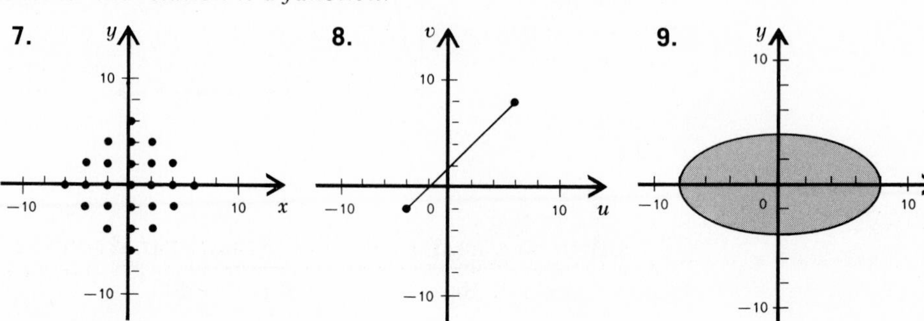

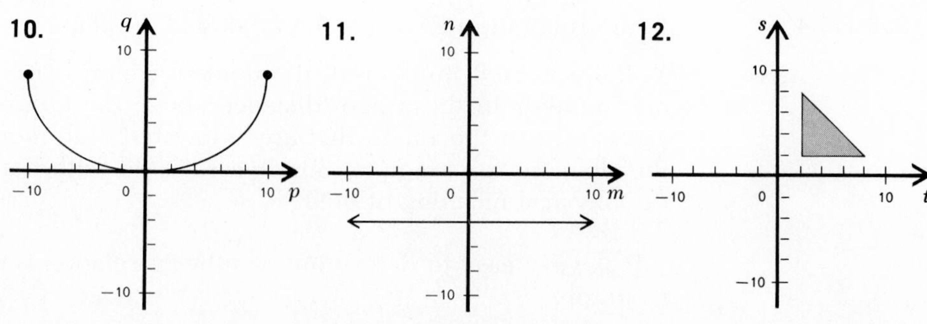

Graph each relation for the domain {0, 1, 4}. What is the range of each and which is a function?

13. $F: y = |x|$

14. $G: y = x^2$

15. $f: |y| = x$

16. $g: y^2 = x$

17. $H: x^2 + y^2 = 4,\ x \in \{-2, 0, 2\}$

18. $h: x^2 + y^2 = 9,\ x \in \{-3, 0, 3\}$

C *Graph each relation, state its domain and range, and indicate which are functions.*

19. $H = \{(x, y)\,|\,y = \frac{x}{2},\ x \in \{-4, -2, 0, 2, 4\}\}$

20. $h = \{(x, y)\,|\,y = x + 3,\ x \in \{-3, -1, 0, 2\}\}$

21. $F = \{(x, y)\,|\,0 \le y \le x,\ 0 \le x \le 3;\ x, y \in N\}$

22. $G = \{(x, y)\,|\,0 \le y < |x|,\ -2 \le x \le 2;\ x, y \in I\}$

10.2 Function Notation

We have just seen that a function involves two sets of elements, a domain and a range, and a rule that enables one to assign each element in the domain to exactly one element in the range. We use different letters to denote names for numbers; in essentially the same way, we will now use different letters to denote names for functions. For example, f and g may be used to name the two functions

$$f: y = 2x + 1$$

$$g: y = x^2$$

Unless otherwise stated, the functions we are considering are real functions; that is, functions whose domains and ranges are sets of real numbers.

If x represents an element in the domain of a function f, then we will often use the symbol $f(x)$ in place of y to designate the number in the range of f to which x is paired. It is important not to confuse this new symbol and think of it as the product of f and x. The symbol $f(x)$ is read "f of x," or "the value of f at x."

This new function notation is extremely important, and its correct use should be mastered as early as possible. For example, in place of the more formal representation of the functions f and g above, we can now write

$$f(x) = 2x + 1 \qquad \text{and} \qquad g(x) = x^2$$

Thus $f(x)$ and $2x + 1$ name the same number in the range of f for each replacement of the variable x from the domain of f; $g(x)$ and x^2 name the same number in the range of g for each replacement of the variable x from the domain of g. In particular,

$$f(3) = 2(3) + 1 = 7 \qquad \text{and} \qquad g(5) = 5^2 = 25$$

That is, the function f assigns 7 to the number 3, and g assigns 25 to 5. The ordered pair (3, 7) belongs to the function f, and (5, 25) belongs to the function g.

EXAMPLE 5 Let $f(x) = \frac{x}{2} + 1$ and $g(x) = 1 - x^2$. Then,

(A) $f(6) = \frac{6}{2} + 1 = 3 + 1 = 4$

(B) $g(-2) = 1 - (-2)^2 = 1 - 4 = -3$

(C) $f(4) + g(0) = \left(\frac{4}{2} + 1\right) + (1 - 0^2) = 3 + 1 = 4$

(D) $\dfrac{2g(-3) + 6}{f(8)} = \dfrac{2[1 - (-3)^2] + 6}{\frac{8}{2} + 1} = \dfrac{2(-8) + 6}{5} = \dfrac{-10}{5} = -2$

PROBLEM 5 If $f(x) = \frac{x}{3} - 2$ and $g(x) = 4 - x^2$, find

(A) $f(9)$ (B) $g(-2)$

(C) $f(0) + g(2)$ (D) $\dfrac{4g(-1) - 4}{f(12)}$

ANSWER (A) 1 (B) 0 (C) −2 (D) 4

EXAMPLE 6 For f and g in Example 5,

(A) $g(2 + h) = 1 - (2 + h)^2 = 1 - (4 + 4h + h^2)$

$$= -3 - 4h - h^2$$

(B) $\dfrac{g(2 + h) - g(2)}{h} = \dfrac{[1 - (2 + h)^2] - [1 - 2^2]}{h}$

$$= \frac{-3 - 4h - h^2 + 3}{h} = \frac{-4h - h^2}{h} = -4 - h$$

(C) $f[g(3)] = f[1 - 3^2]$

$$= f(-8) = \frac{-8}{2} + 1 = -3$$

PROBLEM 6 For f and g in problem 5, find

(A) $g(3+h)$ (B) $\dfrac{g(3+h)-g(3)}{h}$ (C) $g[f(3)]$

ANSWER (A) $-5-6h-h^2$ (B) $-6-h$ (C) 3

Exercise 57

A *If* $f(x)=3x-2$, *find:*

1. $f(2)$ **2.** $f(1)$ **3.** $f(-2)$

4. $f(-1)$ **5.** $f(0)$ **6.** $f(4)$

If $g(x)=x-x^2$, *find:*

7. $g(2)$ **8.** $g(1)$ **9.** $g(4)$

10. $g(5)$ **11.** $g(-2)$ **12.** $g(-1)$

B *For* $f(x)=10x-7$

$g(t)=6-2t$

$F(u)=3u^2$

$G(v)=v-v^2$, *find:*

13. $f(-2)$ **14.** $F(-1)$ **15.** $g(2)$

16. $G(-3)$ **17.** $g(0)$ **18.** $G(0)$

19. $f(3)+g(2)$ **20.** $F(2)+G(3)$

21. $2g(-1)-3G(-1)$ **22.** $4G(-2)-g(-3)$

23. $\dfrac{f(2)\cdot g(-4)}{G(-1)}$ **24.** $\dfrac{F(-1)\cdot G(2)}{g(-1)}$

25. $g(2+h)$ **26.** $F(2+h)$

27. $\dfrac{g(2+h)-g(2)}{h}$ **28.** $\dfrac{F(2+h)-F(2)}{h}$

29. $\dfrac{f(3+h)-f(3)}{h}$ **30.** $\dfrac{G(2+h)-G(2)}{h}$

31. $F[g(1)]$ **32.** $G[F(1)]$ **33.** $g[f(1)]$

34. $g[G(0)]$ **35.** $f[G(1)]$ **36.** $G[g(2)]$

37. If $A(w)=\dfrac{w-3}{w+5}$, find $A(5)$, $A(0)$, and $A(-5)$.

38. If $h(s) = \dfrac{s}{s-2}$, find $h(3)$, $h(0)$, and $h(2)$.

C **39.** Let the distance that a car travels at 30 miles/hour in t hours be given by $d(t) = 30t$. Find

(A) $d(1)$, $d(10)$; (B) $\dfrac{d(2+h) - d(2)}{h}$

40. The distance that an object falls in a vacuum is given by $s(t) = 16t^2$. Find

(A) $s(0)$, $s(1)$, $s(2)$, and $s(3)$; (B) $\dfrac{s(2+h) - s(2)}{h}$

What happens as h tends to 0? Interpret physically.

Each of the relationships in Problems 41 and 42 can be described by a function. Write an equation that specifies each function.

41. The cost per day $C(x)$ for renting a car at \$10 per day plus 10 cents a mile for x miles. (The rental cost per day depends on the mileage per day.)

42. The cost $C(x)$ of manufacturing x pairs of skis if fixed costs are \$300 per day and the variable costs are \$50 per pair of skis. (The cost per day depends on the number of pairs of skis manufactured per day.)

43. For $f(x) = 5x$ (A) Does $f(at) = af(t)$? (B) Does $f(a+b) = f(a) + f(b)$? (C) Does $f(ab) = f(a) \cdot f(b)$?

44. For $g(x) = x^2$ (A) Does $g(at) = ag(t)$? (B) Does $g(a+b) = g(a) + g(b)$? (C) Does $g(ab) = g(a) \cdot g(b)$?

10.3 Inverse Relations and Functions

In this section we are going to discuss an important method of obtaining new relations and functions from old relations and functions. We will use this method in Chap. 11 to obtain the logarithmic functions from the exponential functions.

Given a relation R, if we interchange the order of the components in each ordered pair belonging to R, we obtain a new relation R^{-1}, called the *inverse of* R. For example, if

$$R = \{(3, 5), (5, -1), (7, 0)\}$$

then

$$R^{-1} = \{(5, 3), (-1, 5), (0, 7)\}$$

It follows from the definition (and is evident from the example) that the domain of R is the *range of* R^{-1} and the range of R is the *domain of* R^{-1}.

If a relation R is specified by an equation, say

$$R\colon y = 2x - 1 \tag{1}$$

then how do we find R^{-1}? The answer is easy: We interchange the variables in (1). Thus,

$$R^{-1}\colon x = 2y - 1 \tag{2}$$

or, solving for y,

$$R^{-1}\colon y = \frac{x+1}{2} \tag{3}$$

Any ordered pair of numbers that satisfies (1), when reversed in order, will satisfy (2) and (3). For example, (3, 5) satisfies (1) and (5, 3) satisfies (2) and (3), as can easily be checked. The graphs of R and R^{-1} are given in Fig. 1.

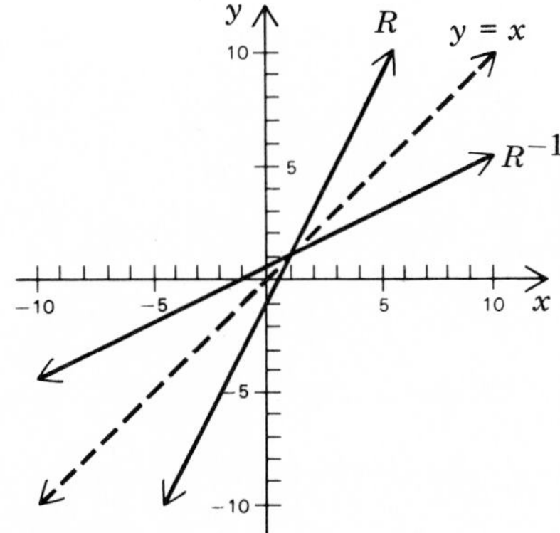

Figure 1

It is useful to sketch in the line $y = x$ and observe that if we fold the paper along this line, then R and R^{-1} will match. Actually, we can graph R^{-1} by drawing R with wet ink and fold the paper along $y = x$ before the ink dries; R will then print R^{-1}. [To prove this, one has to show that the line $y = x$ is the perpendicular bisector of the line segment joining (a, b) to (b, a).] Knowing that the graph of R and R^{-1} are symmetric relative to the line $y = x$ makes it easy to graph R^{-1} if R is known and vice versa.

In Fig. 1 observe that R and R^{-1} are both functions. This is not always the case, however. Inverses of some functions may not be functions and inverses of some relations that are not functions may be functions. We must check each case.

EXAMPLE 7 If R is given by $y = x^2$,

(A) Find R^{-1}.

(*B*) Graph *R* and R^{-1}.

(*C*) Indicate which are functions.

SOLUTION (*A*) $R: y = x^2$

$R^{-1}: x = y^2$ or $y = \pm\sqrt{x}$

(*B*)

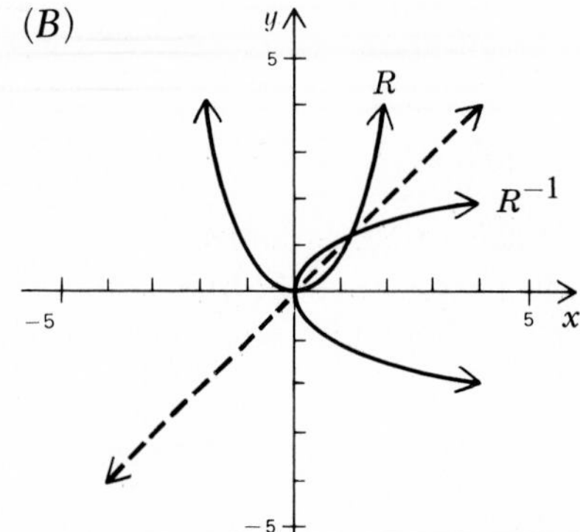

(*C*) *R* is a function; R^{-1} is not.

PROBLEM 7 Repeat Example 7 for *R* given by $|y| = x$.

ANSWER (*A*) $R^{-1}: |x| = y$

(*B*)

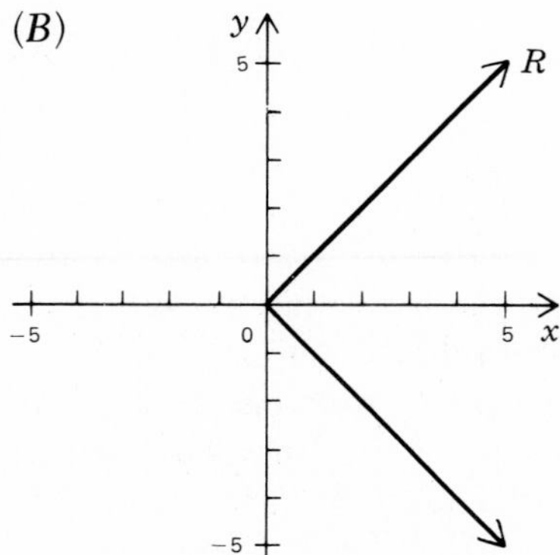

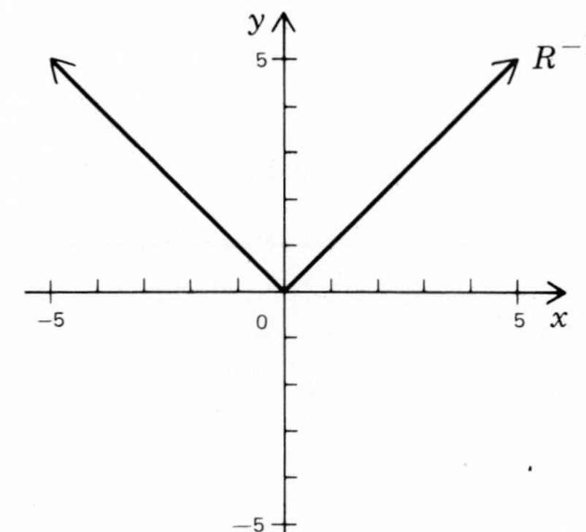

(*C*) *R* is not a function
R^{-1} is a function

EXAMPLE 8 Given $f(x) = 2x - 1$, find

(*A*) $f^{-1}(x)$ (*B*) $f^{-1}(2)$ (*C*) $f[f^{-1}(3)]$

SOLUTION (A) First write f in the form

f: $y = 2x - 1$

Interchange variables to obtain f^{-1}

f^{-1}: $x = 2y - 1$

Solve for y to obtain $f^{-1}(x)$

$$y = \frac{x+1}{2}$$

or

$$f^{-1}(x) = \frac{x+1}{2}$$

(B) $$f^{-1}(2) = \frac{2+1}{2} = \frac{3}{2}$$

(C) $$f[f^{-1}(3)] = f\left[\frac{3+1}{2}\right] = f(2) = 2 \cdot 2 - 1 = 3$$

PROBLEM 8 Given $g(x) = \frac{x}{2} + 1$, find

(A) $g^{-1}(x)$ (B) $g^{-1}(3)$ (C) $g^{-1}[g(2)]$

ANSWER (A) $g^{-1}(x) = 2(x - 1)$ (B) 4 (C) 2

We conclude this section with the following important theorem:

THEOREM If f and f^{-1} are both functions, then

$$f[f^{-1}(x)] = x \qquad \text{and} \qquad f^{-1}[f(x)] = x$$

For illustrations of this theorem, see part C in Example 8 and problem 8 above.

Exercise 58

A *Find the inverse for each of the following relations.*

1. $R = \{(-2, 1), (0, 3), (2, 2)\}$

2. $F = \{(-3, -1), (0, 1), (3, 2)\}$

3. $G = \{(-2, 4), (-1, 1), (0, 0), (1, 1), (2, 4)\}$

4. $H = \{(-5, 0), (-2, 1), (0, 0), (2, 1), (5, 0)\}$

Graph on the same coordinate system along with $y = x$.

5. R and R^{-1} in Problem 1
6. F and F^{-1} in Problem 2
7. G and G^{-1} in Problem 3
8. H and H^{-1} in Problem 4

Indicate which are functions.

9. R or R^{-1} in Problem 1
10. F or F^{-1} in Problem 2
11. G or G^{-1} in Problem 3
12. H or H^{-1} in Problem 4

B *Find the inverse for each of the following relations in the form of an equation.*

13. $f\colon y = 3x - 2$
14. $g\colon y = 2x + 3$
15. $F\colon y = \frac{x}{3} - 2$
16. $G\colon y = \frac{x}{2} + 5$
17. $h\colon y = \frac{x^2}{2}$
18. $H\colon y = |2x|$

Graph on the same coordinate system along with $y = x$. Indicate which are functions.

19. f and f^{-1} in Problem 13
20. g and g^{-1} in Problem 14
21. h and h^{-1} in Problem 17
22. H and H^{-1} in Problem 18
23. For $f(x) = 3x - 2$, find (A) $f^{-1}(x)$; (B) $f^{-1}(2)$; (C) $f[f^{-1}(3)]$.
24. For $g(x) = 2x + 3$, find (A) $g^{-1}(x)$; (B) $g^{-1}(5)$; (C) $g[g^{-1}(4)]$.
25. For $F(x) = \frac{x}{3} - 2$; find (A) $F^{-1}(x)$; (B) $F^{-1}(-1)$; (C) $F^{-1}[F(4)]$.
26. For $G(x) = \frac{x}{2} + 5$, find (A) $G^{-1}(x)$; (B) $G^{-1}(8)$; (C) $G^{-1}[G(-4)]$.

C 27. For $f(x) = \frac{x}{3} + 2$, find (A) $f^{-1}(x)$; (B) $f[f^{-1}(a)]$.

28. For $g(x) = 4x + 2$, find (A) $g^{-1}(x)$; (B) $g^{-1}[g(a)]$.
29. Find $F[F^{-1}(x)]$ for Problem 15.
30. Find $G^{-1}[G(x)]$ for Problem 16.

10.4 Linear and Quadratic Functions

We have studied linear and quadratic forms in some detail. These forms (with certain restrictions) define linear and quadratic functions.

LINEAR FUNCTIONS

Any function defined by an equation of the form

$$f(x) = ax + b \qquad a \neq 0$$

where a and b are constants, and x is a variable, is called a *linear function*. The graph, of course, is a nonvertical straight line with slope a and y intercept b.

EXAMPLE 9 Graph the linear function defined by $f(x) = x/3 + 1$, and indicate its slope and y intercept.

SOLUTION

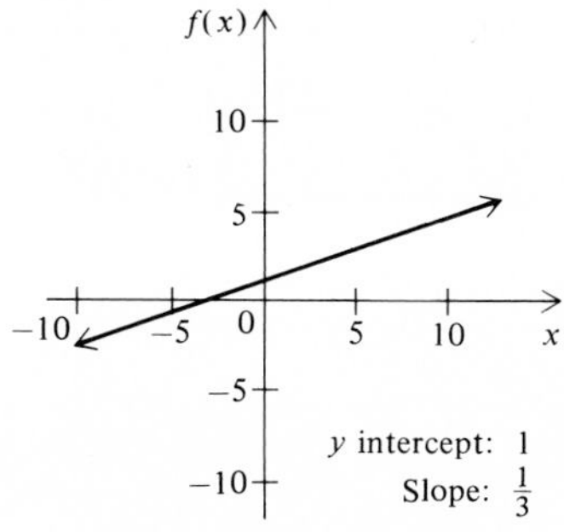

PROBLEM 9 Graph the linear function defined by $f(x) = -x/2 + 3$, and indicate its slope and y intercept.

ANSWER

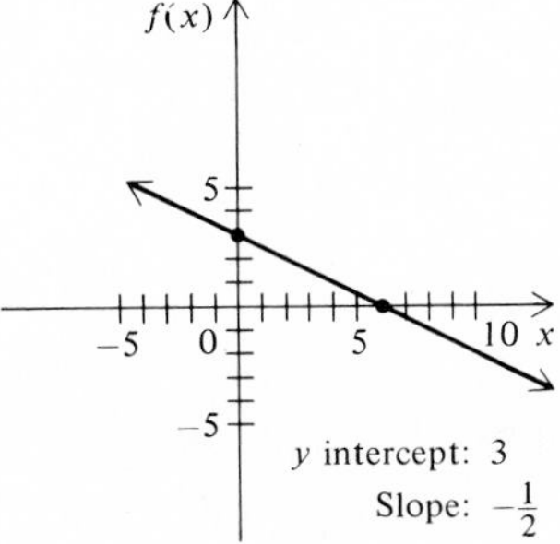

QUADRATIC FUNCTIONS

Any function defined by an equation of the form

$$\boxed{f(x) = ax^2 + bx + c \qquad a \neq 0} \tag{4}$$

where a, b, and c are constants, and x is a variable, is called a *quadratic function*. In earlier sections you graphed some of these functions and found that they are not linear. In more advanced courses, it is shown that the graph of (4) is a *parabola* which has its axis (line of symmetry) parallel to the y axis, and which opens upward if $a > 0$ and downward if $a < 0$. If you point a water hose up into the air at an angle, then the water forms an arc that is very close to a parabola.

To graph a quadratic function, plot enough points so that when they are joined by a smooth curve the resulting figure will look like a parabola (or part of a parabola if the domain of the function is restricted). The following example illustrates the process.

EXAMPLE 10 Graph: $f(x) = x^2 - 2x - 3 \qquad x \geq 0$

SOLUTION

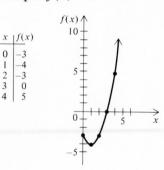

x	$f(x)$
0	−3
1	−4
2	−3
3	0
4	5

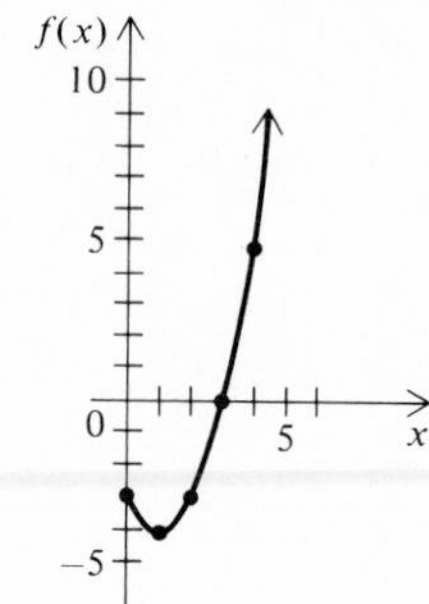

PROBLEM 10 Graph: $g(t) = 12 - 2t^2 \qquad 0 \leq t \leq 6$

ANSWER

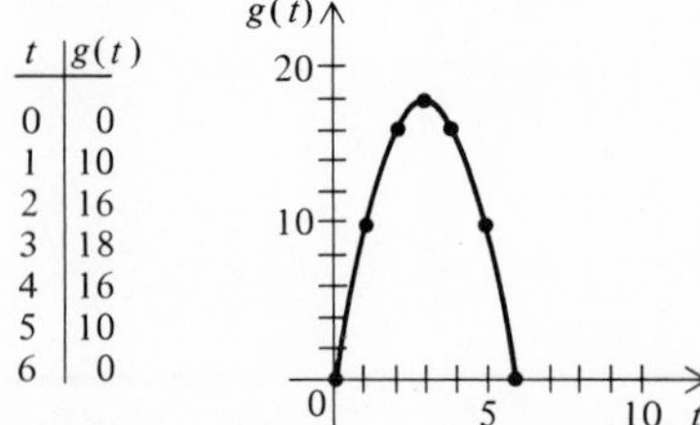

t	$g(t)$
0	0
1	10
2	16
3	18
4	16
5	10
6	0

QUADRATIC FUNCTIONS AND QUADRATIC EQUATIONS

In Chap. 9 we discussed the quadratic equation $ax^2 + bx + c = 0$ in detail. We found, among other things, that this equation has two real, one real, or no real roots, depending on whether $b^2 - 4ac$ is positive, 0, or negative. Now that we have taken a brief informal look at quadratic functions and their graphs, we have another way of looking at real solutions of quadratic equations.

For a given function f, if r is a number such that

$$f(r) = 0$$

then r is called a *zero of f*. Thus if r is a zero of a quadratic function $f(x) =$

$ax^2 + bx + c$, then r is a solution of the corresponding quadratic equation $ax^2 + bx + c = 0$, and vice versa.

To solve a quadratic equation by graphing, one forms the related quadratic function, graphs the function, and estimates the x coordinates of the points where the graph crosses the x axis, if it crosses.

EXAMPLE 11 Solve $x^2 - 4x + 3 = 0$, $x^2 - 6x + 9 = 0$, and $x^2 - 4x + 5 = 0$ by graphing.

SOLUTION

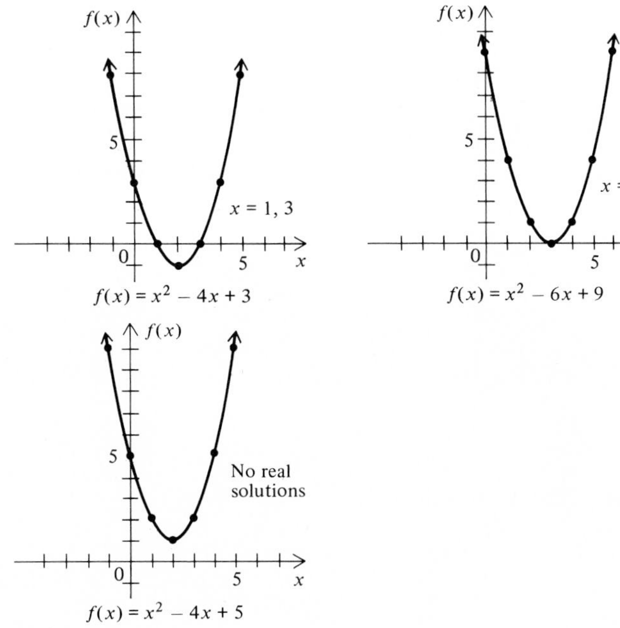

PROBLEM 11 Solve $x^2 + x - 2 = 0$, $x^2 + 4x + 4 = 0$, and $x^2 + 4 = 0$ by graphing.

ANSWER

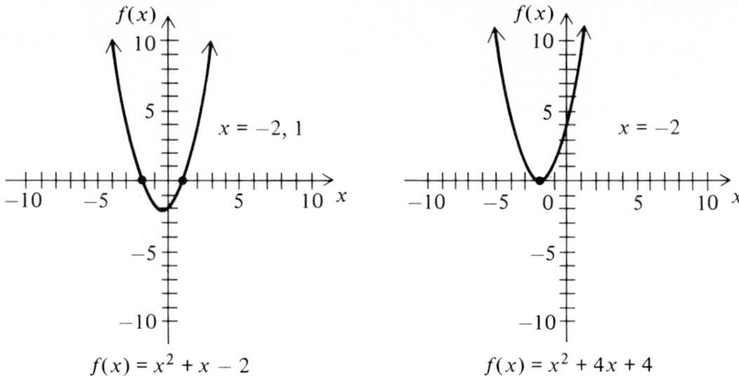

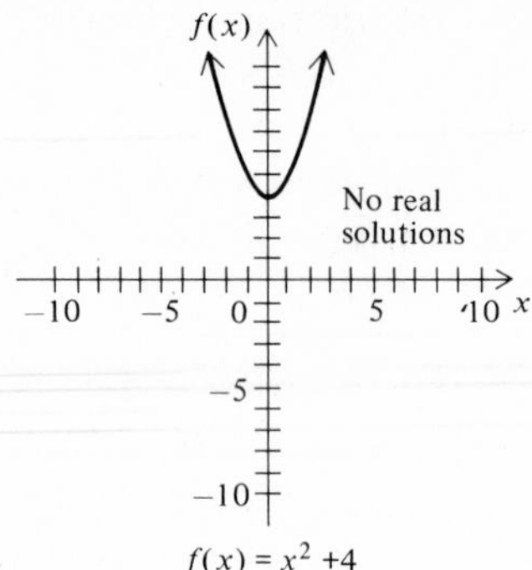

Exercise 59

A *Graph each of the following linear functions. Indicate the slope and y intercept of each.*

1. $f(x) = 2x - 4$

2. $g(x) = \frac{x}{2}$

3. $h(x) = 4 - 2x$

4. $y = -\frac{x}{2} + 1$

Graph each of the following quadratic functions and estimate the zeros of each function from its graph.

5. $f(x) = x^2$

6. $g(x) = -x^2$

7. $F(x) = 4 - x^2, -2 \leq x \leq 2$

8. $G(x) = 9 - x^2, -3 \leq x \leq 3$

B *Graph each of the following functions. Indicate whether they are linear, quadratic, or neither.*

9. $g(x) = -3, -1 \leq x \leq 5$

10. $f(x) = 4, -6 \leq x \leq 6$

11. $P(x) = x^3, -3 \leq x \leq 3$

12. $Q(x) = -x^3, -3 \leq x \leq 3$

13. $g(t) = 96t - 16t^2, 0 \leq t \leq 6$

14. $G(x) = 16x - 2x^2, 0 \leq x \leq 8$

Find all real solutions of the quadratic equations in the following problems by graphing associated quadratic functions.

15. $x^2 - 4 = 0$

16. $x^2 - 16 = 0$

17. $x^2 + 9 = 0$

18. $x^2 + 1 = 0$

19. $x^2 + 8x + 16 = 0$

20. $x^2 - 0x + 25 = 0$

21. $6x - x^2 = 0$

22. $4 + 3x - x^2 = 0$

23. $x^2 - 2x + 4 = 0$

24. $x^2 + x + 1 = 0$

C **25.** If an object is projected vertically from the ground at 96 feet/second, then (neglecting air resistance) its velocity $v(t)$ at time t is given by the equation $v(t) = 96 - 32t$. (*A*) Graph the function v for $0 \leq t \leq 6$. (*B*) Find the zero(s) of the function from the graph and interpret physically.

26. If an object is projected vertically from the ground at 96 feet/second, then (neglecting air resistance) its distance above the ground $d(t)$ at time t is given by the equation $d(t) = 96t - 16t^2$. (*A*) Graph the function d for a suitable set of values of t. (*B*) Determine from the graph the zero(s) of the function (interpret physically), the domain of the function, and the maximum height of the object.

27. A rectangular dog pen is to be made with 100 feet of fence wire. (*A*) If x represents the length of the pen, express its area $A(x)$ in terms of x. (*B*) What is the domain of the resulting function? (*C*) Graph the function for this domain. (*D*) From the graph estimate the dimension of the rectangle that will make the area maximum.

28. Work the preceding problem with the added assumption that an existing property fence will be used for one side of the pen. (Let x represent the width of the pen.)

10.5 Direct and Inverse Variation

In reading scientific material, one is likely to come across statements such as "The pressure of an enclosed gas varies directly as the absolute temperature," or, "The frequency of vibration of air in an organ pipe varies inversely as the length of the pipe," or even more complicated statements such as "The force of attraction between two bodies varies jointly as their masses and inversely as the square of the distance between the two bodies." These statements have precise mathematical meaning in that they represent particular types of functions. The purpose of this section is to investigate these special functions.

The statement *y varies directly as x* means

$$\boxed{y = kx \qquad k \neq 0}$$

where k is a constant called the *constant of variation*. Similarly, the statement "y varies directly as the square of x" means

$$y = kx^2 \qquad k \neq 0$$

and so on. The first equation defines a linear function, and the second a quadratic function.

Direct variation is illustrated by the familiar formulas

$$C = \pi D \qquad \text{and} \qquad A = \pi r^2$$

where the first formula asserts that the circumference of a circle varies directly as the diameter, and the second that the area of a circle varies directly as the square of the radius. In both cases, π is the constant of variation.

The statement y *varies inversely as* x means

$$y = \frac{k}{x} \qquad k \neq 0$$

where k is a constant (the constant of variation). As in the case of direct variation, we also discuss y varying inversely as the square of x, and so on.

An illustration of inverse variation is given in the distance-rate-time formula. In driving a fixed distance, time varies inversely as the rate. Thus,

$$t = \frac{d}{r}$$

where d is the constant of variation—as the rate increases, the time decreases, and vice versa.

EXAMPLE 12 Translate each of the following statements into appropriate equations, and find the constants of variation if $y = 16$ when $x = 4$.

(A) y varies directly as the cube of x

SOLUTION $y = kx^3$

To find the constant of variation k, substitute in $x = 4$ and $y = 16$.

$$16 = k \cdot 4^3$$

Thus $k = \frac{1}{4}$ and the equation of variation is

$$y = \tfrac{1}{4}x^3$$

(B) y varies inversely as the square root of x

SOLUTION $y = \dfrac{k}{\sqrt{x}}$

To find k, substitute in $x = 4$ and $y = 16$. Thus, $16 = k/\sqrt{4}$ and $k = 32$. The equation of variation is

$$y = \frac{32}{\sqrt{x}}$$

PROBLEM 12 If $y = 4$ when $x = 8$, find the equation of variation for each of the following statements:

(A) y varies directly as the cube root of x

(B) y varies inversely as the square of x

ANSWER (A) $y = 2\sqrt[3]{x}$ (B) $y = 256/x^2$

The statement *w varies jointly as x and y* means

$$w = kxy \qquad k \neq 0$$

where k is a constant (the constant of variation). Similarly, if

$$w = kxyz^2 \qquad k \neq 0$$

we would say that "w varies jointly as x, y, and the square of z," and so on. For example, the area of a rectangle varies jointly as its length and width (recall $A = lw$), and the volume of a right circular cylinder varies jointly as the square of its radius and its height (recall $V = \pi r^2 h$). What is the constant of variation in each case?

The above types of variation are often combined. For example, the statement "w varies jointly as x and y, and inversely as the square of z" means

$$w = k\frac{xy}{z^2} \qquad k \neq 0$$

Thus the statement, "The force of attraction F between two bodies varies jointly as their masses m_1 and m_2, and inversely as the square of the distance d between the two bodies," means

$$F = k\frac{m_1 m_2}{d^2} \qquad k \neq 0$$

If (assuming k is positive) either of the two masses is increased, the force of attraction increases; on the other hand, if the distance is increased, the force of attraction decreases.

EXAMPLE 13 The pressure P of enclosed gas varies directly as the absolute temperature T, and inversely as the volume V. If 500 cubic feet of gas yields a pressure of 10 pounds per square feet at a temperature of 300 K (absolute temperature[†]), what will be the pressure of the same gas if the volume is decreased to 300 cubic feet and the temperature increased to 360 K?

[†]A Kelvin (absolute) and Celsius degree are the same size, but 0 on the Kelvin scale is $-273°$ on the Celsius scale. This is the point at which molecular action is supposed to stop.

SOLUTION METHOD 1 Write the equation of variation $P = k\frac{T}{V}$, and find k using the first set of values:

$$10 = k\frac{300}{500}$$

$$k = \frac{50}{3}$$

Hence, the equation of variation for this particular gas is $P = \frac{50}{3}\frac{T}{V}$.

Now find the new pressure P, using the second set of values:

$$P = \frac{50}{3}\frac{360}{300} = 20 \text{ pounds per square feet}$$

METHOD 2 (generally faster than Method 1) Write the equation of variation $P = k\frac{T}{V}$ then convert to the equivalent form:

$$\frac{PV}{T} = k$$

If P_1, V_1, and T_1 are one set of values for the gas, and P_2, V_2, and T_2 are another set, then

$$\frac{P_1V_1}{T_1} = k \quad \text{and} \quad \frac{P_2V_2}{T_2} = k$$

Hence

$$\frac{P_1V_1}{T_1} = \frac{P_2V_2}{T_2}$$

Since all values are known except P_2, substitute and solve. Thus

$$\frac{(10)(500)}{300} = \frac{P_2(300)}{360}$$

$$P_2 = 20 \text{ pounds per square feet}$$

PROBLEM 13 The length L of skid marks of a car's tires (when brakes are applied) varies directly as the square of the speed v of the car. If skid marks of 20 feet are produced at 30 miles/hour how fast would the same car be going if it produced skid marks of 80 feet? Solve two ways (see Example 13).

ANSWER 60 miles/hour

Exercise 60

A *Translate each statement into an equation using k as the constant of variation.*

1. F varies directly as the square of v

2. u varies directly as v

3. The pitch or frequency of a guitar string f of a given length varies directly as the square root of the tension T of the string.

4. Geologists have found in studies of earth erosion that the erosive force (sediment carrying power) P of a swiftly flowing stream varies directly as the sixth power of the velocity v of the water.

5. y varies inversely as the square root of x

6. I varies inversely as t

7. The biologist Reaumur suggested in 1735 that the length of time t that it takes fruit to ripen during the growing season varies inversely as the sum of the average daily temperatures T during the growing season.

8. In a study on urban concentration, F. Auerbach discovered an interesting law. After arranging all the cities of a given country according to their population size, starting with the largest, it was found that the population p of a city varied inversely as the number n indicating its position in the ordering.

9. R varies jointly as S, T, and V

10. g varies jointly as x and the square of y

11. The volume of a cone v varies jointly as its height h and the square of the radius r of its base.

12. The amount of heat Q put out by an electrical appliance (in calories) varies jointly as time t, resistance R in the circuit, and the square of the current I.

Solve using either of the two methods illustrated in Example 13.

13. u varies directly as the square root of v. If $u = 2$ when $v = 2$, find u when $v = 8$.

14. y varies directly as the square of x. If $y = 20$ when $x = 2$, find y when $x = 5$.

15. L varies inversely as the square root of m. If $L = 9$ when $M = 9$, find L when $M = 3$.

16. I varies inversely as the cube of t. If $I = 4$ when $t = 2$, find I when $t = 4$.

B *Translate each equation into an equation using k as the constant of variation.*

17. U varies jointly as a and b, and inversely as the cube of c.

18. w varies directly as the square of x and inversely as the square root of y.

19. The maximum safe load L for a horizontal beam varies jointly as its width w and the square of its height h, and inversely as its length l.

20. Joseph Cavanaugh, a sociologist, found that the number of long-distance phone calls n between pairs of cities in a given time period varied (approximately) jointly as the populations P_1 and P_2 of the two cities, and inversely as the distance d between the two cities.

Solve using either of the two methods illustrated in Example 13.

21. Q varies jointly as m and the square of n, and inversely as P. If $Q = -4$ when $m = 6$, $n = 2$, and $P = 12$, find Q when $m = 4$, $n = 3$, and $P = 6$.

22. w varies jointly as x, y, and z, and inversely as the square of t. If $w = 2$ when $x = 2$, $y = 3$, $z = 6$, and $t = 3$, find w when $x = 3$, $y = 4$, $z = 2$, and $t = 2$.

23. The weight w of an object on or above the surface of the earth varies inversely as the square of the distance d between the object and the center of the earth. If a girl weighs 100 pounds on the surface of the earth, how much would she weigh (to the nearest pound) 400 miles above the earth's surface? (Assume the radius of the earth is 4,000 miles.)

24. A child was struck by a car in a crosswalk. The driver of the car had slammed on the brakes and left skid marks 160 feet long. The driver told the police that the car was traveling 30 miles/hour. The police know that the length of skid marks L (when brakes are applied) varies directly as the square of the speed of the car v, and that at 30 miles/hour (under ideal conditions) skid marks would be 40 feet long. How fast was the driver actually going before the brakes were applied?

25. Ohm's law states that the current I in a wire varies directly as the electromotive force E and inversely as the resistance R. If $I = 22$ amperes when $E = 110$ volts and $R = 5$ ohms, find I if $E = 220$ volts and $R = 11$ ohms.

26. Anthropologists, in their study of race and human genetic groupings, often use an index called the cephalic index. The cephalic index C varies directly as the width w of the head and inversely as the length l of the head (both looking down from the top). If an Indian in Baja, California (Mexico) has $C = 75$, $w = 6$ inches, and $l = 8$ inches, then what would C be for an Indian in Northern California with $w = 8.1$ in. and $l = 9$ in.?

C **27.** If the horsepower P required to drive a speedboat through water varies directly as the cube of the speed v of the boat, what change in horsepower is required to double the speed of a boat?

28. The intensity of illumination E on a surface varies inversely as the square of its distance d from a light source. What is the effect on the total illumination on a book if the distance between the light source and the book is doubled?

29. The frequency of vibration f of a musical string varies directly with the square root of the tension T and inversely as the length L of the string. If the tension of a string is increased by a factor of 4 and the length of a string is doubled, what is the effect on the frequency?

30. In an automobile accident the destructive force F of a car varies (approximately) jointly as the weight w of the car and the square of the speed v of the car. (This is why accidents at high speed are generally so severe.) What would be the effect on the destructive force of a car if its weight were doubled and its speed were doubled?

Exercise 61 Chapter Review

A **1.** Given the relation $M = \{(2, 3), (3, 3), (4, 4), (5, 4)\}$: (*A*) State its domain and range. (*B*) Graph M. (*C*) Is M a function?

2. If $f(x) = 6 - x$, find $f(6)$, $f(0)$, $f(-3)$, and $f(m)$.

3. If $G(z) = z - 2z^2$, find $G(2)$, $G(0)$, $G(-1)$, and $G(c)$.

4. Graph the relation $M = \{(0, 5), (2, 7), (2, 3)\}$, its inverse M^{-1}, and $y = x$, all on the same coordinate system. Indicate whether M or M^{-1} is a function.

5. What is the domain and range of M^{-1} in the preceding problem?

6. Graph $f(x) = 1 - 2x$. Is this function linear, quadratic, or neither?

7. Solve $x^2 - 9 = 0$ graphically.

Translate each statement into an equation using k as the constant of variation.

8. m varies directly as the square of n

9. P varies inversely as the square of Q

10. A varies jointly as a and b

11. y varies directly with the cube of x and inversely with the square root of z

B **12.** Is every relation a function? Explain.

13. If $f(t) = 4 - t^2$ and $g(t) = t - 3$, find:

(*A*) $f(0) - g(0)$ (*B*) $\dfrac{g(6)}{f(-1)}$ (*C*) $g(x) - f(x)$ (*D*) $f[g(2)]$

14. If $g(t) = 1 - t^2$, find $g(1 + h)$ and $\dfrac{g(1+h) - g(1)}{h}$

15. The cost $C(x)$ for renting a business copying machine is \$200 per month plus 5 cents a copy for x copies. Express this functional relationship in terms of an equation and graph it for $0 \le x \le 3{,}000$.

16. Let $M(x) = \dfrac{x + 3}{2}$

(A) Find $M^{-1}(x)$. (B) Are both M and M^{-1} functions? (C) Find $M^{-1}[M(3)]$.

17. Graph $g(t) = 8t - t^2$, $0 \leq t \leq 10$. Is this function linear, quadratic, or neither?

18. Solve the quadratic equation $x^2 + 6x + 5 = 0$ graphically.

19. The time t required for an elevator to lift a weight varies jointly as the weight w and the distance d through which it is lifted, and inversely as the power P of the motor. Write the equation of variation using k as the constant of variation.

20. If y varies directly as x and inversely as z, (A) write the equation of variation; (B) if $y = 4$ when $x = 6$ and $z = 2$, find y when $x = 4$ and $z = 4$.

C **21.** Graph $E(x) = 2^x$, $x \in \{-2, -1, 0, 1, 2\}$; E^{-1}; and $y = x$ on the same coordinate system.

22. For which of the following functions does $f(a + b) = f(a) + f(b)$: $f(x) = x^2$, $f(x) = 2x - 3$, $f(x) = 3x$?

23. For which of the following functions does $f(kx) = kf(x)$: $f(x) = x^2$, $f(x) = 2x - 3$, $f(x) = 3x$?

24. Which of the following functions have inverses that are functions: $f(x) = x^3$; $g(x) = x^2$; $h(x) = 2x - 3$, $F(x) = x^2$, $x \geq 0$?

25. The intensity of illumination E at a given point varies inversely as the square of the distance d that the point is from a light source. If the illumination is 50 footcandles at a point 2 feet from a light, what will be the illumination at a point 5 feet from the same light? At what distance will the illumination be 100 footcandles?

26. The total force F of a wind on a wall varies jointly as the area of the wall A and the square of the velocity of the wind v. How is the total force on a wall affected if the area is cut in half and the velocity is doubled?

Chapter 11

EXPONENTIAL AND LOGARITHMIC FUNCTIONS

Most of the functions we have considered have been algebraic functions, that is, functions defined by means of the basic algebraic operations on variables and constants. (By algebraic operations we mean addition, subtraction, multiplication, division, powers, and roots.) In this chapter we will define and investigate the properties of two new and important classes of functions called exponential and logarithmic functions.

11.1 Exponential Functions

What kind of function does

$$g(x) = 2^x \tag{1}$$

define? Since a variable cannot appear as an exponent in an algebraic function, g must be a new kind of function. The function g is a particular example of a general class of functions called exponential functions. An *exponential function* is a function defined by the equation

$$f(x) = b^x \qquad b > 0, b \neq 1$$

where b is a constant, called the base, and the exponent x is a variable. The replacement set for the exponent, the domain of f, is the set of real numbers R.

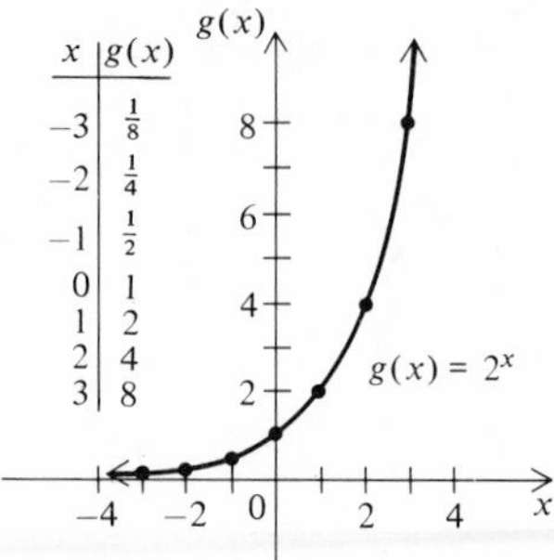

Figure 1

Most students, if asked to graph an exponential function such as that in (1), would not hesitate at all. They would likely make up a table by assigning integers to x, plot the resulting set of ordered pairs of numbers, and then join these points with a smooth curve (see Fig. 1). The only catch is that 2^x has not been defined for all real numbers. We know what 2^5, 2^{-3}, $2^{2/3}$, $2^{-3/5}$, $2^{1.4}$, and $2^{-3.15}$ mean (that is, 2^p, where p is a rational number—see Chap. 7), but what does

$$2^{\sqrt{2}}$$

mean? The question is not easy to answer at this time. In fact, a precise definition of $2^{\sqrt{2}}$ must wait for more advanced courses, where we can show that

$$b^x$$

names a real number for b a positive real number and x any real number, and that the graph of $g(x) = 2^x$ is as indicated in Fig. 1. We can also show that, for x irrational, b^x can be approximated as closely as we like by using rational number approximations for x. For example, each of the terms in the sequence

$$2^{1.4}, \qquad 2^{1.41}, \qquad 2^{1.414}, \qquad \ldots$$

approximates $2^{\sqrt{2}}$, and as we move to the right the approximation improves.

Finally, we can show that the laws of exponents continue to hold for irrational exponents.

All these statements about irrational exponents (which can be justified in a course on calculus) will be assumed true for the rest of this book.

It is useful to compare the graphs of $y = 2^x$ and $y = (\frac{1}{2})^x = 2^{-x}$ by plotting them on the same coordinate system (Fig. 2). The graph of

$$f(x) = b^x \qquad b > 1$$

will look very much like $y = 2^x$, and the graph of

$$f(x) = b^x \qquad 0 < b < 1$$

will look very much like $y = (\frac{1}{2})^x$.

Exponential functions are often referred to as growth functions because of their widespread use in describing different kinds of growth phenomena. These functions are used to describe population growth of people, animals, and bacteria; radioactive decay (negative growth); growth of a new chemical substance in a chemical reaction; increase or decline in the temperature of a substance being heated or cooled; growth of money at compound interest; light absorption (negative growth) as it passes through air, water, or glass; decline of atmospheric pressure as altitude is increased; and growth of learning a skill such as swimming or typing relative to practice.

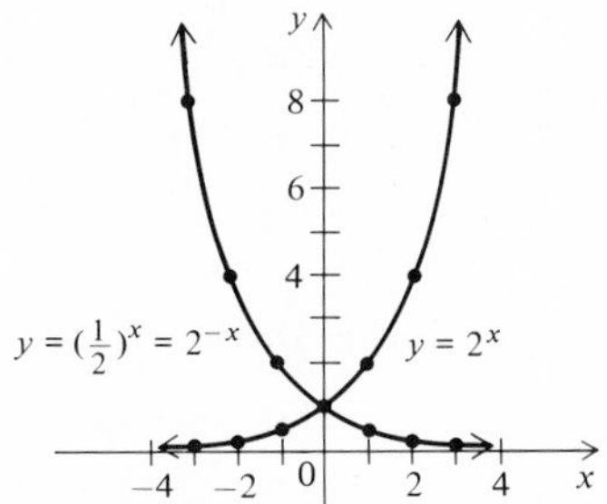

Figure 2

For introductory purposes, the bases 2 and $\frac{1}{2}$ were convenient choices; however, a certain irrational number, denoted by e, is by far the most frequently used exponential base for both theoretical and practical purposes. In fact,

$$f(x) = e^x$$

is often referred to as *the* exponential function because of its widespread use. The reasons for the preference for e as a base are made clear in more advanced courses. And at that time, it is shown that e is approximated by

$(1 + 1/n)^n$ to any decimal accuracy desired by taking n (an integer) sufficiently large. The irrational number e to five decimal places is

$$e \approx 2.71828$$

Similarly, e^x can be approximated to any decimal-place accuracy desired by using $[1 + (1/n)]^{nx}$ for sufficiently large n. Because of the importance of e^x, tables for its evaluation are readily available in almost any mathematical handbook. A short table for e^x and e^{-x} is included in the Appendix for convenient reference.

Finally, we state without proof that for $b > 0$, $b \neq 1$

$$\boxed{b^m = b^n \qquad \text{if and only if} \qquad m = n}$$

Thus, if $2^{15} = 2^{3x}$, then $3x = 15$ and $x = 5$.

Exercise 62

A *Graph each function for $-3 \le x \le 3$. Plot points using integers for x, then join the points with a smooth curve.*

1. $y = 3^x$

2. $y = 2^x$

3. $y = (\frac{1}{3})^x = 3^{-x}$

4. $y = (\frac{1}{2})^x = 2^{-x}$

5. $y = 4 \cdot 3^x$

6. $y = 5 \cdot 2^x$

7. $y = 3(\frac{1}{3})^x = 3 \cdot 3^{-x}$

8. $y = 4(\frac{1}{2})^x = 4 \cdot 2^{-x}$

B *Graph for $-3 \le x \le 3$. Use Table 2 in the Appendix if base is e.*

9. $y = 7(\frac{1}{2})^{2x} = 7 \cdot 2^{-2x}$

10. $y = 11 \cdot 2^{-2x}$

11. $y = e^x$

12. $y = e^{-x}$

13. $y = 10e^{-0.12x}$

14. $y = 100e^{0.25x}$

15. If we start with 2 cents and double up each day, at the end of n days we would have 2^n cents. Graph $f(n) = 2^n$ for $1 \le n \le 10$. (Pick the scale on the vertical axis so that the graph will not go off the paper.)

16. Radioactive strontium 90 has a half-life of 28 years; that is, in 28 years one-half of any amount of strontium 90 will change to another substance because of radioactive decay. If we place a bar containing 100 milligrams of strontium 90 in a nuclear reactor, then the amount of strontium 90 that will be left after t years is given by $A = 100(\frac{1}{2})^{t/28}$. Graph this exponential function for $t = 0$, 28, 2(28), 3(28), 4(28), 5(28), and 6(28), and then join these points with a smooth curve.

C **17.** Graph $y = e^{-x^2}$ for $x = -1.5, -1.0, -0.5, 0, 0.5, 1.0$, and 1.5, and then join these points with a smooth curve. (Use Table 2.)

18. For $f = \{(x, y) \mid y = 10^x\}$, graph f and f^{-1} using the same coordinate axes.

19. For $f = \{(x, y) \mid y = 2^x\}$, graph f and f^{-1} using the same coordinate axes.

20. Graph $y = y_0 e^{-0.22x}$, y_0 constant.

21. Graph $y = y_0 2^x$, y_0 constant.

22. If bacteria in a certain culture doubles every hour, write a formula that gives the number of bacteria N in the culture after n hours, assuming that the culture has N_0 bacteria to start with.

23. Daniel Lowenthal, a sociologist at Columbia University, made a study of the sale of popular records relative to their positions in the top 20. He found, over a 5-year period, that the average number of sales $N(i)$ of the ith-ranking record is given approximately by the exponential function

$$N(i) = N_1 e^{-0.09(i-1)} \qquad 1 \le i \le 20$$

where N_1 is the number of sales of the top-ranking record on the list. Graph the function using Table 2 in the Appendix.

24. Sociologists Stephan and Mischler found that if the members of a discussion group of 10 were ranked according to the number of times each participated, then the number of times $N(i)$ the ith-ranked person participated was given approximately by the exponential function

$$N(i) = N_1 e^{-0.11(i-1)} \qquad 1 \le i \le 10$$

where N_1 was the number of times the top-ranked person participated in the discussion. Graph the exponential function using N_1 as the basic unit on the vertical scale and Table 2 in the Appendix.

11.2 Logarithmic Functions

Now we will define a new class of functions, called *logarithmic functions*, as inverses of exponential functions. Here you will see why we placed special emphasis on the general concept of inverse function. If we start with the exponential function

$$f\colon y = 2^x$$

and interchange the variables x and y, we obtain the inverse of f

$$f^{-1}\colon x = 2^y$$

Graphing f and f^{-1} on the same coordinate system (Fig. 3), we see that

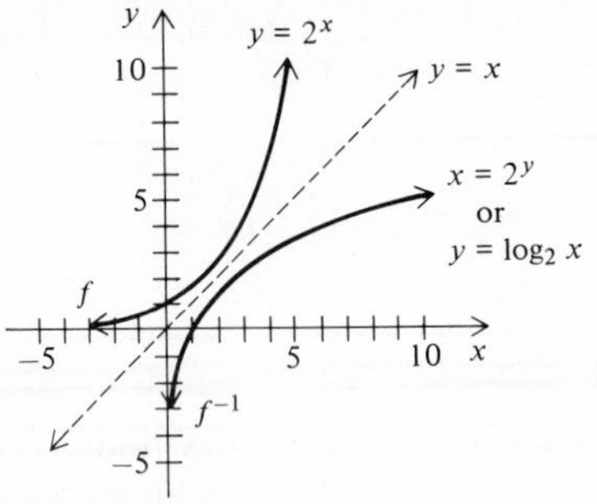

Figure 3

f^{-1} is also a function. This new function is given the name *logarithmic function with base* 2, and we write

$$y = \log_2 x \qquad \text{if and only if} \qquad x = 2^y$$

In general, we define the logarithmic function with base b to be the inverse of the exponential function with base b ($b > 0$, $b \neq 1$). Symbolically,

$$\boxed{y = \log_b x \qquad \text{if and only if} \qquad x = b^y} \tag{2}$$

In words, (2) states that the log to the base b of x is the power y to which b must be raised to equal x. It is very important to remember that $y = \log_b x$ and $x = b^y$ define the same function, and as such can be used interchangeably.

Since the domain of an exponential function includes all real numbers and its range is the set of positive real numbers, the domain of a logarithmic function is the set of all positive real numbers and its range is the set of all real numbers.

EXAMPLE 1 **FROM LOGARITHMIC FORM TO EXPONENTIAL FORM**

(A) $\log_2 8 = 3$ is equivalent to $8 = 2^3$

(B) $\log_{25} 5 = \frac{1}{2}$ is equivalent to $5 = 25^{1/2}$

(C) $\log_2 \frac{1}{4} = -2$ is equivalent to $\frac{1}{4} = 2^{-2}$

PROBLEM 1 Change to an equivalent exponential form.

(A) $\log_3 27 = 3$ (B) $\log_{36} 6 = \frac{1}{2}$ (C) $\log_3 \frac{1}{9} = -2$

ANSWER (A) $27 = 3^3$ (B) $6 = 36^{1/2}$ (C) $\frac{1}{9} = 3^{-2}$

EXAMPLE 2 **FROM EXPONENTIAL FORM TO LOGARITHMIC FORM**

(A) $49 = 7^2$ is equivalent to $\log_7 49 = 2$

(*B*) $3 = \sqrt{9}$ is equivalent to $\log_9 3 = \frac{1}{2}$

(*C*) $\frac{1}{5} = 5^{-1}$ is equivalent to $\log_5 \frac{1}{5} = -1$

PROBLEM 2 Change to an equivalent logarithmic form.

(*A*) $64 = 4^3$ (*B*) $2 = \sqrt[3]{8}$ (*C*) $\frac{1}{16} = 4^{-2}$

ANSWER (*A*) $\log_4 64 = 3$ (*B*) $\log_8 2 = \frac{1}{3}$ (*C*) $\log_4 \frac{1}{16} = -2$

EXAMPLE 3 Find x, b, or y as indicated.

(*A*) $y = \log_4 8$

SOLUTION If $y = \log_4 8$, then $8 = 4^y$.

Write each member to the same base 2:

$$2^3 = 2^{2y}$$

Thus,

$$2y = 3 \qquad \text{and} \qquad y = \frac{3}{2}$$

(*B*) $\log_3 x = -2$

SOLUTION Write in equivalent exponential form

$$x = 3^{-2}$$

Thus,

$$x = \frac{1}{9}$$

(*C*) $\log_b 1{,}000 = 3$

SOLUTION Write in equivalent exponential form

$$1{,}000 = b^3$$

and write the left member as a power of 10

$$10^3 = b^3$$

Thus,

$$b = 10$$

PROBLEM 3 Find x, b, or y as indicated.

(*A*) $y = \log_9 27$

(*B*) $\log_2 x = -3$

(*C*) $\log_b 100 = 2$

ANSWER (A) $y = \frac{3}{2}$ (B) $x = \frac{1}{8}$ (C) $b = 10$

Recall from the last chapter that if f and f^{-1} are both functions, then

$$f[f^{-1}(x)] = x \qquad \text{and} \qquad f^{-1}[f(x)] = x$$

Since the logarithmic and exponential functions are inverses of each other for the same base, it follows that

$$\log_b b^x = x$$
$$b^{\log_b x} = x \qquad x > 0$$

Both of these equations are very useful, and are encountered in many mathematics courses.

EXAMPLE 4

(A) $\log_{10} 10^5 = 5$

(B) $\log_2 8 = \log_2 2^3 = 3$

(C) $10^{\log_{10} 7} = 7$

PROBLEM 4 Evaluate each of the following.

(A) $\log_{10} 10^{-2}$ (B) $\log_5 25$ (C) $2^{\log_2 9}$

ANSWER (A) -2 (B) 2 (C) 9

Exercise 63

A *Rewrite in exponential form.*

1. $\log_3 9 = 2$ **2.** $\log_2 4 = 2$

3. $\log_3 81 = 4$ **4.** $\log_5 125 = 3$

5. $\log_{10} 1{,}000 = 3$ **6.** $\log_{10} 100 = 2$

7. $\log_e 1 = 0$ **8.** $\log_8 1 = 0$

Rewrite in logarithmic form.

9. $64 = 8^2$ **10.** $25 = 5^2$

11. $10{,}000 = 10^4$ **12.** $1{,}000 = 10^3$

13. $u = v^x$ **14.** $a = b^c$

15. $9 = 27^{2/3}$ **16.** $8 = 4^{3/2}$

Find each of the following.

17. $\log_{10} 10^5$

18. $\log_5 5^3$

19. $\log_2 2^{-4}$

20. $\log_{10} 10^{-7}$

21. $\log_6 36$

22. $\log_3 9$

23. $\log_{10} 1{,}000$

24. $\log_{10} 0.001$

Find x, y, or b as indicated.

25. $\log_2 x = 2$

26. $\log_3 x = 2$

27. $\log_4 16 = y$

28. $\log_8 64 = y$

29. $\log_b 16 = 2$

30. $\log_b 10^{-3} = -3$

B *Rewrite in exponential form.*

31. $\log_{10} 0.001 = -3$

32. $\log_{10} 0.01 = -2$

33. $\log_{81} 3 = \frac{1}{4}$

34. $\log_4 2 = \frac{1}{2}$

35. $\log_{1/2} 16 = -4$

36. $\log_{1/3} 27 = -3$

37. $\log_a N = e$

38. $\log_k u = v$

Rewrite in logarithmic form.

39. $0.01 = 10^{-2}$

40. $0.001 = 10^{-3}$

41. $1 = e^0$

42. $1 = (\frac{1}{2})^0$

43. $\frac{1}{8} = 2^{-3}$

44. $\frac{1}{8} = (\frac{1}{2})^3$

45. $\frac{1}{3} = 81^{-1/4}$

46. $\frac{1}{2} = 32^{-1/5}$

47. $7 = \sqrt{49}$

48. $11 = \sqrt{132}$

Find each of the following.

49. $\log_b b^u$

50. $\log_b b^{uv}$

51. $\log_e e^{1/2}$

52. $\log_e e^{-3}$

53. $\log_2 \sqrt{8}$

54. $\log_5 \sqrt[3]{5}$

55. $\log_{23} 1$

56. $\log_{17} 1$

57. $\log_4 8$

58. $\log_4 \frac{1}{4}$

Find x, y, or b as indicated.

59. $\log_4 x = \frac{1}{2}$

60. $\log_{25} x = \frac{1}{2}$

61. $\log_{1/3} 9 = y$

62. $\log_{49} \frac{1}{7} = y$

63. $\log_b 1{,}000 = \frac{3}{2}$

64. $\log_b 4 = \frac{2}{3}$

C **65.** $\log_b 1 = 0$ **66.** $\log_b b = 1$

67. For $f = \{(x, y) \mid y = 1^x\}$ discuss the domain and range for f and f^{-1}. Are both relations functions?

68. Why is 1 not a suitable logarithmic base? (HINT: Try to find $\log_1 5$.)

69. (A) For $f = \{(x\ y) \mid y = 10^x\}$, graph f and f^{-1} using the same coordinate axes.

(B) Discuss the domain and range of f and f^{-1}.

(C) What other name could you use for the inverse of f?

70. Prove that $\log_b (1/x) = -\log_b x$.

71. If $\log_b x = 3$, find $\log_b (1/x)$.

11.3 Properties of Logarithmic Functions

Logarithmic functions have several very useful properties that follow directly from the fact that they are inverses of exponential functions. These properties will enable us to convert multiplication problems into addition problems, division problems into subtraction problems, and power and root problems into multiplication problems. In addition, we will be able to solve exponential equations such as $2 = 10^x$.

THEOREM 1 (*Properties of logarithmic functions.*) If b, M, and N are positive real numbers, $b \neq 1$, and p is a real number, then

1 $\log_b MN = \log_b M + \log_b N$
2 $\log_b \dfrac{M}{N} = \log_b M - \log_b N$
3 $\log_b M^p = p \log_b M$

The proof of the first property (as well as the proof of the other two) is based on the laws of exponents. To bring exponents into the proof, we let

$$u = \log_b M \qquad \text{and} \qquad v = \log_b N$$

and convert these to the equivalent exponential forms

$$M = b^u \qquad \text{and} \qquad N = b^v$$

Now, see if you can provide the reasons for each of the following steps:

$$\log_b MN = \log_b b^u b^v = \log_b b^{u+v} = u + v = \log_b M + \log_b N$$

The proofs of the other two properties are left as exercises.

EXAMPLE 5 (*A*)

$$\log_b \frac{mn}{pq} = \log_b mn - \log_b pq$$
$$= \log_b m + \log_b n - (\log_b p + \log_b q)$$
$$= \log_b m + \log_b n - \log_b p - \log_b q$$

(*B*) $\log_b (mn)^{2/3} = \frac{2}{3} \log_b mn = \frac{2}{3}(\log_b m + \log_b n)$

(*C*) $\log_b \dfrac{x^8}{y^{1/5}} = \log_b x^8 - \log_b y^{1/5} = 8 \log_b x - \frac{1}{5} \log_b y$

PROBLEM 5 Write in terms of simpler logarithmic forms, as in Example 5.

(*A*) $\log_b(r/uv)$ (*B*) $\log_b(m/n)^{3/5}$ (*C*) $\log_b(u^{1/3}/v^5)$

ANSWER (*A*) $\log_b r - \log_b u - \log_b v$ (*B*) $\frac{3}{5}(\log_b m - \log_b n)$

(*C*) $\frac{1}{3} \log_b u - 5 \log_b v$

EXAMPLE 6 If $\log_e 3 = 1.10$ and $\log_e 7 = 1.95$, find

(*A*) $\log_e \frac{7}{3}$

(*B*) $\log_e \sqrt[3]{21}$

SOLUTION (*A*) $\log_e \frac{7}{3} = \log_e 7 - \log_e 3 = 1.95 - 1.10 = 0.85$

(*B*) $\log_e \sqrt[3]{21} = \log_e (21)^{1/3} = \frac{1}{3} \log_e (3 \cdot 7) = \frac{1}{3}(\log_e 3 + \log_e 7)$

$$= \tfrac{1}{3}(1.10 + 1.95) = 1.02$$

PROBLEM 6 If $\log_e 5 = 1.609$ and $\log_e 8 = 2.079$, find

(*A*) $\log_e(5^{10}/8)$ (*B*) $\log_e \sqrt[4]{8/5}$

ANSWER (*A*) 14.011 (*B*) 0.118

Finally, we state without proof that for n and m any positive real numbers

$$\boxed{\log_b m = \log_b n \quad \text{if and only if} \quad m = n}$$

Thus, if $\log_{10} x = \log_{10} 32.15$, then $x = 32.15$.

The following example and problem, though somewhat artificial, will give you additional practice in using the properties in Theorem 1.

EXAMPLE 7 Find x so that $\log_b x = \frac{2}{3}\log_b 27 + 2\log_b 2 - \log_b 3$.

SOLUTION

$$\begin{aligned}\log_b x &= \tfrac{2}{3}\log_b 27 + 2\log_b 2 - \log_b 3\\ &= \log_b 27^{2/3} + \log_b 2^2 - \log_b 3\\ &= \log_b 9 + \log_b 4 - \log_b 3\\ &= \log_b \frac{9\cdot 4}{3} = \log_b 12\end{aligned}$$

Thus

$$\log_b x = \log_b 12$$

hence

$$x = 12$$

PROBLEM 7 Find x so that $\log_b x = \frac{2}{3}\log_b 8 + \frac{1}{2}\log_b 9 - \log_b 6$.

ANSWER $x = 2$

Exercise 64

A *Write in terms of simpler logarithmic forms (going as far as you can with logarithmic properties—see Example 5).*

1. $\log_b uv$ **2.** $\log_b rt$ **3.** $\log_b \dfrac{A}{B}$

4. $\log_b \dfrac{p}{q}$ **5.** $\log_b u^5$ **6.** $\log_b w^{25}$

7. $\log_b N^{3/5}$ **8.** $\log_b u^{-2/3}$ **9.** $\log_b \sqrt{Q}$

10. $\log_b \sqrt[5]{M}$ **11.** $\log_b uvw$ **12.** $\log_b \dfrac{u}{vw}$

Write each expression in terms of a single logarithm with a coefficient of 1. Example: $\log_b u^2 - \log_b v = \log_b (u^2/v)$.

13. $\log_b A + \log_b B$ **14.** $\log_b P + \log_b Q + \log_b R$

15. $\log_b X - \log_b Y$ **16.** $\log_b x^2 - \log_b y^3$

17. $\log_b w + \log_b x - \log_b y$ **18.** $\log_b w - \log_b x - \log_b y$

If $\log_e 2 = 0.69$, $\log_e 3 = 1.10$, and $\log_e 5 = 1.61$, find the logarithm to the base e of each of the following numbers.

19. $\log_e 30$ **20.** $\log_e 6$

21. $\log_e \frac{2}{5}$ **22.** $\log_e \frac{5}{3}$

23. $\log_e 27$ **24.** $\log_e 16$

B *Write in terms of simpler logarithmic forms (going as far as you can with logarithmic properties—see Example 5).*

25. $\log_b u^2v^7$ **26.** $\log_b u^{1/2}v^{1/3}$

27. $\log_b \frac{1}{a}$ **28.** $\log_b \frac{1}{M^3}$

29. $\log_b \frac{\sqrt[3]{N}}{p^2q^3}$ **30.** $\log_b \frac{m^5n^3}{\sqrt{p}}$

31. $\log_b \sqrt[4]{\frac{x^2y^3}{\sqrt{z}}}$ **32.** $\log_b \sqrt[5]{\left(\frac{x}{y^4z^9}\right)^3}$

Write each expression in terms of a single logarithm with a coefficient of 1.

33. $2 \log_b x - \log_b y$ **34.** $\log_b m - \frac{1}{2} \log n$

35. $3 \log_b x + 2 \log_b y - 4 \log_b z$ **36.** $\frac{1}{3} \log_b w - 3 \log x - 5 \log y$

37. $\frac{1}{5}(2 \log_b x + 3 \log_b y)$ **38.** $\frac{1}{3}(\log_b x - \log_b y)$

If $\log_b 2 = 0.69$, $\log_b 3 = 1.10$, and $\log_b 5 = 1.61$, find the logarithm to the base b of the following numbers.

39. $\log_b 7.5$ **40.** $\log_b 1.5$

41. $\log_b \sqrt[3]{2}$ **42.** $\log_b \sqrt{3}$

43. $\log_b \sqrt{0.9}$ **44.** $\log_b \sqrt{\frac{3}{2}}$

45. Find x so that $\frac{3}{2} \log_b 4 - \frac{2}{3} \log_b 8 + 2 \log_b 2 = \log_b x$.

46. Find x so that $3 \log_b 2 + \frac{1}{2} \log_b 25 - \log_b 20 = \log_b x$.

C **47.** Write $\log_b y - \log_b c + kt = 0$ in exponential form free of logarithms.

48. Write $\log_e x - \log_e 100 = -0.08t$ in exponential form free of logarithms.

49. Prove that $\log_b (M/N) = \log_b M - \log_b N$ under the hypotheses of Theorem 1.

50. Prove that $\log_b M^p = p \log_b M$ under the hypotheses of Theorem 1.

51. Prove that $\log_b MN = \log_b M + \log_b N$ by starting with $M = b^{\log_b M}$ and $N = b^{\log_b N}$.

52. Prove that $\log_b (M/N) = \log_b M - \log_b N$ by starting with $M = b^{\log_b M}$ and $N = b^{\log_b N}$.

11.4 Determining Common Logarithms

Of all possible logarithmic bases, the base e and the base 10 are used almost exclusively. Logarithmic functions with base e are of great value in pure and applied mathematics from calculus onward. Logarithmic functions with base 10 (because of our base-10 decimal system) find widespread use as computational tools for evaluating difficult problems such as

$$\frac{(538{,}000)^3(0.002374)^2}{\sqrt[5]{0.1083}}$$

or solving exponential equations such as

$$2 = (1.05)^x$$

Logarithms of numbers to the base 10 are called *common logarithms*. Since common logarithms will be used extensively in the work that follows, we do not indicate the base 10 unless special emphasis is desired. Thus, we define

$$\log x = \log_{10} x$$

To find the logarithm of any power of 10 is easy. Since $\log_b b^x = x$, we can write

$$\log 10^k = k$$

But how do we find logarithms of numbers such as 53,200 or 0.00876? Recalling that any positive decimal fraction can be written in scientific notation (see Chap. 7), we see that

$$\log 53{,}200 = \log (5.32 \times 10^4) = \log 5.32 + \log 10^4 = \log 5.32 + 4$$

and that

$$\log 0.00876 = \log(8.76 \times 10^{-3}) = \log 8.76 + \log 10^{-3} = \log 8.76 - 3$$

In general, any positive decimal fraction N can be written as the product of a number between 1 and 10 and a power of 10. Hence, $\log N$ involves two parts: the logarithm of a number between 1 and 10 and an integer.

$$N = r \times 10^k \qquad 1 \le r < 10,\ k \text{ an integer}$$

$$\log N = \log (r \times 10^k) = \log r + \log 10^k = \log r + k$$

What we now need is a way of finding $\log r$, where $1 \le r < 10$. Using methods of advanced mathematics, tables can be constructed for this purpose to any decimal accuracy desired. Table 3 in the Appendix is

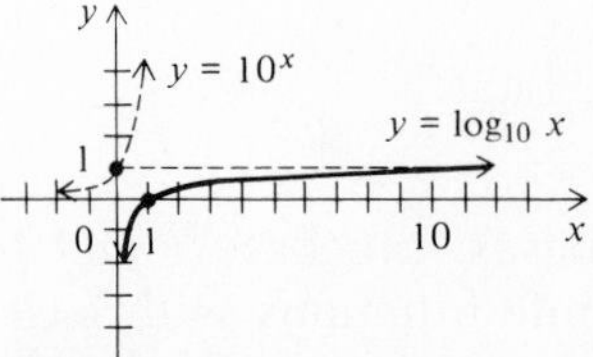

Figure 4

such a table to four decimal places. Looking at Fig. 4 we note that if x is between 1 and 10, then $\log x$ is between 0 and 1.

To illustrate the use of Table 3, a small portion of it is reproduced in Fig. 5. To find log 3.47, for example, we first locate 3.4 under the x heading; then we move across to the column headed 7, where we find .5403. Thus $\log 3.47 = 0.5403$.

x	0	1	2	3	4	5	6	7	8	9
3.2	.5051	.5065	.5079	.5092	.5105	.5119	.5132	.5145	.5159	.5172
3.3	.5185	.5198	.5211	.5224	.5237	.5250	.5263	.5276	.5289	.5302
3.4	.5315	.5328	.5340	.5353	.5366	.5378	.5391	**.5403**	.5416	.5428
3.5	.5441	.5453	.5465	.5478	.5490	.5502	.5514	.5527	.5539	.5551

$\log 3.47 = 0.5403$

Figure 5

EXAMPLE 8 (*A*)

$$\log 33{,}800 = \log (3.38 \times 10^4) = \log 3.38 + \log 10^4$$
$$= 0.5289 + 4$$
$$= 4.5289$$

(*B*)

$$\log 0.00351 = \log (3.51 \times 10^{-3}) = \log 3.51 + \log 10^{-3}$$
$$= 0.5453 - 3 \quad \text{or} \quad -2.4547$$

NOTE: The form $0.5453 - 3$ is usually preferred to -2.4547 for reasons that will be made clear later.

PROBLEM 8 Use the table in Fig. 5 to find

(*A*) log 328,000 (*B*) log 0.000342

ANSWER (*A*) 5.5159 (*B*) $0.5340 - 4$

We see that the log of a positive decimal fraction has two main parts: a nonnegative decimal fraction between 0 and 1 (called the *mantissa*) and an integer part (called the *characteristic*). With a little practice you will be able to write down the characteristic and mantissa directly, doing the second and third steps in Example 8 mentally.

Now we will reverse the problem; that is, given the log of a number, find the number. To find the number, we first write the log of the number in the form $m + c$, where m (the mantissa) is a nonnegative number between 0 and 1, and c (the characteristic) is an integer; then reverse the process illustrated in Example 8.

EXAMPLE 9 (*A*)

$$\log x = 2.5224$$
$$= 0.5224 + 2$$
$$x = 3.33 \times 10^2 \quad \text{or} \quad 333 \qquad \text{From table in Fig. 5}$$

$(B) \qquad \log x = 0.5172 - 4$

$x = 3.29 \times 10^{-4} \quad \text{or} \quad 0.000329$

$(C) \qquad \log x = -4.4685$

$= 0.5315 - 5$

$x = 3.4 \times 10^{-5} \quad \text{or} \quad 0.000034$

Add and subtract 5 to obtain a positive mantissa. NOTE: $-4.4685 \neq .4685 - 4$

PROBLEM 9 Use Fig. 5 to find x if

(A) $\log x = 5.5378$ (B) $\log x = 0.5289 - 3$ (C) $\log x = -2.4921$

ANSWER (A) 3.45×10^5 (B) 3.38×10^{-3} (C) 3.22×10^{-3}

LINEAR INTERPOLATION

Any printed table is necessarily limited to a finite number of entries. The logarithmic function

$$y = \log x \leq 1 \quad x < 10$$

is defined for an infinite number of values—all real numbers between 1 and 10. What, then, do we do about values of x not in a table? How do we find, for example, log 3.276? If a certain amount of accuracy can be sacrificed, we can round 3.276 to the closest entry in the table and proceed as before. Thus,

$$\log 3.276 \approx \log 3.28 = 0.5159$$

We can do better than this, however, without too much additional work, by using a process called *linear interpolation* (see Fig. 6).

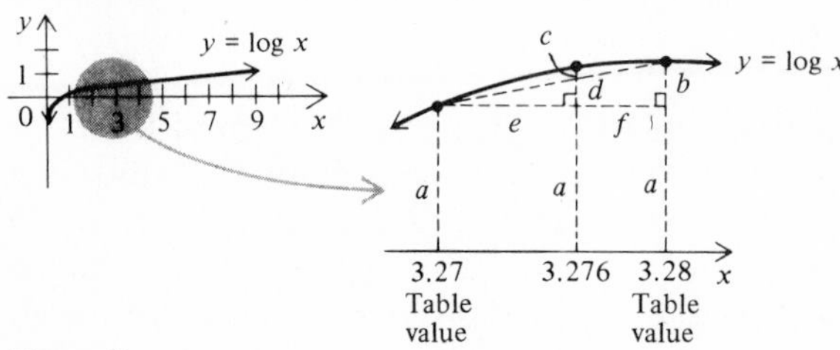

Figure 6

In Fig. 6, we would like to find $a + d + c$. Using proportional parts of similar triangles, we will be able to find d without difficulty, since a and $a + b$ can be determined from the table. Hence, we will settle for $a + d$ as an approximation for $a + d + c$. This approximation is better than what appears in the figure, since the curve is distorted in the drawing for clarity. The logarithmic curve is actually much flatter.

Now to the process of linear interpolation. We find d using the proportion

$$\frac{d}{b} = \frac{e}{e+f}$$

We organize the work as follows for convenient computation.

$$\begin{array}{ccc} & x & \log x \\ 0.010\left\{\begin{array}{l} \phantom{0.006\{}3.280 \\ 0.006\left\{\begin{array}{l}3.276\\3.270\end{array}\right.\end{array}\right. & & \left.\begin{array}{r}0.5159d} \\ \left.\begin{array}{r}n\\0.5145\end{array}\right\}d\end{array}\right\}0.0014 \end{array}$$

Here $b = 0.0014$, $e = 0.006$, and $e + f = 0.010$, thus

$$\frac{d}{0.0014} = \frac{0.006}{0.010}$$

$$d = 0.0008$$

Hence,

$$\log 3.276 = 0.5145 + 0.0008 = 0.5153$$

In practice, the linear interpolation process is carried out by means of a few key operational steps (often done mentally), as indicated in the next example. Notice how decimal points have been dropped to simplify the arithmetic. This is convenient, and no harm is done as long as the decimal points are properly reintroduced at the end of the calculation. If in doubt, proceed with all the decimal points, as in the example above.

EXAMPLE 10 Use linear interpolation to find log 3,514.

SOLUTION $\log 3{,}514 = \log (3.514 \times 10^3) = \log 3.514 + 3$

$$\begin{array}{ccc} & x & \log x \\ 10\left\{\begin{array}{l} \phantom{4\{}3.520 \\ 4\left\{\begin{array}{l}3.514\\3.510\end{array}\right.\end{array}\right. & & \left.\begin{array}{r}0.5465d} \\ \left.\begin{array}{r}n\\0.5453\end{array}\right\}d\end{array}\right\}12 \end{array}$$

$$\frac{4}{10} = \frac{d}{12}$$

$$d = 5$$

Thus,

$$\log 3.514 = 0.5453 + 0.0005 = 0.5458$$

and

$$\log 3{,}514 = 0.5458 + 3 = 3.5458$$

PROBLEM 10 Use linear interpolation to find log 326.6.

ANSWER 2.5140

The linear interpolation process just described is also used to find x, given $\log x$. The following example illustrates the procedure.

EXAMPLE 11 Use linear interpolation to find x, given $\log x = 2.5333$.

SOLUTION

$$\log x = 2.5333$$
$$= 0.5333 + 2$$

$$\begin{array}{cc} x & \log x \\ 10\left\{\begin{array}{l} \quad 3.420 \\ d\left\{\begin{array}{l} \; n \\ 3.410 \end{array}\right. \end{array}\right. & \left.\begin{array}{l} 0.5340 \\ \left.\begin{array}{l} 0.5333 \\ 0.5328 \end{array}\right\}5 \end{array}\right\}12 \end{array}$$

$$\frac{d}{10} = \frac{5}{12}$$

$$d = 4$$

$$n = 3.410 + 0.004 = 3.414$$

Thus,

$$x = 3.414 \times 10^2 \qquad \text{or} \qquad 341.4$$

PROBLEM 11 Use linear interpolation to find x, given $\log x = 7.5230$.

ANSWER 3.335×10^7

Exercise 65

A *Use Table 3 to find logarithm.*

1. log 2.35
2. log 7.82
3. log 5.03
4. log 9.39
5. log 74
6. log 123
7. log 3,100,000
8. log 48,700
9. log 0.00636
10. log 0.0398
11. log 0.000049
12. log 0.72

Use Table 3 to find x, given log x as indicated.

13. $\log x = 0.7226$
14. $\log x = 0.9289$
15. $\log x = 0.9713$
16. $\log x = 0.4594$
17. $\log x = 4.9196$
18. $\log x = 2.8331$
19. $\log x = 0.6096 - 1$
20. $\log x = 0.6998 - 3$
21. $\log x = 1.4440$

22. $\log x = 1.3160$ **23.** $\log x = 0.9609 - 4$ **24.** $\log x = 0.0086 - 2$

25. $\log x = 0.6749 - 3$ **26.** $\log x = 3.9523 - 5$ **27.** $\log x = -0.1387$

28. $\log x = -0.3696$ **29.** $\log x = -3.1675$ **30.** $\log x = -2.2958$

B *Use linear interpolation to find each logarithm.*

31. $\log 2.317$ **32.** $\log 5.143$ **33.** $\log 703{,}400$

34. $\log 28{,}430$ **35.** $\log 65.03$ **36.** $\log 20.35$

37. $\log 0.004006$ **38.** $\log 0.03713$ **39.** $\log 0.9008$

40. $\log 0.6413$ **41.** $\log 692{,}300$ **42.** $\log 84{,}660$

Use linear interpolation to find x.

43. $\log x = 0.7163$ **44.** $\log x = 0.4085$ **45.** $\log x = 5.5458$

46. $\log x = 2.4735$ **47.** $\log x = 3.4303$ **48.** $\log x = 1.9141$

49. $\log x = 0.6038 - 3$ **50.** $\log x = 0.2177 - 1$ **51.** $\log x = 0.8392 - 1$

52. $\log x = 0.8509 - 4$ **53.** $\log x = -0.8315$ **54.** $\log x = -2.6651$

Find each logarithm.

55. $\log(42.2)(0.0038)$ **56.** $\log(352)(0.0218)$ **57.** $\log(3870/0.05)$

58. $\log(34{,}000/45.2)$ **59.** $\log 2^{20}$ **60.** $\log(39.8)^5$

61. $\log(0.627)^7$ **62.** $\log(0.00243)^3$ **63.** $\log \sqrt[5]{0.000053}$

64. $\log \sqrt[3]{(431)^2}$

C **65.** Find y if $y = 10^{1.6138}$.

66. Find y if $y = 10^{3.8627}$.

Evaluate by finding log x, and then x. Use linear interpolation where necessary.

67. $x = \dfrac{78.12}{0.00386}$ **68.** $x = (3.718)(43.8)$

69. $(0.004153)^3$ **70.** $\sqrt[3]{43.67}$

11.5 Computation with Logarithms†

The ability to find common logarithms, together with our knowledge of the properties of logarithms discussed earlier, provides us with a new

†Optional. Even though small hand calculators are rapidly replacing logarithms as a computational tool, the logarithmic functions, along with their properties, are still very important. Logarithmic forms appear in many formulas and logarithmic properties are indispensable to the solution of some types of equations.

computational tool. Several detailed examples will illustrate how logarithms are used in this regard.

EXAMPLE 12 Find the product (4,325)(32.94) using logarithms.

SOLUTION Let

$$N = (4{,}325)(32.94)$$

then

$$\log N = \log\,(4{,}325)(32.94)$$
$$= \log 4{,}325 + \log 32.94$$

The last member of this equation tells us how to proceed. We arrange our work as follows for ease and accuracy of computation:

$$\begin{aligned} \log 4{,}325 &= 3.6360 \\ \log 32.94 &= \underline{1.5177} \\ & \ 5.1537 \end{aligned}$$

$$\log N = 5.1537$$
$$= 0.1537 + 5$$
$$N = 1.427 \times 10^5 \quad \text{or} \quad 142{,}700$$ to four significant figures using linear interpolation

NOTE: By direct multiplication we obtain 142,465.5, which when rounded to four significant figures is 142,500. The use of logarithms produced an error of two units in the fourth significant figure.

PROBLEM 12 Find the product (873.2)(21,030) using logarithms. Also compute the product directly and compare results.

ANSWER By logarithms: 1.836×10^7; by direct computation: 18,363,396

EXAMPLE 13 Find the value of $\sqrt[3]{263.8}/(4.37)^4$ using logarithms.

SOLUTION Let

$$N = \frac{(263.8)^{1/3}}{(4.37)^4}$$

then

$$\log N = \log\,(263.8)^{1/3} - \log\,(4.37)^4 = \tfrac{1}{3}\log 263.8 - 4\log 4.37$$

$$\tfrac{1}{3}\log 263.8 = \tfrac{1}{3}(2.4213) = 0.8071$$

$$4\log 4.37 = 4(0.6405) = 2.5620$$

Before subtracting, we note that to be able to find N we must have log N in the form $m + c$, where m is a mantissa (a positive number between 0 and 1), and c is a characteristic (an integer). To obtain the desired $m + c$ form directly, we simply add, and subtract 2 from 0.8071 before we subtract 2.5620. Thus,

$$\begin{array}{rl} & 2.8071 - 2 \\ & -2.5620 \\ \hline \log N = & 0.2451 - 2 \end{array}$$

$$N = 1.758 \times 10^{-2} \quad \text{or} \quad 0.01758$$

PROBLEM 13 Find $\sqrt[4]{88.18}/(2.09)^6$ using logarithms.

ANSWER 3.678×10^{-2}

EXAMPLE 14 Find: $\sqrt[5]{-0.00627}$

SOLUTION Since log x is defined only for positive x, we first determine the sign of the answer, and then proceed with positive quantities only. In this case, the answer will be negative. Let

$$N = (0.00627)^{1/5}$$

then

$$\log N = \tfrac{1}{5} \log (6.27 \times 10^{-3})$$
$$= \tfrac{1}{5}(0.7973 - 3)$$

To keep the arithmetic as short and as simple as possible, we add and subtract 2 to the second factor, so that the integral part of that factor will be exactly divisible by 5. Thus,

$$\log N = \tfrac{1}{5}(2.7973 - 5)$$
$$= 0.5595 - 1$$
$$N = 3.627 \times 10^{-1} \quad \text{or} \quad 0.3627$$

Hence,

$$\sqrt[5]{-0.00627} = -0.3627$$

PROBLEM 14 Find $\sqrt[7]{-0.000418}$ using logarithms.

ANSWER -0.3291

Logarithms were first conceived by the Scotsman John Napier (1550–1617) as a computational device to ease the backbreaking calculations required in astronomy studies. The new computational tool was immediately welcomed by the mathematical and scientific world. Now, with the increased availability of electronic calculating devices, logarithms have lost some of their importance as a computational tool. However, the logarithmic concept has been greatly generalized since its first introduction, and logarithmic functions are of great importance in both pure and applied mathematics.

Exercise 66

Compute, using common logarithms.

A

1. $(92.4)(837)$

2. $(4.53)(3{,}620)$

3. $\dfrac{452}{32.7}$

4. $\dfrac{65.4}{8.03}$

5. $(5{,}172)(32.6)$

6. $(6{,}632{,}000)(0.0198)$

7. $\dfrac{43.2}{6{,}405}$

8. $\dfrac{0.813}{25.13}$

9. $(0.064)(382.1)(4.07)$

10. $(45)(3{,}680)(0.5173)$

11. $(5.138)^5$

12. $(32.72)^4$

B

13. $(0.0304)^7$

14. $(0.143)^6$

15. $\sqrt[4]{408.2}$

16. $\sqrt[3]{32.63}$

17. $\sqrt[6]{0.05903}$

18. $\sqrt[7]{0.0004177}$

19. $\dfrac{(0.0517)^3(0.253)}{0.00183}$

20. $\dfrac{(238.1)(0.704)^2}{4.82}$

21. $\dfrac{(0.03039)^3}{(0.00792)(-444)}$

22. $\dfrac{(6{,}114{,}000)^2}{(-738{,}000)(0.0395)}$

23. $\dfrac{(4.63)^3}{\sqrt{0.431}}$

24. $\dfrac{\sqrt[3]{62.5}}{(0.038)^4}$

25. $\sqrt[3]{\dfrac{42.52}{(0.047)(0.964)}}$

26. $\sqrt{\dfrac{(212)(13.72)}{(0.019)}}$

27. $\dfrac{(-0.0732)^{1/5}}{(6430)^2(0.003214)}$

28. $\dfrac{(-547)\sqrt[3]{0.000365}}{(-3691)}$

29. $\dfrac{(538{,}000)^3(0.002374)^2}{\sqrt[5]{0.1083}}$

30. $\dfrac{(87.64)^{1/3}(-0.0008240)}{(7.923)(0.6273)}$

31. $18^{-1.27}$

32. $(4.63)^{2.08}$

33. $8 - \sqrt[5]{0.00348}$

34. $5 + \sqrt[3]{(7.320)^2}$

35. The time t it takes a simple pendulum L feet long to complete one oscillation (through a small arc) is given by the formula

$$t = 2\pi\sqrt{\frac{L}{g}}$$

Find t for a pendulum 5 feet long. (Use $\pi = 3.14$ and $g = 32.2$.)

36. Find the radius of a sphere whose volume is 25 cubic feet. [$V = \frac{4}{3}\pi r^3$; use $\pi = 3.142$.]

37. Suppose a distant ancestor of yours at the time of the birth of Christ deposited $1 in a bank that paid 3 per cent interest compounded annually. How much could you have collected in 1970 if you had been the designated heir? (Now you should understand why banks do not allow deposits to remain in an inactive account indefinitely.) [$A = A_0(1 + r)^n$]

38. If when you were born your parents had deposited $2,000 in a bank for your college education, how much would you have at age 17 if the bank paid 5 percent interest compounded annually? [$A = A_0(1 + r)^n$.]

C **39.** $\sqrt[5]{\sqrt[4]{13.2} - \sqrt[3]{435}} = ?$

40. $(3.08)^{2.32}(42.5)^{0.03} + 4 \log 33.04 = ?$

41. In a study on learning, the psychologist Thurstone found that the time t required for learning a list of symbols varied directly as the $\frac{3}{2}$ power of the number n of items on the list. (*A*) Write the equation of variation. (*B*) If it takes a person 2 hours to learn a list of 25 symbols, approximately how many symbols should the student be able to learn in 4 hours? (HINT: Obtain $\frac{2}{25^{3/2}} = \frac{4}{n^{3/2}}$, and solve for n.)

42. The German astronomer J. Kepler (1571–1630) discovered that the time t it takes a planet to travel once around the sun varies directly as the $\frac{3}{2}$ power of its mean distance d from the sun. Using 93 million miles as the mean distance between the earth and the sun, and 141.5 million miles as the mean distance between Mars and the sun, compute the number of days it takes Mars to complete one revolution about the sun.

11.6 Exponential and Logarithmic Equations

Often, when dealing with exponential functions, one must solve exponential equations in which the variable appears in the exponent. Logarithms play an important role in the solution of such equations.

EXAMPLE 15 Solve $2^{3x-2} = 5$ for x.

SOLUTION

$$2^{3x-2} = 5$$

$$\log 2^{3x-2} = \log 5$$

$$(3x - 2) \log 2 = \log 5$$

$$(3x - 2) = \frac{\log 5}{\log 2}$$

$$x = \frac{1}{3}\left(2 + \frac{\log 5}{\log 2}\right)$$

$$= \frac{1}{3}\left(2 + \frac{0.6990}{0.3010}\right)$$

$$= 1.441$$

NOTE: $\dfrac{\log A}{\log B} \neq \log A - \log B$

PROBLEM 15 Solve $35^{1-2x} = 7$ for x.

ANSWER 0.2263

The next example illustrates an approach to solving some types of logarithmic equations.

EXAMPLE 16 Solve $\log (x + 3) + \log x = 1$

SOLUTION

$$\log(x + 3) + \log x = 1$$

$$\log[x(x + 3)] = 1$$

$$x(x + 3) = 10^1$$

$$x^2 + 3x - 10 = 0$$

$$(x + 5)(x - 2) = 0$$

$$x = -5, 2$$

-5 is not a solution (why?). Thus 2 is the only solution.

PROBLEM 16 Solve: $\log (x - 15) = 2 - \log x$

ANSWER $x = 20$

Exercise 67

A *Solve to two significant figures.*

1. $23 = 3^x$

2. $5 = 2^x$

3. $10^{5x-2} = 348$

4. $10^{3x} = 92$

5. $3^{x-2} = 78.4$

6. $2^{2x-3} = 435$

7. $0.074 = 3^{-x}$

8. $0.238 = 2^{-x}$

9. $\log x - \log 8 = 1$

10. $\log 5 + \log x = 2$

B *Solve to two significant figures (use $e = 2.718$ and $\log e = 0.4343$).*

11. $3 = (1.06)^x$

12. $2 = (1.05)^x$

13. $12 = e^{-x}$

14. $42.1 = e^x$

15. $438 = 100e^{0.25x}$

16. $123 = 500e^{-0.12x}$

17. $3 = (1 + x)^{18}$

18. $2 = (1 + x)^{12}$

19. $\log (x - 9) + \log 100x = 3$

20. $\log x + \log (x - 3) = 1$

21. $e^{\log_e 4x} = 16$

22. $\log_e e^{3x} = 12$

23. Radioactive strontium 90 is used in nuclear reactors and decays according to the equation

$$A = Pe^{-0.0248t}$$

where P is the amount present at $t = 0$ and A is the amount remaining after t years. If 500 milligrams of strontium 90 are placed in a nuclear reactor, how much will be left after 10 years?

24. If 500 milligrams of strontium 90 are placed in a nuclear reactor, how much will be left after 100 years?

25. Find the half-life (to the nearest year) of strontium 90; that is, find t so that $A = P/2$.

26. How long (to the nearest year) will it take for strontium 90 to decompose so that only 1 percent of the original amount remains. HINT: Find t so that $A = 0.01P$.

27. If you start with 2 cents and double it each day, at the end of n days you would have 2^n cents. How many dollars would you have at the end of 31 days?

C **28.** Show that $\log_b N = \dfrac{\log_a N}{\log_a b}$

Find each to four significant figures (use $e = 2.718$ and $\log e = 0.4343$). (Use Problem 28 with $a = 10$.)

29. $\log_{12} 42.8$

30. $\log_2 12$

31. $\log_e 235$

32. $\log_e 10$

33. Show that $(\log_e 10)(\log_{10} e) = 1$.

34. Show that $(\log_a b)(\log_b a) = 1$ for $a, b > 0$; $a, b \neq 1$.

35. Earlier we mentioned that e could be approximated to any decimal accuracy desired by taking the integer n sufficiently large in $(1 + 1/n)^n$. Use common logs to approximate e for $n = 10$ and $n = 100$.

11.7 Additional Applications

Exercise 68

Logarithmic and exponential functions are widely used in both applied and pure mathematics. The following additional applications illustrate the use of these functions in the social, natural, and physical sciences. The most difficult problems are double-starred (★★), those of moderate difficulty single-starred (★), and the easier problems are not marked.

ARCHAEOLOGY CARBON 14 DATING

Cosmic-ray bombardment of the atmosphere produces neutrons, which in turn react with nitrogen to produce radioactive carbon 14. In 1946, Willard Libby, the Nobel prize-winning chemist, reasoned that carbon 14 entered all living tissues through the carbon dioxide which is first absorbed by plants. As long as a plant or animal is alive, he found, carbon 14 is maintained at a constant level in its tissues. Once dead, however, it ceases taking in carbon and, to the slow beat of time, the carbon 14 diminishes by radioactive decay at a rate determined by its half-life of 5,600 years. Thus, a piece of old bone or a piece of charcoal from an ancient campfire can be dated by measuring the amount of carbon 14 left. The method is reasonably accurate up to about 40 or 50 thousand years, covering ancient civilization back to Cro-Magnon man.

Mathematically, if at a given time there is A_0 milligrams of carbon 14 in a nonliving substance, at the end of t years there will be

$$A = A_0\left(\frac{1}{2}\right)^{t/5{,}600} \qquad \text{milligrams}$$

of carbon 14 in that substance.

1. Graph the exponential equation above for $t = 0$, 5,600, 2(5,600), 3(5,600), 4(5,600), and 5(5,600).

★ **2.** The remains of an ancient campfire were discovered in a cave near the site of the discovery of Cro-Magnon man in France. Charcoal from the campfire was tested for its carbon-14 content. Estimate the age of the campfire if it was found that 10 percent of the original amount of carbon 14 was still present.

ASTRONOMY
BRIGHTNESS
OF STARS

Ever since the time of the Greek astronomer Hipparchus (second century B.C.), the brightness of stars has been measured in terms of *magnitude*. The brightest stars (excluding the sun) are classed as magnitude 1, and the dimmest visible by the eye are classed as magnitude 6. In 1856, the English astronomer N. R. Pegson showed that first-magnitude stars are 100 times brighter than sixth-magnitude stars. And he concluded that the ratio of brightness between any two consecutive magnitudes is $0.01^{1/5}$.

3. Find the ratio of brightness between two consecutive magnitudes, $0.01^{1/5}$, to three decimal places.

4. An optical instrument is required to observe stars beyond the sixth magnitude, the limit of ordinary vision. However, even optical instruments have their limitations. The limiting magnitude LM of any optical telescope with lens diameter D is given by the formula

$$LM = 8.8 + 51 \log_{10} D$$

Find the limiting magnitude of (A) a homemade 6-inch reflecting telescope; (B) the 200-inch Mount Palomar telescope in California.

BIOLOGY
BACTERIA
GROWTH

⋆ **5.** At a given temperature, the number of bacteria in milk double every 3 hours. If we start with A_0 bacteria in a batch of milk, after t hours there will be $A = A_0 2^{t/3}$ bacteria in the milk. Graph this exponential function for $t = 0, 3, 6, 9, 12$, and 15; then join these points with a smooth curve.

6. A single cholera bacterium divides every $\frac{1}{2}$ hour to produce two complete cholera bacteria. If we start with a colony of A_0 bacteria, in t hours (assuming adequate food supply) we will have $A = A_0 2^{2t}$ bacteria. Find A for $A_0 = 5{,}000$ and $t = 12.6$.

ECOLOGY
HUMAN
POPULATION
GROWTH

If a population of A_0 people grows at the constant rate of r percent per year, in t years there will be $A = A_0(1 + r)^t$ people in the population. Use this formula for the following two problems.

⋆ **7.** It is estimated that the population of the world is increasing at the rate of 2 percent per year. At this rate how long will it take the world population to double? To triple?

⋆⋆ **8.** If the world population continues to grow at its present growth rate of 2 percent a year, approximately how many square yards of land will there be for each person 565 years from now? (The earth has approximately 1.7×10^{14} square yards of land, and the present population is approximately 4×10^9.)

BIOLOGY
ANCESTORS
AND
HEREDITY

⋆ **9.** Galton's law of heredity states that the influence of ancestors on an individual is as follows: each parent, $\frac{1}{2}$; each grandparent, $(\frac{1}{2})^2$; each great-grandparent, $(\frac{1}{2})^3$; and so on. How much hereditary influence would an ancestor 20 generations back (approximately 600 years) have had on you?

⋆⋆ **10.** For each generation back, the number of our direct ancestors (parents, grandparents, and so forth) doubles. If we figure 30 years to a generation, then how many ancestors did you have 600 years ago? 1,980 years ago? (Note that at the time of Christ there were an estimated 3.5×10^8 people on the earth.)

BUSINESS

11. If a certain amount of money P (called the principal) is invested at r percent interest compounded annually, the amount of money A after t years is given by the equation

$$A = P(1 + r)^t$$

How much will $1,000 be worth in 5 years at 6 percent interest compounded annually?

★12. How long (to the nearest year) will it take for a sum of money to double if invested at 5 percent interest compounded annually?

★13. At what rate of interest (compounded annually) should a sum of money be invested if it is to double in 10 years?

★14. Monthly payments p on a loan of L dollars for n months at i percent interest per month on the unpaid balance are determined by the formula

$$p = \frac{Li}{1 - (1 + i)^{-n}}$$

What are the monthly payments on a 12-month loan of $500 at 1 percent interest per month on the unpaid balance?

EARTH SCIENCE

15. For relatively clear bodies of fresh water or saltwater, light intensity is reduced according to the exponential function

$$I = I_0 e^{-kd}$$

where I is the intensity at d feet below the surface, and I_0 is the intensity at the surface; k is called the coefficient of extinction. Until recently two of the clearest bodies of water in the world were the fresh-water Crystal Lake in Wisconsin ($k = 0.0485$) and the saltwater Sargasso Sea off the West Indies ($k = 0.00942$). Find the depths (to the nearest foot) in these two bodies of water at which the light was reduced to 1 percent of that at the surface.

16. An approximation of atmospheric pressure in pounds per square inch may be calculated from the formula

$$P = 14.7e^{-0.21h}$$

where h is altitude above sea level in miles. (*A*) What is the pressure 2 miles above sea level? (*B*) What is the pressure at the bottom of a mine shaft $\frac{1}{3}$ mile below sea level? (Use Table 2 in the Appendix.)

GEOMETRY

17. Find the radius of a sphere with surface area 160 square inches. ($A = 4\pi r^2$; use $\pi = 3.142$.)

18. It is possible to show that for a triangle with sides a, b, and c, the area is given by

$$A = \sqrt{s(s-a)(s-b)(s-c)}$$

where s is the semiperimeter $\frac{1}{2}(a + b + c)$. Find the area of a triangle with sides 16, 28.3, and 23.5 feet.

PHYSICS
SOUND

Because of the extraordinary range of sensitivity of the human ear (a range of over 1,000 million million to 1), it is helpful to use a logarithmic scale to measure sound intensity over this range rather than an absolute scale. The unit of measure is called the *decibel*, after the inventor of the telephone, Alexander Graham Bell. If we let N be the number of decibels, I the power of the sound in question in watts per cubic centimeter, and I_0 the power of sound just below the threshold of hearing (approximately 10^{-16} watt per square centimeter),

$$I = I_0 10^{N/10}$$

19. Starting with the decibel formula above, show that $N = 10 \log_{10}(I/I_0)$.

20. Use the formula in Problem 19 (with $I_0 = 10^{-16}$ watt per square centimeter) to find the decibel ratings of the following sounds:
(*A*) Whisper (10^{-13} watt per square centimeter)
(*B*) Normal conversation (3.16×10^{-10} watt per square centimeter)
(*C*) Heavy traffic (10^{-8} watt per square centimeter)
(*D*) Jet plane with afterburner (10^{-1} watt per square centimeter)

PSYCHOLOGY
LEARNING

In learning a particular task, such as typing or swimming, one progresses faster at the beginning and then levels off. If you plot the level of performance against time, you will obtain a curve of the type shown in Fig. 7. This is called a learning curve and can be very closely approximated by an exponential equation of the form $y = a(1 - e^{-cx})$, where a and c are positive constants. Curves of this type have applications in psychology, education, and industry.

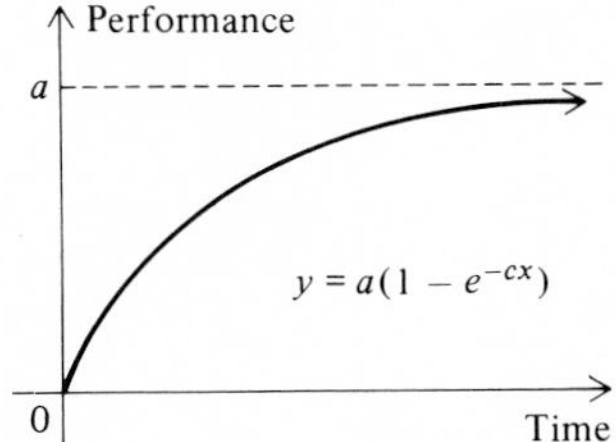

Figure 7

21. A particular person's history of learning to type is given by the exponential equation $N = 80(1 - e^{-0.08n})$, where N is the number of words per minute this person was able to type after n weeks of instruction. Graph this equation for $n = 0, 5, 10, 15, 20, 25, 30$ (using Table 2 in the Appendix), and join the points with a smooth curve.

22. Approximately how many weeks did it take the person in Problem 21 to learn to type 60 words per minute?

PSYCHOLOGY
SENSATION

Professor S. S. Stevens of Harvard University discovered that sensation magnitude y grows as a power function of the stimulus magnitude x, in terms of the formula $y = kx^p$. A law of this form seems to govern our reactions to light, sound, smell, taste, vibration, shock, warmth, and cold.

23. For brightness of light, $y = kx^{0.33}$, where y is the sensed brightness of a light of x candlepower. Approximately what increase in candlepower will double the sensed brightness of a light?

24. For electric shocks, $y = kx^{3.5}$, where y is the sensed magnitude of the shock produced by a current of x amperes. What increase in current will double the sensed magnitude of an electric shock?

PUZZLES

25. If a very thin piece of paper 0.001 inches thick is torn in half, and each half is again torn in half, and this process is repeated for a total of 50 times, how high will the paper be if the pieces are stacked one on top of the other? Give the answer to two significant figures in miles (1 mile = 63,360 inches).

26. Find the fallacy.

$$3 > 2$$
$$3 \log \tfrac{1}{2} > 2 \log \tfrac{1}{2}$$
$$\log \left(\tfrac{1}{2}\right)^3 > \log \left(\tfrac{1}{2}\right)^2$$
$$\left(\tfrac{1}{2}\right)^3 > \left(\tfrac{1}{2}\right)^2$$
$$\tfrac{1}{8} > \tfrac{1}{4}$$

Exercise 69 Chapter Review

A

1. Write $\log_{10} x = y$ in exponential form.

2. Write $m = 10^n$ in logarithmic form with base 10.

Solve for x without the use of a table.

3. $\log_2 x = 2$ **4.** $\log_x 25 = 2$ **5.** $\log_2 8 = x$

Use common logarithms to evaluate each problem.

6. $(27{,}300)(0.00418)$ **7.** $\dfrac{5.987}{0.7904}$ **8.** $\dfrac{(52.5)(1.33)}{0.079}$

Solve for x to two significant figures.

9. $10^x = 17.5$ **10.** $\log 4 + \log x = 2$

B

11. For $f: y = \log_2 x$, graph f and f^{-1} using the same coordinate axes. What are the domains and range of f and f^{-1}?

12. Write $\log_e y = x$ in exponential form.

13. Write $y = e^x$ in logarithmic form with base e.

Solve for x without the use of a table.

14. $\log_{1/4} 16 = x$ **15.** $\log_x 9 = -2$ **16.** $\log_{16} x = \frac{3}{2}$

17. $\log_x e^5 = 5$ **18.** $10^{\log_{10} x} = 33$

Use common logarithms to evaluate each problem.

19. $\dfrac{(0.318)^3}{(0.00457)(328)}$ **20.** $\sqrt[7]{0.00004803}$

21. $\sqrt[5]{\dfrac{(0.315)^3}{0.075}}$ **22.** $(6.07)^{1.35}$

23. $\sqrt[3]{(752)^2} - 40.32$ **24.** If $\log x = -2.6073$, find x.

Solve for x to two significant figures (e = 2.718).

25. $25 = 5(2)^x$ **26.** $0.01 = e^{-0.05x}$

27. $\log_e x = 0$ **28.** $\log (3x^2) - \log 9x = 1$

29. $\log_e e^{x^2} = 4$ **30.** $\log_5 23 = x$

31. Many countries in the world have a population growth rate of 3 percent (or more) per year. At this rate how long, to the nearest year, will it take a population to double? $[P = P_0(1.03)^t]$

C **32.** Explain why 1 cannot be used as a logarithmic base.

33. Prove that $\log_b (M/N) = \log_b M - \log_b N$.

34. Write $\log_e y = -5t + \log_e c$ in exponential form free of logarithms.

35. $\log_e 100 = ?$ (NOTE: $e \approx 2.718$)

36. Radioactive argon 39 has a half-life of 4 minutes. After t minutes the amount A of argon 39 left after starting with A_0 is given by the formula

$$A = A_0\left(\tfrac{1}{2}\right)^{t/4}$$

Graph this exponential function for $t = 0, 4, 8, 12, 16$, and 20, and then join these points with a smooth curve.

37. How long, to three significant figures, will it take for the carbon 14 to diminish to 1 percent of the original amount after the death of a plant or animal? $[A = A_0\left(\tfrac{1}{2}\right)^{t/5{,}600}]$

Chapter 12

SEQUENCES AND SERIES

In this chapter we are going to consider functions whose domains are the set of natural numbers or particular subsets of the natural numbers. These special functions, called sequences, are encountered with increased frequency as one progresses in mathematics.

12.1 Sequences and Series

An *infinite sequence* is a function a whose domain is the set of all natural numbers $N = \{1, 2, 3, \ldots, n, \ldots\}$. The range of the function is $a(1)$, $a(2)$, $a(3)$, . . . , $a(n)$, . . . , which is usually written

$$a_1, a_2, a_3, \ldots, a_n, \ldots$$

where

$$a_n = a(n)$$

The elements in the range are called the *terms of the sequence*; a_1 is the first term, a_2 the second term, and a_n the nth term. For example, if

$$a_n = \frac{n-1}{n} \qquad n \in N$$

the function a is a sequence with terms

$$0, \frac{1}{2}, \frac{2}{3}, \frac{3}{4}, \ldots, \frac{n-1}{n}, \ldots$$

If the domain of a function a is the set of positive integers $\{1, 2, 3, \ldots, n\}$ for some fixed n, then a is called a *finite sequence*. Thus, if

$$a_1 = 5$$

and

$$a_n = a_{n-1} + 2 \qquad n \in \{2, 3, 4\}$$

the function a is a finite sequence with terms

5, 7, 9, 11

The two examples above illustrate two common ways in which sequences are specified:

1. The nth term a_n is expressed by means of n.
2. One or more terms are given, and the nth term is expressed by means of preceding terms.

EXAMPLE 1 Find the first four terms of a sequence whose nth term is

$$a_n = \frac{1}{2^n}$$

SOLUTION $a_1 = \frac{1}{2^1} = \frac{1}{2} \qquad a_2 = \frac{1}{2^2} = \frac{1}{4} \qquad a_3 = \frac{1}{2^3} = \frac{1}{8} \qquad a_4 = \frac{1}{2^4} = \frac{1}{16}$

PROBLEM 1 Find the first four terms of a sequence whose nth term is

$$a_n = \frac{n}{n^2+1}$$

ANSWER $\frac{1}{2}, \frac{2}{5}, \frac{3}{10}, \frac{4}{17}$

EXAMPLE 2 Find the first five terms of a sequence specified by

$$a_1 = 5$$

$$a_n = \tfrac{1}{2}a_{n-1} \qquad n \geq 2$$

SOLUTION

$$a_1 = 5$$
$$a_2 = \tfrac{1}{2}a_{2-1} = \tfrac{1}{2}a_1 = \tfrac{5}{2}$$
$$a_3 = \tfrac{1}{2}a_{3-1} = \tfrac{1}{2}a_2 = \tfrac{5}{4}$$
$$a_4 = \tfrac{1}{2}a_{4-1} = \tfrac{1}{2}a_3 = \tfrac{5}{8}$$
$$a_5 = \tfrac{1}{2}a_{5-1} = \tfrac{1}{2}a_4 = \tfrac{5}{16}$$

PROBLEM 2 Find the first five terms of a sequence specified by

$$a_1 = 3$$
$$a_n = a_{n-1} + 4 \qquad n \geq 2$$

ANSWER 3, 7, 11, 15, 19

Now let us look at the problem in reverse. That is, given the first few terms of a sequence (and assuming the sequence continues in the indicated pattern), find a_n in terms of n.

EXAMPLE 3 Find a_n in terms of n for the given sequences.

(*A*) 5, 6, 7, 8, . . .

SOLUTION $a_n = n + 4$

(*B*) 2, −4, 8, −16, . . .

SOLUTION $a_n = (-1)^{n+1}2^n$ Note how $(-1)^{n+1}$ functions as a sign alternator.

NOTE: These representations are not unique.

PROBLEM 3 Find a_n in terms of n for

(*A*) 3, 5, 7, 9, . . . (*B*) $1, -\tfrac{1}{2}, \tfrac{1}{4}, -\tfrac{1}{8}, \ldots$

ANSWER (*A*) $a_n = 2n + 1$ (*B*) $a_n = (-1)^{n+1}/2^{n-1}$

The sum of the terms of a sequence is called a *series*. Thus, if a_1, a_2, a_3, . . . , a_n are the terms of a sequence,

$$S_n = a_1 + a_2 + a_3 + \cdots + a_n$$

is called a series. If the sequence is infinite, the corresponding series is called an *infinite series*. We will restrict our attention to finite series in this section.

A series is often represented in a compact form using *sigma notation*, as follows:

$$S_n = \sum_{k=1}^{n} a_k = a_1 + a_2 + a_3 + \cdots + a_n$$

where the terms of the series on the right are obtained from the middle expression by successively replacing k in a_k with integers, starting with 1 and ending with n. Thus, if a sequence is given by

$$\frac{1}{2}, \frac{1}{4}, \frac{1}{8}, \ldots, \frac{1}{2^n}$$

the corresponding series is given by

$$S_n = \frac{1}{2} + \frac{1}{4} + \frac{1}{8} + \cdots + \frac{1}{2^n}$$

or

$$S_n = \sum_{k=1}^{n} \frac{1}{2^k}$$

EXAMPLE 4 Write $S_5 = \sum_{k=1}^{5} \frac{k-1}{k}$ without sigma notation.

SOLUTION

$$S_5 = \sum_{k=1}^{5} \frac{k-1}{k} \qquad \text{Replace } k \text{ in } \frac{k-1}{k} \text{ successively with 1, 2, 3, 4, and 5, then add.}$$

$$= \frac{1-1}{1} + \frac{2-1}{2} + \frac{3-1}{3} + \frac{4-1}{4} + \frac{5-1}{5}$$

$$= 0 + \tfrac{1}{2} + \tfrac{2}{3} + \tfrac{3}{4} + \tfrac{4}{5}$$

PROBLEM 4 Write $S_6 = \sum_{k=1}^{6} \frac{(-1)^{k+1}}{2k-1}$ without sigma notation.

ANSWER $S_6 = 1 - \tfrac{1}{3} + \tfrac{1}{5} - \tfrac{1}{7} + \tfrac{1}{9} - \tfrac{1}{11}$.

EXAMPLE 5 Write the following series using sigma notation.

$$S_6 = 1 - \tfrac{1}{2} + \tfrac{1}{3} - \tfrac{1}{4} + \tfrac{1}{5} - \tfrac{1}{6}$$

SOLUTION We first note that the nth term of the series is given by

$$a_n = (-1)^{n+1} \frac{1}{n}$$

hence,

$$S_6 = \sum_{k=1}^{6} (-1)^{k+1} \frac{1}{k}$$

PROBLEM 5 Write the following series using sigma notation

$$S_5 = 1 - \tfrac{2}{3} + \tfrac{4}{9} - \tfrac{8}{27} + \tfrac{16}{81}$$

ANSWER $S_5 = \sum_{k=1}^{5} \left(-\tfrac{2}{3}\right)^{k-1}$

Exercise 70

A *Write the first four terms for each sequence.*

1. $a_n = n - 2$

2. $a_n = n + 3$

3. $a_n = \dfrac{n-1}{n+1}$

4. $a_n = \left(1 + \dfrac{1}{n}\right)^n$

5. $a_n = (-2)^{n+1}$

6. $a_n = \dfrac{(-1)^{n+1}}{n^2}$

7. Write the 8th term of the sequence in Problem 1.

8. Write the 10th term of the sequence in Problem 2.

9. Write the 100th term of the sequence in Problem 3.

10. Write the 200th term of the sequence in Problem 4.

Write each series in expanded form without sigma notation.

11. $S_5 = \sum_{k=1}^{5} k$

12. $S_4 = \sum_{k=1}^{4} k^2$

13. $S_3 = \sum_{k=1}^{3} \dfrac{1}{10^k}$

14. $S_5 = \sum_{k=1}^{5} \left(\tfrac{1}{3}\right)^k$

15. $S_4 = \sum_{k=1}^{4} (-1)^k$

16. $S_6 = \sum_{k=1}^{6} (-1)^{k+1}k$

B *Write the first five terms of each sequence.*

17. $a_n = (-1)^{n+1}n^2$

18. $a_n = (-1)^{n+1}\dfrac{1}{2^n}$

19. $a_n = \dfrac{1}{3}\left(1 - \dfrac{1}{10^n}\right)$

20. $a_n = n[1 - (-1)^n]$

21. $a_1 = 7;\ a_n = a_{n-1} - 4,\ n \geq 2$

22. $a_1 = a_2 = 1;\ a_n = a_{n-1} + a_{n-2},\ n \geq 3$ (Fibonacci sequence)

23. $a_1 = 4;\ a_n = \tfrac{1}{4}a_{n-1},\ n \geq 2$

24. $a_1 = 2;\ a_n = 2a_{n-1},\ n \geq 2$

Find a_n in terms of n.

25. 4, 5, 6, 7, . . .

26. −2, −1, 0, 1, . . .

27. 3, 6, 9, 12, . . .

28. −2, −4, −6, −8, . . .

29. $\frac{1}{2}, \frac{2}{3}, \frac{3}{4}, \frac{4}{5}, \ldots$

30. $\frac{1}{2}, \frac{3}{4}, \frac{5}{6}, \frac{7}{8}, \ldots$

31. 1, −1, 1, −1, . . .

32. 1, −2, 3, −4, . . .

33. −2, 4, −8, 16, . . .

34. 1, −3, 5, −1, . . .

35. $x, \frac{x^2}{2}, \frac{x^3}{3}, \frac{x^4}{4}, \ldots$

36. $x, -x^3, x^5, -x^7, \ldots$

Write each series in expanded form without sigma notation.

37. $\sum_{k=1}^{4} \frac{(-2)^{k+1}}{k}$

38. $\sum_{k=1}^{5} (-1)^{k+1}(2k-1)^2$

39. $S_3 = \sum_{k=1}^{3} \frac{1}{k} x^{k+1}$

40. $S_5 = \sum_{k=1}^{5} x^{k-1}$

41. $\sum_{k=1}^{5} \frac{(-1)^{k+1}}{k} x^k$

42. $\sum_{k=0}^{4} \frac{(-1)^k x^{2k+1}}{2k+1}$

Write each series using sigma notation.

43. $S_4 = 1^2 + 2^2 + 3^2 + 4^2$

44. $S_5 = 2 + 3 + 4 + 5 + 6$

45. $S_5 = \frac{1}{2} + \frac{1}{2^2} + \frac{1}{2^3} + \frac{1}{2^4} + \frac{1}{2^5}$

46. $S_4 = 1 - \frac{1}{2} + \frac{1}{3} - \frac{1}{4}$

47. $S_n = 1 + \frac{1}{2^2} + \frac{1}{3^2} + \cdots + \frac{1}{n^2}$

48. $S_n = 2 + \frac{3}{2} + \frac{4}{3} + \cdots + \frac{n+1}{n}$

49. $S_n = 1 - 4 + 9 + \cdots + (-1)^{n+1}n^2$

50. $S_n = \frac{1}{2} - \frac{1}{4} + \frac{1}{8} + \cdots + \frac{(-1)^{n+1}}{2n}$

C **51.** Show that $\sum_{k=1}^{n} ca_k = c \sum_{k=1}^{n} a_k$

52. Show that $\sum_{k=1}^{n} (a_k + b_k) = \sum_{k=1}^{n} a_k + \sum_{k=1}^{n} b_k$

The sequence $a_n = \frac{a_{n-1}^2 + N}{2a_{n-1}}$, $n \geq 2$, $N \in R^+$, *can be used to find* $\sqrt{N}$ *to any decimal accuracy desired. To start the sequence, choose* a_1 *arbitrarily from the positive real numbers.*

53. (A) Find the first four terms of the sequence $a_1 = 3$;

$$a_n = \frac{a_{n-1}^2 + 2}{2a_{n-1}}, \qquad n \geq 2$$

(A small hand calculator is useful, but not necessary.)

(*B*) Compare terms with decimal approximation of $\sqrt{2}$ from table.

(*C*) Repeat (*A*) and (*B*) by letting a_1 be any other positive number, say, 1.

54. (*A*) Find the first four terms of the sequence $a_1 = 2$;

$$a_n = \frac{a_{n-1}^2 + 5}{2a_{n-1}}, \qquad n \geq 3$$

(*B*) Find $\sqrt{5}$ in table and compare with (*A*).

(*C*) Repeat (*A*) and (*B*) by letting a_1 be any other positive number say, 3.

12.2 Arithmetic Sequences and Series

Consider the sequence

5, 9, 13, 17, . . .

Can you guess what the fifth term is? If you guessed 21, you have observed that each term after the first can be obtained from the preceding one by adding 4 to it. This is an example of an arithmetic sequence. In general, a sequence

$$a_1, a_2, a_3, \ldots, a_n, \ldots$$

is called an *arithmetic sequence* (or an *arithmetic progression*) if there exists a constant d, called the *common difference*, such that

$$a_n = a_{n-1} + d \qquad \text{for every } n > 1$$

EXAMPLE 6 Which sequence is an arithmetic sequence and what is its common difference?

(*A*) 1, 2, 3, 5, . . . (*B*) 3, 5, 7, 9, . . .

SOLUTION (*B*) is an arithmetic sequence with $d = 2$

PROBLEM 6 Repeat Example 6 with

(*A*) −4, −1, 2, 5, . . . (*B*) 2, 4, 8, 16, . . .

ANSWER (*A*) with $d = 3$

Arithmetic sequences have several convenient properties. For example, it is easy to derive formulas for the nth term in terms of n and the sum of any number of consecutive terms. To obtain an nth-term formula, we note that if a is an arithmetic sequence, then

$$a_2 = a_1 + d$$

$$a_3 = a_2 + d = a_1 + 2d$$

$$a_4 = a_3 + d = a_1 + 3d$$

which suggests

$$a_n = a_1 + (n - 1)d \qquad \text{for every } n > 1$$

EXAMPLE 7 If the 1st and 10th terms of an arithmetic sequence are 3 and 30, respectively, find the 50th term of the sequence.

SOLUTION First find d:

$$a_n = a_1 + (n - 1)d$$

$$a_{10} = a_1 + (10 - 1)d$$

$$30 = 3 + 9d$$

$$d = 3$$

Now find a_{50}:

$$a_{50} = a_1 + (50 - 1)3$$

$$= 3 + 49 \cdot 3$$

$$= 150$$

PROBLEM 7 If the 1st and 15th terms of an arithmetic sequence are -5 and 23, respectively, find the 73d term of the sequence.

ANSWER 139

The sum of the terms of an arithmetic sequence is called an *arithmetic series*. We will derive two simple and very useful formulas for finding the *sum of the first n terms of an arithmetic series*. Let

$$S_n = a_1 + (a_1 + d) + \cdots + [a_1 + (n - 2)d] + [a_1 + (n - 1)d]$$

Reversing the order of the sum, we obtain

$$S_n = [a_1 + (n - 1)d] + [a_1 + (n - 2)d] + \cdots + (a_1 + d) + a_1$$

Adding left members and corresponding elements of the right members of the two equations, we see that

$$2S_n = [2a_1 + (n-1)d] + [2a_1 + (n-1)d] + \cdots + [2a_1 + (n-1)d]$$
$$= n[2a_1 + (n-1)d]$$

or

$$\boxed{S_n = \frac{n}{2}[2a_1 + (n-1)d]}$$

By replacing $a_1 + (n-1)d$ with a_n, we obtain a second useful formula for the sum:

$$\boxed{S_n = \frac{n}{2}(a_1 + a_n)}$$

EXAMPLE 8 Find the sum of the first 26 terms of an arithmetic series if the first term is -7 and $d = 3$.

SOLUTION
$$S_n = \frac{n}{2}[2a_1 + (n-1)d]$$
$$S_{26} = \frac{26}{2}[2(-7) + (26-1)3]$$
$$= 793$$

PROBLEM 8 Find the sum of the first 52 terms of an arithmetic series if the first term is 23 and $d = -2$.

ANSWER $-1{,}456$

EXAMPLE 9 Find the sum of all the odd numbers between 51 and 99, inclusive.

SOLUTION First find n:

$$a_n = a_1 + (n-1)d$$
$$99 = 51 + (n-1)2$$
$$n = 25$$

Now find S_{25}:

$$S_n = \frac{n}{2}(a_1 + a_n)$$
$$S_{25} = \frac{25}{2}(51 + 99)$$
$$= 1{,}875$$

PROBLEM 9 Find the sum of all the even numbers between -22 and 52, inclusive.

ANSWER 570

Exercise 71

A **1.** Determine which of the following are arithmetic sequences. Find d and the next two terms for those that are.

(A) 2, 4, 8, . . . (B) 7, 6.5, 6, . . .

(C) −11, −16, −21, . . . (D) $\frac{1}{2}, \frac{1}{6}, \frac{1}{18}, \ldots$

2. Repeat Problem 1 for

(A) 5, −1, −7, . . . (B) 12, 4, $\frac{4}{3}$, . . .

(C) $\frac{1}{2}, \frac{2}{3}, \frac{3}{4}, \ldots$ (D) 16, 48, 80, . . .

Let $a_1, a_2, a_3, \ldots, a_n, \ldots$ be an arithmetic sequence. In Problems 3 to 18 find the indicated quantities.

3. $a_1 = -5$, $d = 4$, $a_2 = ?$, $a_3 = ?$, $a_4 = ?$

4. $a_1 = -18$, $d = 3$, $a_2 = ?$, $a_3 = ?$, $a_4 = ?$

5. $a_1 = -3$, $d = 5$, $a_{15} = ?$, $S_{11} = ?$

6. $a_1 = 3$, $d = 4$, $a_{22} = ?$, $S_{21} = ?$

7. $a_1 = 1$, $a_2 = 5$, $S_{21} = ?$

8. $a_1 = 5$, $a_2 = 11$, $S_{11} = ?$

B **9.** $a_1 = 7$, $a_2 = 5$, $a_{15} = ?$

10. $a_1 = -3$, $d = -4$, $a_{10} = ?$

11. $a_1 = 3$, $a_{20} = 117$, $d = ?$, $a_{101} = ?$

12. $a_1 = 7$, $a_8 = 28$, $d = ?$, $a_{25} = ?$

13. $a_1 = -12$, $a_{40} = 22$, $S_{40} = ?$

14. $a_1 = 24$, $a_{24} = -28$; $S_{24} = ?$

15. $a_1 = \frac{1}{3}$, $a_2 = \frac{1}{2}$, $a_{11} = ?$, $S_{11} = ?$

16. $a_1 = \frac{1}{6}$, $a_2 = \frac{1}{4}$, $a_{19} = ?$, $S_{19} = ?$

17. $a_3 = 13$, $a_{10} = 55$, $a_1 = ?$

18. $a_9 = -12$, $a_{13} = 3$, $a_1 = ?$

19. $S_{51} = \sum_{k=1}^{51} (3k + 3) = ?$

20. $S_{40} = \sum_{k=1}^{40} (2k - 3) = ?$

C **21.** Find $g(1) + g(2) + g(3) + \cdots + g(51)$ if $g(t) = 5 - t$.

22. Find $f(1) + f(2) + f(3) + \cdots + f(20)$ if $f(x) = 2x - 5$.

23. Find the sum of all the even integers between 21 and 135.

24. Find the sum of all the odd integers between 100 and 500.

25. Show that the sum of the first n odd natural numbers is n^2, using appropriate formulas from this section.

26. Show that the sum of the first n even natural numbers is $n + n^2$, using appropriate formulas from this section.

27. If in a given sequence $a_1 = -3$ and $a_n = a_{n-1} + 3$, $n > 1$, find a_n in terms of n.

28. For the sequence in Problem 27 find $S_n = \sum_{k=1}^{n} a_k$ in terms of n.

29. An object falling from rest in a vacuum near the surface of the earth falls 16 feet during the 1st second, 48 feet during the 2d second, and 80 feet during the 3d second, and so on. (*A*) How far will the object fall during the 11th second? (*B*) How far will the object fall in 11 seconds? (*C*) How far will the object fall in t seconds?

30. In investigating different job opportunities, you find that firm A will start you at $6,000 per year and guarantee you a raise of $300 each year, while firm B will start you at $7,000 per year, but will only guarantee you a raise of $200 each year. Over a 15-year period which firm will pay the greatest total amount?

12.3 Geometric Sequences and Series

Consider the sequence

$$2, -4, 8, -16, \ldots$$

Can you guess what the fifth and sixth terms are? If you guessed 32 and −64, respectively, you have observed that each term after the first can be obtained from the preceding one by multiplying it by −2. This is an example of a geometric sequence. In general, a sequence

$$a_1, a_2, a_3, \ldots, a_n, \ldots$$

is called a *geometric sequence* (or a *geometric progression*) if there exists a nonzero constant r, called the *common ratio*, such that

$$a_n = ra_{n-1} \qquad \text{for every } n > 1$$

EXAMPLE 10 Which sequence is a geometric sequence and what is its common ratio?

(*A*) 2, 6, 8, 10, . . . (*B*) −1, 3, −9, 27, . . .

SOLUTION (*B*) is a geometric sequence with $r = -3$

PROBLEM 10 Repeat Example 10 with

(A) $\frac{1}{4}, \frac{1}{2}, 1, 2, \ldots$ (B) $\frac{1}{2}, \frac{1}{4}, \frac{1}{16}, \frac{1}{256}, \ldots$

ANSWER (A) with $r = 2$

Just as with arithmetic sequences, geometric sequences have several convenient properties. It is easy to derive formulas for the nth term in terms of n and the sum of any number of consecutive terms. To obtain an nth-term formula, we note that if a is a geometric sequence, then

$$a_2 = ra_1$$

$$a_3 = ra_2 = r^2a_1$$

$$a_4 = ra_3 = r^3a_1$$

which suggests that

$$a_n = a_1r^{n-1} \qquad \text{for every } n > 1$$

EXAMPLE 11 Find the seventh term of the geometric sequence $1, \frac{1}{2}, \frac{1}{4}, \ldots$

SOLUTION

$$r = \tfrac{1}{2}$$

$$a_n = a_1r^{n-1}$$

$$a_7 = 1\left(\tfrac{1}{2}\right)^{7-1} = \tfrac{1}{64}$$

PROBLEM 11 Find the eighth term of the geometric sequence $\frac{1}{64}, -\frac{1}{32}, \frac{1}{16}, \ldots$

ANSWER -2

EXAMPLE 12 If the 1st and 10th terms of a geometric sequence are 1 and 2, respectively, find the common ratio r.

SOLUTION

$$a_n = a_1r^{n-1}$$

$$2 = 1r^{10-1}$$

$$r = 2^{1/9} = 1.08 \qquad \text{calculation by logarithms or calculator}$$

PROBLEM 12 If the first and eighth terms of a geometric sequence are 2 and 16, respectively, find the common ratio r.

ANSWER $r = 1.346$

A *geometric series* is any series whose terms form a geometric sequence. As was the case with an arithmetic series, we can derive two simple and very useful formulas for finding the *sum of the first n terms of a geometric series.* Let

$$S_n = a_1 + a_1r + a_1r^2 + a_1r^3 + \cdots + a_1r^{n-1}$$

and multiply both members by r to obtain

$$rS_n = a_1r + a_1r^2 + a_1r^3 + \cdots + a_1r^{n-1} + a_1r^n$$

Now subtract the left member of the second equation from the left member of the first, and the right member of the second equation from the right member of the first to obtain

$$S_n - rS_n = a_1 - a_1r^n$$

$$S_n(1 - r) = a_1 - a_1r^n$$

Thus,

$$S_n = \frac{a_1 - a_1r^n}{1 - r} \qquad r \neq 1$$

Since $a_n = a_1r^{n-1}$, or $ra_n = a_1r^n$, the sum formula can also be written as

$$S_n = \frac{a_1 - ra_n}{1 - r} \qquad r \neq 1$$

EXAMPLE 13 Find the sum of the first 20 terms of a geometric series if the first term is 1 and $r = 2$.

SOLUTION

$$S_n = \frac{a_1 - a_1r^n}{1 - r}$$

$$S_n = \frac{1 - 1 \cdot 2^{20}}{1 - 2} \approx 1{,}050{,}000 \qquad \text{calculation by logarithms or calculator}$$

PROBLEM 13 Find the sum of the first 14 terms of a geometric series if the first term is $\frac{1}{64}$ and $r = -2$.

ANSWER -85.33

INFINITE GEOMETRIC SERIES

Given a geometric series, what happens to the sum S_n as n increases? To answer this question, we first write the sum formula in the more convenient form

$$S_n = \frac{a_1 - a_1r^n}{1 - r} = \frac{a_1}{1 - r} - \frac{a_1r^n}{1 - r} \tag{1}$$

It is possible to show that if $|r| < 1$, that is, if $-1 < r < 1$, then r^n will tend to 0 as n increases. (See what happens, for example, if $r = \frac{1}{2}$.) Thus, S_n can be made as close to

$$\frac{a_1}{1-r}$$

as we wish by taking n sufficiently large. Thus, we define

$$\boxed{S_\infty = \frac{a_1}{1-r} \qquad |r| < 1}$$

and call this the *sum of an infinite geometric series.* If $|r| \geq 1$, an infinite geometric series has no sum.

EXAMPLE 14 Represent the repeating decimal $0.4545\overline{45}$ as the quotient of two integers. (The bar over the last two digits indicates that these digits repeat indefinitely.)

SOLUTION $0.4545\overline{45} = 0.45 + 0.0045 + 0.000045 + \cdots$

The right member of the equation is an infinite geometric series with $a_1 = 0.45$ and $r = 0.01$. Thus,

$$S_\infty = \frac{a_1}{1-r} = \frac{0.45}{1-0.01} = \frac{0.45}{0.99} = \frac{5}{11}$$

Hence, $0.4545\overline{45}$ and $\frac{5}{11}$ name the same rational number. Check the result by dividing 5 by 11.

PROBLEM 14 Repeat Example 14 for $0.8181\overline{81}$.

ANSWER $\frac{9}{11}$

Exercise 72

A **1.** Determine which of the following are geometric sequences. Find r and the next two terms for those that are.

(*A*) $2, -4, 8, \ldots$ (*B*) $7, 6.5, 6, \ldots$

(*C*) $-11, -16, -21, \ldots$ (*D*) $\frac{1}{2}, \frac{1}{6}, \frac{1}{18}, \ldots$

2. Repeat Problem 1 for:

(*A*) $5, -1, -7, \ldots$ (*B*) $12, 4, \frac{4}{3}, \ldots$

(*C*) $\frac{1}{2}, \frac{2}{3}, \frac{3}{4}, \ldots$ (*D*) $16, 48, 80, \ldots$

Let $a_1, a_2, a_3, \ldots, a_n, \ldots$ be a geometric sequence. Find each of the indicated quantities.

3. $a_1 = -6$, $r = -\frac{1}{2}$, $a_2 = ?$, $a_3 = ?$, $a_4 = ?$

4. $a_1 = 12$, $r = \frac{2}{3}$, $a_2 = ?$, $a_3 = ?$, $a_4 = ?$

5. $a_1 = 81$, $r = \frac{1}{3}$, $a_{10} = ?$

6. $a_1 = 64$, $r = \frac{1}{2}$, $a_{13} = ?$

7. $a_1 = 3$, $a_7 = 2{,}187$, $r = 3$, $S_7 = ?$

8. $a_1 = 1$, $a_7 = 729$, $r = -3$, $S_7 = ?$

B

9. $a_1 = 100$, $a_6 = 1$, $r = ?$

10. $a_1 = 10$, $a_{10} = 30$, $r = ?$

11. $a_1 = 5$, $r = -2$, $S_{10} = ?$

12. $a_1 = 3$, $r = 2$, $S_{10} = ?$

13. $a_1 = 9$, $a_4 = \frac{8}{3}$, $a_2 = ?$, $a_3 = ?$

14. $a_1 = 12$, $a_4 = -\frac{4}{9}$, $a_2 = ?$, $a_3 = ?$

15. $S_7 = \sum_{k=1}^{7} (-3)^{k-1} = ?$

16. $S_7 = \sum_{k=1}^{7} 3^k = ?$

17. Find $g(1) + g(2) + \cdots + g(10)$ if $g(x) = (\frac{1}{2})^x$.

18. Find $f(1) + f(2) + \cdots + f(10)$ if $f(x) = 2^x$.

19. Find a positive number x so that $-2 + x - 6$ is a geometric series.

20. Find a positive number x so that $6 + x + 8$ is a geometric series.

Find the sum of each infinite geometric series that has a sum.

21. $3 + 1 + \frac{1}{3} + \cdots$

22. $16 + 4 + 1 + \cdots$

23. $2 + 4 + 8 + \cdots$

24. $4 + 6 + 9 + \cdots$

25. $2 - \frac{1}{2} + \frac{1}{8} + \cdots$

26. $21 - 3 + \frac{3}{7} + \cdots$

C *Represent each repeating decimal fraction as the quotient of two integers.*

27. $0.777\overline{7}$

28. $0.555\overline{5}$

29. $0.5454\overline{54}$

30. $0.2727\overline{27}$

31. $3.216216\overline{216}$

32. $5.6363\overline{63}$

33. If P dollars is invested at r percent compounded annually, the amount A present after n years forms a geometric progression with a constant ratio $(1 + r)$.

Write a formula for the amount present after n years. How long will it take for a sum of money P to double if invested at 6 percent interest compounded annually?

34. If a population of A_0 people grows at the constant rate of r percent per year, the population after t years forms a geometric progression with a constant ratio $(1 + r)$. Write a formula for the total population after t years. If the world's population is increasing at the rate of 2 percent per year, how long will it take to double?

35. A rotating flywheel coming to rest rotates 300 revolutions the first minute. If in each subsequent minute it rotates two-thirds as many times as in the preceding minute, how many revolutions will the wheel make before coming to rest?

36. The first swing of a bob on a pendulum is 10 inches. If on each subsequent swing it travels 0.9 as far as on the preceding swing, how far will the bob travel before coming to rest?

37. The government, through a subsidy program, distributes \$1,000,000. If we assume that each individual or agency spends 0.8 of what is received, and 0.8 of this is spent, and so on, how much total increase in spending results from this government action? (Let $a_1 = \$800{,}000$.)

38. Visualize a hypothetical 440-yard oval race track that has tapes stretched across the track at the halfway point and at each point that marks the halfway point of each remaining distance thereafter. A runner running around the track has to break the first tape before the second, the second before the third, and so on. From this point of view it appears that the runner will never finish the race. (This famous paradox is attributed to the Greek philosopher, Zeno, 495–435 B.C.) If we assume the runner runs at 440 yards per minute, the times between tape breakings form an infinite geometric progression. What is the sum of this progression?

12.4 Binomial Formula

The binomial form

$$(a + b)^n$$

n a natural number, appears more frequently than one might expect. The coefficients in its expansion play an important role in probability studies. In this section we will give an informal derivation of the famous binomial formula, which will enable us to expand $(a + b)^n$ directly for any natural number n, however large. First, we introduce the useful concept of factorial.

FACTORIAL

The product of the first n natural numbers is symbolized by $n!$ and is referred to as *n factorial*. Symbolically,

$$1! = 1 \qquad n! = n \cdot (n-1)! \qquad \text{or} \qquad n! = n(n-1)(n-2)\cdots 3\cdot 2\cdot 1$$

In addition, we define

$$0! = 1$$

EXAMPLE 15 Evaluate each.

(A) $4! = 4 \cdot 3! = 4 \cdot 3 \cdot 2! = 4 \cdot 3 \cdot 2 \cdot 1! = 4 \cdot 3 \cdot 2 \cdot 1 = 24$

(B) $5! = 5 \cdot 4 \cdot 3 \cdot 2 \cdot 1 = 120$

(C) $\dfrac{7!}{6!} = \dfrac{7 \cdot \cancel{6!}}{\cancel{6!}} = 7$

(D) $\dfrac{8!}{5!} = \dfrac{8 \cdot 7 \cdot 6 \cdot \cancel{5!}}{\cancel{5!}} = 336$

PROBLEM 15 Evaluate:

(A) $7!$ (B) $\dfrac{8!}{7!}$ (C) $\dfrac{6!}{3!}$

ANSWER (A) 5,040 (B) 8 (C) 120

A form involving factorials that is very useful, is given by the formula

$$\binom{n}{r} = \frac{n!}{(n-r)!r!} \qquad n \geq r$$

EXAMPLE 16 Find:

(A) $\dbinom{5}{2}$ (B) $\dbinom{4}{4}$

SOLUTION (A) $\dbinom{5}{2} = \dfrac{5!}{(5-2)!2!} = \dfrac{5!}{3!2!} = 10$

(B) $\dbinom{4}{4} = \dfrac{4!}{(4-4)!4!} = \dfrac{4!}{0!4!} = 1$

PROBLEM 16 Find:

(A) $\dbinom{9}{2}$ (B) $\dbinom{5}{5}$

ANSWER (A) 36 (B) 1

BINOMIAL FORMULA

Let us try to discover a formula for the expansion of $(a + b)^n$, n a natural number.

$$(a + b)^1 = a + b$$

$$(a + b)^2 = a^2 + 2ab + b^2$$

$$(a + b)^3 = a^3 + 3a^2b + 3ab^2 + b^3$$

$$(a + b)^4 = a^4 + 4a^3b + 6a^2b^2 + 4ab^3 + b^4$$

$$(a + b)^5 = a^5 + 5a^4b + 10a^3b^2 + 10a^2b^3 + 5ab^4 + b^5$$

OBSERVATIONS

1 The expansion of $(a + b)^n$ has $n + 1$ terms.
2 The power of a starts at n and decreases for each term until it is 0 in the last term.
3 The power of b starts at 0 in the first term and increases 1 for each term until it is n in the last term.
4 The sum of the powers of a and b in each term is the constant n.
5 The coefficient of any term after the first can be obtained from the preceding term as follows. In the preceding term multiply the coefficient by the exponent of a, and then divide this product by the number representing the position of the preceding term in the series.

We now postulate these same properties for the general case:

$$(a + b)^n = a^n + \frac{n}{1}a^{n-1}b + \frac{n(n-1)}{1 \cdot 2}a^{n-2}b^2 + \frac{n(n-1)(n-2)}{1 \cdot 2 \cdot 3}a^{n-3}b^3 + \cdots + b^n$$

$$= \frac{n!}{0!(n-0)!}a^n + \frac{n!}{1!(n-1)!}a^{n-1}b + \frac{n!}{2!(n-2)!}a^{n-2}b^2 + \frac{n!}{3!(n-3)!}a^{n-3}b^3 + \cdots + \frac{n!}{n!(n-n)!}b^n$$

$$= \binom{n}{0}a^n + \binom{n}{1}a^{n-1}b + \binom{n}{2}a^{n-2}b^2 + \binom{n}{3}a^{n-3}b^3 + \cdots + \binom{n}{n}b^n$$

Thus, it appears that

BINOMIAL FORMULA

$$(a + b)^n = \sum_{k=0}^{n} \binom{n}{k}a^{n-k}b^k \qquad n \geq 1$$

This result is known as the *binomial formula*, and its general proof requires a method of proof called mathematical induction that is considered in more advanced courses.

EXAMPLE 17 Use the binomial formula to expand $(x+y)^6$.

SOLUTION

$$(x+y)^6 = \sum_{k=0}^{6} \binom{6}{k} x^{6-k}y^k$$
$$= \binom{6}{0}x^6 + \binom{6}{1}x^5y + \binom{6}{2}x^4y^2 + \binom{6}{3}x^3y^3 + \binom{6}{4}x^2y^4 + \binom{6}{5}xy^5 + \binom{6}{6}y^6$$
$$= x^6 + 6x^5y + 15x^4y^2 + 20x^3y^3 + 15x^2y^4 + 6xy^5 + y^6$$

PROBLEM 17 Use the binomial formula to expand $(x+1)^5$.

ANSWER $x^5 + 5x^4 + 10x^3 + 10x^2 + 5x + 1$

EXAMPLE 18 Use the binomial formula to find the fourth term in the expansion of $(x-2)^{20}$.

SOLUTION

$$\text{Fourth term} = \binom{20}{3}x^{17}(-2)^3$$
$$= \frac{20 \cdot 19 \cdot 18}{3 \cdot 2 \cdot 1}x^{17}(-8)$$
$$= -9{,}120x^{17}$$

PROBLEM 18 Use the binomial formula to find the fifth term in the expansion of $(u-1)^{18}$.

ANSWER $3{,}060u^{14}$

Exercise 73

A *Evaluate:*

1. $6!$ **2.** $4!$ **3.** $\frac{20!}{19!}$ **4.** $\frac{5!}{4!}$

5. $\frac{10!}{7!}$ **6.** $\frac{9!}{6!}$ **7.** $\frac{6!}{4!2!}$ **8.** $\frac{5!}{2!3!}$

9. $\frac{9!}{0!(9-0)!}$ **10.** $\frac{8!}{8!(8-8)!}$ **11.** $\frac{8!}{2!(8-2)!}$ **12.** $\frac{7!}{3!(7-3)!}$

Write as the quotient of two factorials.

13. 9 **14.** 12 **15.** $6 \cdot 7 \cdot 8$ **16.** $9 \cdot 10 \cdot 11 \cdot 12$

B *Evaluate:*

17. $\binom{9}{5}$ **18.** $\binom{5}{2}$ **19.** $\binom{6}{5}$ **20.** $\binom{7}{1}$

21. $\binom{9}{9}$ **22.** $\binom{5}{0}$ **23.** $\binom{17}{13}$ **24.** $\binom{20}{16}$

Expand, using the binomial formula.

25. $(u + v)^5$ **26.** $(x + y)^4$ **27.** $(y - 1)^4$

28. $(x - 2)^5$ **29.** $(2x - y)^5$ **30.** $(m + 2n)^6$

Find the indicated term in each expansion.

31. $(u + v)^{15}$; 7th term **32.** $(a + b)^{12}$; 5th term

33. $(2m + n)^{12}$; 11th term **34.** $(x + 2y)^{20}$; 3d term

35. $[(w/2) - 2]^{12}$; 7th term **36.** $(x - 3)^{10}$; 4th term

C **37.** Evaluate $(1.01)^{10}$ to four decimal places, using the binomial formula. (HINT: Let $1.01 = 1 + 0.01$.)

38. Evaluate $(0.99)^6$ to four decimal places, using the binomial formula.

39. Show that

$$\binom{n}{r} = \binom{n}{n - r}$$

40. Can you guess what the next two rows in Pascal's triangle are? Compare the numbers in the triangle with the binomial coefficients obtained with the binomial formula.

```
    1
   1 1
  1 2 1
 1 3 3 1
1 4 6 4 1
```

Exercise 74 Chapter Review

A **1.** Identify all arithmetic and all geometric sequences from the following list of sequences:

(*A*) 16, −8, 4, . . . (*B*) 5, 7, 9, . . .

(*C*) $-8, -5, -2, \ldots$ (*D*) $2, 3, 5, 8, \ldots$

(*E*) $-1, 2, -4, \ldots$

In Problems 2 to 5

(A) Write the first four terms of each sequence.

(B) Find a_{10}.

(C) Find S_{10}.

2. $a_n = 2n + 3$

3. $a_n = 32(\frac{1}{2})^n$

4. $a_1 = -8$; $a_n = a_{n-1} + 3$, $n \geq 2$

5. $a_1 = -1$; $a_n = (-2)a_{n-1}$, $n \geq 2$

6. Find S_∞ in Problem 3.

Evaluate:

7. $6!$

8. $\dfrac{22!}{19!}$

9. $\dfrac{7!}{2!(7-2)!}$

B *Write Problems 10 and 11 without sigma notation and find the sums.*

10. $S_{10} = \sum_{k=1}^{10} (2k - 8)$

11. $S_7 = \sum_{k=1}^{7} \dfrac{16}{2^k}$

12. $S_\infty = 27 - 18 + 12 - \cdots = ?$

13. Write $S_n = \dfrac{1}{3} - \dfrac{1}{9} + \dfrac{1}{27} + \cdots + \dfrac{(-1)^{n+1}}{3^n}$ using sigma notation and find S_∞.

14. If in an arithmetic sequence $a_1 = 13$ and $a_7 = 31$, find the common difference d and the fifth term a_5.

Evaluate:

15. $\dfrac{20!}{18!(20-18)!}$

16. $\dbinom{16}{12}$

17. $\dbinom{11}{11}$

18. Expand $(x - y)^5$ using the binomial formula.

19. Find the 10th term in the expansion of $(2x - y)^{12}$.

C **20.** Write $0.727\overline{272}$ as the quotient of two integers.

21. A free-falling body travels $g/2$ feet in the first second, $3g/2$ feet during the next second, $5g/2$ feet the next, and so on, where g is the gravitational constant. Find the distance fallen during the 25th second and the total distance fallen from the start to the end of the 25th second. Express answers in terms of g.

22. Expand $(x + i)^6$, i the complex unit, using the binomial formula.

TABLES

TABLE 1 Squares and Square Roots (0 to 199)

n	n^2	$\sqrt{n}$	n	n^2	$\sqrt{n}$	n	n^2	$\sqrt{n}$	n	n^2	$\sqrt{n}$
0	0	0.000	50	2,500	7.071	100	10,000	10.000	150	22,500	12.247
1	1	1.000	51	2,601	7.141	101	10,201	10.050	151	22,801	12.288
2	4	1.414	52	2,704	7.211	102	10,404	10.100	152	23,104	12.329
3	9	1.732	53	2,809	7.280	103	10,609	10.149	153	23,409	12.369
4	16	2.000	54	2,916	7.348	104	10,816	10.198	154	23,716	12.410
5	25	2.236	55	3,025	7.416	105	11,025	10.247	155	24,025	12.450
6	36	2.449	56	3,136	7.483	106	11,236	10.296	156	24,336	12.490
7	49	2.646	57	3,249	7.550	107	11,449	10.344	157	24,649	12.530
8	64	2.828	58	3,364	7.616	108	11,664	10.392	158	24,964	12.570
9	81	3.000	59	3,481	7.681	109	11,881	10.440	159	25,281	12.610
10	100	3.162	60	3,600	7.746	110	12,100	10.488	160	25,600	12.649
11	121	3.317	61	3,721	7.810	111	12,321	10.536	161	25,921	12.689
12	144	3.464	62	3,844	7.874	112	12,544	10.583	162	26,244	12.728
13	169	3.606	63	3,969	7.937	113	12,769	10.630	163	26,569	12.767
14	196	3.742	64	4,096	8.000	114	12,996	10.677	164	26,896	12.806
15	225	3.873	65	4,225	8.062	115	13,225	10.724	165	27,225	12.845
16	256	4.000	66	4,356	8.124	116	13,456	10.770	166	27,556	12.884
17	289	4.123	67	4,489	8.185	117	13,689	10.817	167	27,889	12.923
18	324	4.243	68	4,624	8.246	118	13,924	10.863	168	28,224	12.961
19	361	4.359	69	4,761	8.307	119	14,161	10.909	169	28,561	13.000
20	400	4.472	70	4,900	8.367	120	14,400	10.954	170	28,900	13.038
21	441	4.583	71	5,041	8.426	121	14,641	11.000	171	29,241	13.077
22	484	4.690	72	5,184	8.485	122	14,884	11.045	172	29,584	13.115
23	529	4.796	73	5,329	8.544	123	15,129	11.091	173	29,929	13.153
24	576	4.899	74	5,476	8.602	124	15,376	11.136	174	30,276	13.191
25	625	5.000	75	5,625	8.660	125	15,625	11.180	175	30,625	13.229
26	676	5.099	76	5,776	8.718	126	15,876	11.225	176	30,976	13.266
27	729	5.196	77	5,929	8.775	127	16,129	11.269	177	31,329	13.304
28	784	5.292	78	6,084	8.832	128	16,384	11.314	178	31,684	13.342
29	841	5.385	79	6,241	8.888	129	16,641	11.358	179	32,041	13.379
30	900	5.477	80	6,400	8.944	130	16,900	11.402	180	32,400	13.416
31	961	5.568	81	6,561	9.000	131	17,161	11.446	181	32,761	13.454
32	1,024	5.657	82	6,724	9.055	132	17,424	11.489	182	33,124	13.491
33	1,089	5.745	83	6,889	9.110	133	17,689	11.533	183	33,489	13.528
34	1,156	5.831	84	7,056	9.165	134	17,956	11.576	184	33,856	13.565
35	1,225	5.916	85	7,225	9.220	135	18,225	11.619	185	34,225	13.601
36	1,296	6.000	86	7,396	9.274	136	18,496	11.662	186	34,596	13.638
37	1,369	6.083	87	7,569	9.327	137	18,769	11.705	187	34,969	13.675
38	1,444	6.164	88	7,744	9.381	138	19,044	11.747	188	35,344	13.711
39	1,521	6.245	89	7,921	9.434	139	19,321	11.790	189	35,721	13.748
40	1,600	6.325	90	8,100	9.487	140	19,600	11.832	190	36,100	13.784
41	1,681	6.403	91	8,281	9.539	141	19,881	11.874	191	36,481	13.820
42	1,764	6.481	92	8,464	9.592	142	20,164	11.916	192	36,864	13.856
43	1,849	6.557	93	8,649	9.644	143	20,449	11.958	193	37,249	13.892
44	1,936	6.633	94	8,836	9.659	144	20,736	12.000	194	37,636	13.928
45	2,025	6.708	95	9,025	9.747	145	21,025	12.042	195	38,025	13.964
46	2,116	6.782	96	9,216	9.798	146	21,316	12.083	196	38,416	14.000
47	2,209	6.856	97	9,409	9.849	147	21,609	12.124	197	38,809	14.036
48	2,304	6.928	98	9,604	9.899	148	21,904	12.166	198	39,204	14.071
49	2,401	7.000	99	9,801	9.950	149	22,201	12.207	199	39,601	14.107
n	n^2	$\sqrt{n}$	n	n^2	$\sqrt{n}$	n	n^2	$\sqrt{n}$	n	n^2	$\sqrt{n}$

TABLE 2 Values of e^x and e^{-x} (0.00 to 3.00)

x	e^x	e^{-x}	x	e^x	e^{-x}	x	e^x	e^{-x}
0.00	1.000	1.000	0.50	1.649	0.607	1.00	2.718	0.368
0.01	1.010	0.990	0.51	1.665	0.600	1.01	2.746	0.364
0.02	1.020	0.980	0.52	1.682	0.595	1.02	2.773	0.361
0.03	1.031	0.970	0.53	1.699	0.589	1.03	2.801	0.357
0.04	1.041	0.961	0.54	1.716	0.583	1.04	2.829	0.353
0.05	1.051	0.951	0.55	1.733	0.577	1.05	2.858	0.350
0.06	1.062	0.942	0.56	1.751	0.571	1.06	2.886	0.346
0.07	1.073	0.932	0.57	1.768	0.566	1.07	2.915	0.343
0.08	1.083	0.923	0.58	1.786	0.560	1.08	2.945	0.340
0.09	1.094	0.914	0.59	1.804	0.554	1.09	2.974	0.336
0.10	1.105	0.905	0.60	1.822	0.549	1.10	3.004	0.333
0.11	1.116	0.896	0.61	1.840	0.543	1.11	3.034	0.330
0.12	1.127	0.887	0.62	1.859	0.538	1.12	3.065	0.326
0.13	1.139	0.878	0.63	1.878	0.533	1.13	3.096	0.323
0.14	1.150	0.869	0.64	1.896	0.527	1.14	3.127	0.320
0.15	1.162	0.861	0.65	1.916	0.522	1.15	3.158	0.317
0.16	1.174	0.852	0.66	1.935	0.517	1.16	3.190	0.313
0.17	1.185	0.844	0.67	1.954	0.512	1.17	3.222	0.310
0.18	1.197	0.835	0.68	1.974	0.507	1.18	3.254	0.307
0.19	1.209	0.827	0.69	1.994	0.502	1.19	3.287	0.304
0.20	1.221	0.819	0.70	2.014	0.497	1.20	3.320	0.301
0.21	1.234	0.811	0.71	2.034	0.492	1.21	3.353	0.298
0.22	1.246	0.803	0.72	2.054	0.487	1.22	3.387	0.295
0.23	1.259	0.795	0.73	2.075	0.482	1.23	3.421	0.292
0.24	1.271	0.787	0.74	2.096	0.477	1.24	3.456	0.289
0.25	1.284	0.779	0.75	2.117	0.472	1.25	3.490	0.287
0.26	1.297	0.771	0.76	2.138	0.468	1.26	3.525	0.284
0.27	1.310	0.763	0.77	2.160	0.463	1.27	3.561	0.281
0.28	1.323	0.756	0.78	2.182	0.458	1.28	3.597	0.278
0.29	1.336	0.748	0.79	2.203	0.454	1.29	3.633	0.275
0.30	1.350	0.741	0.80	2.226	0.449	1.30	3.669	0.273
0.31	1.363	0.733	0.81	2.248	0.445	1.31	3.706	0.270
0.32	1.377	0.726	0.82	2.270	0.440	1.32	3.743	0.267
0.33	1.391	0.719	0.83	2.293	0.436	1.33	3.781	0.264
0.34	1.405	0.712	0.84	2.316	0.432	1.34	3.819	0.262
0.35	1.419	0.705	0.85	2.340	0.427	1.35	3.857	0.259
0.36	1.433	0.698	0.86	2.363	0.423	1.36	3.896	0.257
0.37	1.448	0.691	0.87	2.387	0.419	1.37	3.935	0.254
0.38	1.462	0.684	0.88	2.411	0.415	1.38	3.975	0.252
0.39	1.477	0.677	0.89	2.435	0.411	1.39	4.015	0.249
0.40	1.492	0.670	0.90	2.460	0.407	1.40	4.055	0.247
0.41	1.507	0.664	0.91	2.484	0.403	1.41	4.096	0.244
0.42	1.522	0.657	0.92	2.509	0.399	1.42	4.137	0.242
0.43	1.537	0.651	0.93	2.535	0.395	1.43	4.179	0.239
0.44	1.553	0.644	0.94	2.560	0.391	1.44	4.221	0.237
0.45	1.568	0.638	0.95	2.586	0.387	1.45	4.263	0.235
0.46	1.584	0.631	0.96	2.612	0.383	1.46	4.306	0.232
0.47	1.600	0.625	0.97	2.638	0.379	1.47	4.349	0.230
0.48	1.616	0.619	0.98	2.664	0.375	1.48	4.393	0.228
0.49	1.632	0.613	0.99	2.691	0.372	1.49	4.437	0.225

TABLE 2 *(continued)*

x	e^x	e^{-x}	x	e^x	e^{-x}	x	e^x	e^{-x}
1.50	4.482	0.223	2.00	7.389	0.135	2.50	12.182	0.082
1.51	4.527	0.221	2.01	7.463	0.134	2.51	12.305	0.081
1.52	4.572	0.219	2.02	7.538	0.133	2.52	12.429	0.080
1.53	4.618	0.217	2.03	7.614	0.131	2.53	12.554	0.080
1.54	4.665	0.214	2.04	7.691	0.130	2.54	12.680	0.079
1.55	4.712	0.212	2.05	7.768	0.129	2.55	12.807	0.078
1.56	4.759	0.210	2.06	7.846	0.127	2.56	12.936	0.077
1.57	4.807	0.208	2.07	7.925	0.126	2.57	13.066	0.077
1.58	4.855	0.206	2.08	8.004	0.125	2.58	13.197	0.076
1.59	4.904	0.204	2.09	8.085	0.124	2.59	13.330	0.075
1.60	4.953	0.202	2.10	8.166	0.122	2.60	13.464	0.074
1.61	5.003	0.200	2.11	8.248	0.121	2.61	13.599	0.074
1.62	5.053	0.198	2.12	8.331	0.120	2.62	13.736	0.073
1.63	5.104	0.196	2.13	8.415	0.119	2.63	13.874	0.072
1.64	5.155	0.194	2.14	8.499	0.118	2.64	14.013	0.071
1.65	5.207	0.192	2.15	8.585	0.116	2.65	14.154	0.071
1.66	5.259	0.190	2.16	8.671	0.115	2.66	14.296	0.070
1.67	5.312	0.188	2.17	8.758	0.114	2.67	14.440	0.069
1.68	5.366	0.186	2.18	8.846	0.113	2.68	14.585	0.069
1.69	5.420	0.185	2.19	8.935	0.112	2.69	14.732	0.068
1.70	5.474	0.183	2.20	9.025	0.111	2.70	14.880	0.067
1.71	5.529	0.181	2.21	9.116	0.110	2.71	15.029	0.067
1.72	5.585	0.179	2.22	9.207	0.109	2.72	15.180	0.066
1.73	5.641	0.177	2.23	9.300	0.108	2.73	15.333	0.065
1.74	5.697	0.176	2.24	9.393	0.106	2.74	15.487	0.065
1.75	5.755	0.174	2.25	9.488	0.105	2.75	15.643	0.064
1.76	5.812	0.172	2.26	9.583	0.104	2.76	15.800	0.063
1.77	5.871	0.170	2.27	9.679	0.103	2.77	15.959	0.063
1.78	5.930	0.169	2.28	9.777	0.102	2.78	16.119	0.062
1.79	5.989	0.167	2.29	9.875	0.101	2.79	16.281	0.061
1.80	6.050	0.165	2.30	9.974	0.100	2.80	16.445	0.061
1.81	6.110	0.164	2.31	10.074	0.099	2.81	16.610	0.060
1.82	6.172	0.162	2.32	10.176	0.098	2.82	16.777	0.060
1.83	6.234	0.160	2.33	10.278	0.097	2.83	16.945	0.059
1.84	6.297	0.159	2.34	10.381	0.096	2.84	17.116	0.058
1.85	6.360	0.157	2.35	10.486	0.095	2.85	17.288	0.058
1.86	6.424	0.156	2.36	10.591	0.094	2.86	17.462	0.057
1.87	6.488	0.154	2.37	10.697	0.093	2.87	17.637	0.057
1.88	6.553	0.153	2.38	10.805	0.093	2.88	17.814	0.056
1.89	6.619	0.151	2.39	10.913	0.092	2.89	17.993	0.056
1.90	6.686	0.150	2.40	11.023	0.091	2.90	18.174	0.055
1.91	6.753	0.148	2.41	11.134	0.090	2.91	18.357	0.054
1.92	6.821	0.147	2.42	11.246	0.089	2.92	18.541	0.054
1.93	6.890	0.145	2.43	11.359	0.088	2.93	18.728	0.053
1.94	6.959	0.144	2.44	11.473	0.087	2.94	18.916	0.053
1.95	7.029	0.142	2.45	11.588	0.086	2.95	19.106	0.052
1.96	7.099	0.141	2.46	11.705	0.085	2.96	19.298	0.052
1.97	7.171	0.139	2.47	11.822	0.085	2.97	19.492	0.051
1.98	7.243	0.138	2.48	11.941	0.084	2.98	19.688	0.051
1.99	7.316	0.137	2.49	12.061	0.083	2.99	19.886	0.050
						3.00	20.086	0.050

TABLE 3 Common Logarithms

N	*0*	*1*	2	3	*4*	5	6	7	8	9
1.0	0.0000	0.004321	0.008600	0.01284	0.01703	0.02119	0.02531	0.02938	0.03342	0.03743
1.1	0.04139	0.04532	0.04922	0.05308	0.05690	0.06070	0.06446	0.06819	0.07188	0.07555
1.2	0.07918	0.08279	0.08636	0.08991	0.09342	0.09691	0.1004	0.1038	0.1072	0.1106
1.3	0.1139	0.1173	0.1206	0.1239	0.1271	0.1303	0.1335	0.1367	0.1399	0.1430
1.4	0.1461	0.1492	0.1523	0.1553	0.1584	0.1614	0.1644	0.1673	0.1703	0.1732
1.5	0.1761	0.1790	0.1818	0.1847	0.1875	0.1903	0.1931	0.1959	0.1987	0.2014
1.6	0.2041	0.2068	0.2095	0.2122	0.2148	0.2175	0.2201	0.2227	0.2253	0.2279
1.7	0.2304	0.2330	0.2355	0.2380	0.2405	0.2430	0.2455	0.2480	0.2504	0.2529
1.8	0.2553	0.2577	0.2601	0.2625	0.2648	0.2673	0.2695	0.2718	0.2742	0.2765
1.9	0.2788	0.2810	0.2833	0.2856	0.2878	0.2900	0.2923	0.2945	0.2967	0.2989
2.0	0.3010	0.3032	0.3054	0.3075	0.3096	0.3118	0.3139	0.3160	0.3181	0.3201
2.1	0.3222	0.3243	0.3263	0.3284	0.3304	0.3324	0.3345	0.3365	0.3385	0.3404
2.2	0.3424	0.3444	0.3464	0.3483	0.3502	0.3522	0.3541	0.3560	0.3579	0.3598
2.3	0.3617	0.3636	0.3655	0.3674	0.3692	0.3711	0.3729	0.3747	0.3766	0.3784
2.4	0.3802	0.3820	0.3838	0.3856	0.3874	0.3892	0.3909	0.3927	0.3945	0.3962
2.5	0.3979	0.3997	0.4014	0.4031	0.4048	0.4065	0.4082	0.4099	0.4116	0.4133
2.6	0.4150	0.4166	0.4183	0.4200	0.4216	0.4232	0.4249	0.4265	0.4281	0.4298
2.7	0.4314	0.4330	0.4346	0.4362	0.4378	0.4393	0.4409	0.4425	0.4440	0.4456
2.8	0.4472	0.4487	0.4502	0.4518	0.4533	0.4548	0.4564	0.4579	0.4594	0.4609
2.9	0.4624	0.4639	0.4654	0.4669	0.4683	0.4698	0.4713	0.4728	0.4742	0.4757
3.0	0.4771	0.4786	0.4800	0.4814	0.4829	0.4843	0.4857	0.4871	0.4886	0.4900
3.1	0.4914	0.4928	0.4942	0.4955	0.4969	0.4983	0.4997	0.5011	0.5024	0.5038
3.2	0.5051	0.5065	0.5079	0.5092	0.5105	0.5119	0.5132	0.5145	0.5159	0.5172
3.3	0.5185	0.5198	0.5211	0.5224	0.5237	0.5250	0.5263	0.5276	0.5289	0.5302
3.4	0.5315	0.5328	0.5340	0.5353	0.5366	0.5378	0.5391	0.5403	0.5416	0.5428
3.5	0.5441	0.5453	0.5465	0.5478	0.5490	0.5502	0.5514	0.5527	0.5539	0.5551
3.6	0.5563	0.5575	0.5587	0.5599	0.5611	0.5623	0.5635	0.5647	0.5658	0.5670
3.7	0.5682	0.5694	0.5705	0.5717	0.5729	0.5740	0.5752	0.5763	0.5775	0.5786
3.8	0.5798	0.5809	0.5821	0.5832	0.5843	0.5855	0.5866	0.5877	0.5888	0.5899
3.9	0.5911	0.5922	0.5933	0.5944	0.5955	0.5966	0.5977	0.5988	0.5999	0.6010
4.0	0.6021	0.6031	0.6042	0.6053	0.6064	0.6075	0.6085	0.6096	0.6107	0.6117
4.1	0.6128	0.6138	0.6149	0.6160	0.6170	0.6180	0.6191	0.6201	0.6212	0.6222
4.2	0.6232	0.6243	0.6253	0.6263	0.6274	0.6284	0.6294	0.6304	0.6314	0.6325
4.3	0.6335	0.6345	0.6355	0.6365	0.6375	0.6385	0.6395	0.6405	0.6415	0.6425
4.4	0.6435	0.6444	0.6454	0.6464	0.6474	0.6484	0.6493	0.6503	0.6513	0.6522
4.5	0.6532	0.6542	0.6551	0.6561	0.6571	0.6580	0.6590	0.6599	0.6609	0.6618
4.6	0.6628	0.6637	0.6646	0.6656	0.6665	0.6675	0.6684	0.6693	0.6702	0.6712
4.7	0.6721	0.6730	0.6739	0.6749	0.6758	0.6767	0.6776	0.6785	0.6794	0.6803
4.8	0.6812	0.6821	0.6830	0.6839	0.6848	0.6857	0.6866	0.6875	0.6884	0.6893
4.9	0.6902	0.6911	0.6920	0.6928	0.6937	0.6946	0.6955	0.6964	0.6972	0.6981
5.0	0.6990	0.6998	0.7007	0.7016	0.7024	0.7033	0.7042	0.7050	0.7059	0.7067
5.1	0.7076	0.7084	0.7093	0.7101	0.7110	0.7118	0.7126	0.7135	0.7143	0.7152
5.2	0.7160	0.7168	0.7177	0.7185	0.7193	0.7202	0.7210	0.7218	0.7226	0.7235
5.3	0.7243	0.7251	0.7259	0.7267	0.7275	0.7284	0.7292	0.7300	0.7308	0.7316
5.4	0.7324	0.7332	0.7340	0.7348	0.7356	0.7364	0.7372	0.7380	0.7388	0.7396

By permission from Thomas L. Wade and Howard E. Taylor, *Fundamental Mathematics*, McGraw-Hill Book Company, New York, 1960.

TABLE 3 *(continued)*

N	*0*	*1*	*2*	*3*	*4*	*5*	*6*	*7*	*8*	*9*
5.5	0.7404	0.7412	0.7419	0.7427	0.7435	0.7443	0.7451	0.7459	0.7466	0.7474
5.6	0.7482	0.7490	0.7497	0.7505	0.7513	0.7520	0.7528	0.7536	0.7543	0.7551
5.7	0.7559	0.7566	0.7574	0.7582	0.7589	0.7597	0.7604	0.7612	0.7619	0.7627
5.8	0.7634	0.7642	0.7649	0.7657	0.7664	0.7672	0.7679	0.7686	0.7694	0.7701
5.9	0.7709	0.7716	0.7723	0.7731	0.7738	0.7745	0.7752	0.7760	0.7767	0.7774
6.0	0.7782	0.7789	0.7796	0.7803	0.7810	0.7818	0.7825	0.7832	0.7839	0.7846
6.1	0.7853	0.7860	0.7868	0.7875	0.7882	0.7889	0.7896	0.7903	0.7910	0.7917
6.2	0.7924	0.7931	0.7938	0.7945	0.7952	0.7959	0.7966	0.7973	0.7980	0.7987
6.3	0.7993	0.8000	0.8007	0.8014	0.8021	0.8028	0.8035	0.8041	0.8048	0.8055
6.4	0.8062	0.8069	0.8075	0.8082	0.8089	0.8096	0.8102	0.8109	0.8116	0.8122
6.5	0.8129	0.8136	0.8142	0.8149	0.8156	0.8162	0.8169	0.8176	0.8182	0.8189
6.6	0.8195	0.8202	0.8209	0.8215	0.8222	0.8228	0.8235	0.8241	0.8248	0.8254
6.7	0.8261	0.8267	0.8274	0.8280	0.8287	0.8293	0.8299	0.8306	0.8312	0.8319
6.8	0.8325	0.8331	0.8338	0.8344	0.8351	0.8357	0.8363	0.8370	0.8376	0.8382
6.9	0.8388	0.8395	0.8401	0.8407	0.8414	0.8420	0.8426	0.8432	0.8439	0.8445
7.0	0.8451	0.8457	0.8463	0.8470	0.8476	0.8482	0.8488	0.8494	0.8500	0.8506
7.1	0.8513	0.8519	0.8525	0.8531	0.8537	0.8543	0.8549	0.8555	0.8561	0.8567
7.2	0.8573	0.8579	0.8585	0.8591	0.8597	0.8603	0.8609	0.8615	0.8621	0.8627
7.3	0.8633	0.8639	0.8645	0.8651	0.8657	0.8663	0.8669	0.8675	0.8681	0.8686
7.4	0.8692	0.8698	0.8704	0.8710	0.8716	0.8722	0.8727	0.8733	0.8739	0.8745
7.5	0.8751	0.8756	0.8762	0.8768	0.8774	0.8779	0.8785	0.8791	0.8797	0.8802
7.6	0.8808	0.8814	0.8820	0.8825	0.8831	0.8837	0.8842	0.8848	0.8854	0.8859
7.7	0.8865	0.8871	0.8876	0.8882	0.8887	0.8893	0.8899	0.8904	0.8910	0.8915
7.8	0.8921	0.8927	0.8932	0.8938	0.8943	0.8949	0.8954	0.8960	0.8965	0.8971
7.9	0.8976	0.8982	0.8987	0.8993	0.8998	0.9004	0.9009	0.9015	0.9020	0.9025
8.0	0.9031	0.9036	0.9042	0.9047	0.9053	0.9058	0.9063	0.9069	0.9074	0.9079
8.1	0.9085	0.9090	0.9096	0.9101	0.9106	0.9112	0.9117	0.9122	0.9128	0.9133
8.2	0.9138	0.9143	0.9149	0.9154	0.9159	0.9165	0.9170	0.9175	0.9180	0.9186
8.3	0.9191	0.9196	0.9201	0.9206	0.9212	0.9217	0.9222	0.9227	0.9232	0.9238
8.4	0.9243	0.9248	0.9253	0.9258	0.9263	0.9269	0.9274	0.9279	0.9284	0.9289
8.5	0.9294	0.9299	0.9304	0.9309	0.9315	0.9320	0.9325	0.9330	0.9335	0.9340
8.6	0.9345	0.9350	0.9355	0.9360	0.9365	0.9370	0.9375	0.9380	0.9385	0.9390
8.7	0.9395	0.9400	0.9405	0.9410	0.9415	0.9420	0.9425	0.9430	0.9435	0.9440
8.8	0.9445	0.9450	0.9455	0.9460	0.9465	0.9469	0.9474	0.9479	0.9484	0.9489
8.9	0.9494	0.9499	0.9504	0.9509	0.9513	0.9518	0.9523	0.9528	0.9533	0.9538
9.0	0.9542	0.9547	0.9552	0.9557	0.9562	0.9566	0.9571	0.9576	0.9581	0.9586
9.1	0.9590	0.9595	0.9600	0.9605	0.9609	0.9614	0.9619	0.9624	0.9628	0.9633
9.2	0.9638	0.9643	0.9647	0.9652	0.9657	0.9661	0.9666	0.9671	0.9675	0.9680
9.3	0.9685	0.9689	0.9694	0.9699	0.9703	0.9708	0.9713	0.9717	0.9722	0.9727
9.4	0.9731	0.9736	0.9741	0.9745	0.9750	0.9754	0.9759	0.9763	0.9768	0.9773
9.5	0.9777	0.9782	0.9786	0.9791	0.9795	0.9800	0.9805	0.9809	0.9814	0.9818
9.6	0.9823	0.9827	0.9832	0.9836	0.9841	0.9845	0.9850	0.9854	0.9859	0.9863
9.7	0.9868	0.9872	0.9877	0.9881	0.9886	0.9890	0.9894	0.9899	0.9903	0.9908
9.8	0.9912	0.9917	0.9921	0.9926	0.9930	0.9934	0.9939	0.9943	0.9948	0.9952
9.9	0.9956	0.9961	0.9965	0.9969	0.9974	0.9978	0.9983	0.9987	0.9991	0.9996

ANSWERS TO ODD-NUMBERED PROBLEMS AND ALL ANSWERS FOR CHAPTER REVIEWS

EXERCISE 1

1. 1 **3.** 32 **5.** 1 **7.** 8 **9.** 7 **11.** 3 **13.** 2 **15.** 3 **17.** 8 **19.** 8 **21.** 46 **23.** Variables: A, b, h; constants: $\frac{1}{2}$ **25.** Variables: P, s; constant: 4 **27.** Variables: u, v; constants: 3, 2 **29.** 8 **31.** 28 **33.** 6 **35.** $7 + x$ or $x + 7$ **37.** $7x$ **39.** $2x - 8$ **41.** $18 = 3x$ **43.** $80 = 2x + 3$ or $80 = 3 + 2x$ **45.** $4x = 12x - 3$ **47.** 3 **49.** 6 **51.** $2x - 5 = 3(x - 2)$ **53.** (*A*) $d = 1{,}120t$; (*B*) 1120 is a constant, d and t are variables; (*C*) 320 feet more than a mile.

EXERCISE 2

1. Commutative axiom for addition **3.** Associative axiom for addition **5.** Commutative axiom for multiplication **7.** Associative axiom for multiplication **9.** Identity axiom for addition **11.** Identity axiom for multiplication **13.** $x + 9$ **15.** $20y$ **17.** $u + 15$ **19.** $21x$ **21.** x **23.** Commutative axiom for addition **25.** Commutative axiom for multiplication **27.** Commutative axiom for addition **29.** Associative axiom for multiplication **31.** $x + y + z + 12$ **33.** $3x + 4y + 11$ **35.** $x + y + 5$ **37.** $36mnp$ **39.** (*B*) and (*D*) are false, since $12 - 4 \neq 4 - 12$ and $12 \div 4 \neq 4 \div 12$ **41.** *1.* Commutative axiom for addition; *2.* Associative axiom for addition; *3.* Associative axiom for addition; *4.* Substitutive principle for equality; *5.* Commutative axiom for addition; *6.* Associative axiom for addition.

EXERCISE 3

1. -7 **3.** 6 **5.** 2 **7.** 27 9. 0 **11.** -10 **13.** 4 **15.** -6 **17.** 12 **19.** Sometimes **21.** -3 **23.** -2 **25.** -6 **27.** -5 **29.** -5 **31.** -5 **33.** -1 **35.** -6 **37.** 28 **39.** -5 **41.** 7 or -7 **43.** -5 **45.** 5 **47.** -5 **49.** -3 **51.** 6 **53.** \$23 **55.** True **57.** False; $(+7) - (-3) = +10$, $(-3) - (+7) = -10$ **59.** True **61.** False; $|(+9) + (-3)| = +6$, $|+9| + |-3| = +12$ **63.** Commutative axiom for addition, associative axiom for addition, Theorem 2, identity axiom

EXERCISE 4

1. 15 **3.** 3 **5.** -18 **7.** -3 **9.** 0 **11.** 0 **13.** Not defined **15.** Not defined **17.** -7 **19.** -1 **21.** 11 **23.** 9 **25.** Both are 8 **27.** -10 **29.** 3 **31.** -14 **33.** -5 **35.** -70 **37.** 0 **39.** 10 **41.** -12 **43.** Not defined **45.** 56 **47.** 0 **49.** -6 **51.** 8 **53.** Never **55.** Always **57.** -50 **59.** 0 **61.** When x and y are of opposite signs **63.** *1.* Identity axiom for addition; *2.* Distributive axiom; *3.* Addition property for equality; *4.* Inverse and associative axiom for addition; *5.* Inverse axiom for addition; *6.* Identity axiom for addition; *7.* Symmetric property for equality.

EXERCISE 5

1. $18 > 4$ **3.** $-5 < 9$ **5.** $x \geq -8$ **7.** $x \leq 3$ **9.** 12 is greater than 11 **11.** 11 is less than 12 **13.** x is less than or equal to 4 **15.** $6 > 3$ **17.** $-3 > -6$ **19.** $-6 < -3$ **21.** $5 > 0$ **23.** $-5 < 0$ **25.** $-6 < -3$ **27.** $e > a$ **29.** $c > b$ **31.** $0 < d$ **33.** greater than

35. **37.**

39. 41.

43. $<$ 45. $>$ 47. $\{-2, -4, -6\}$ 49. $\{-6, -4\}$ 51. $\{-2, 2, 4, 6\}$
53. (A) $A = x(200 - x)$; (B) $0 < x < 200$

55. 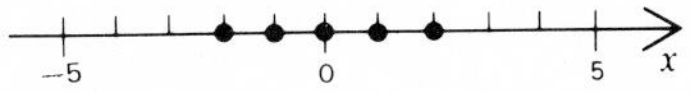 −5 0 5 x

57. 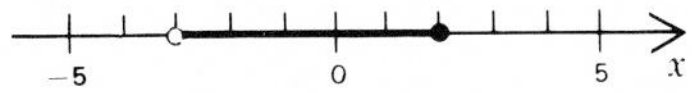 −5 0 5 x

59. If $a < b$, then there exists a positive number p such that $a + p = b$. Subtracting a from each number, we obtain $b - a = p$, a a positive number.

EXERCISE 6

1. (A) F; (B) T 2. (A) $8y$; (B) $y + 8$ 3. (A) Commutative axiom for multiplication; (B) Associative axiom for addition 4. (A) $x + 10$; (B) $10x$; (C) $x + 5$ 5. (A) -4; (B) 8; (C) 2; (D) 9 6. -5 7. -13 8. 6 9. -3 10. 3 11. -12 12. 28 13. -18 14. -4 15. 6 16. Not defined 17. 0 18. -3 19. 4 20. -14 21. -5 22. 4 23. -12 24. $>$ 25. $<$ 26. $>$ 27. $<$ 28. -10 29. -60 30. 6 31. 4 32. 1 33. 10 34. -3 35. 4 36. -26 37. 0 38. 6 39. (A) $+8$; (B) 5 40. -6 41. 15 42. 49 43. -7 44. Commutative axiom for addition 45. Associative axiom for addition 46. Commutative axiom for multiplication 47. Associative axiom for multiplication 48. (A) $m + n + p + 14$; (B) $28xy$ 49. $3x = x + 8$ 50. $p = 2x + 2(2x + 5)$ 51. (A) a; (B) 0 52. (A) F; (B) T 53. (A) $\{-2, -1, 0\}$; (B) $\{-1, 0\}$

54. (A) −5 0 x (B) −5 0 x

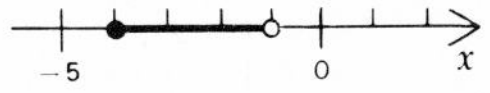

55. Less than 56. (A) -5; $+5$; (B) -7 57. 35 58. -42 59. 6 60. All positive integers and 0 61. $2(x - 3) = 2x + 5$ 62. $C = \{x \in N \mid 3 \le x < 7\}$ 63. *1.* Theorem 5A; *2.* Identity axiom for multiplication; *3.* Distributive axiom; *4.* Inverse axiom for addition; *5.* Theorem 4C

EXERCISE 7

1. -3 3. 3 5. 1 7. $17x$ 9. x 11. $8x$ 13. $-13t$ 15. $5y + 3x$ 17. $4m - 6n$ 19. $9u - 4v$ 21. $-2m - 24n$ 23. $5u - 6v$ 25. $-u + 4v$ 27. $9x - 3$ 29. $-2x - 4$ 31. $7x^2 - x - 12$ 33. $-x + 1$ 35. $-y^2 - 2$ 37. $-3x^2y$ 39. $3y^3 + 4y^2 - y - 3$ 41. $3a^2 - b^2$ 43. $-7x + 9y$ 45. $-5x + 3y$ 47. $4x - 6$ 49. $-8x + 12$ 51. $10t - 18$ 53. $m - 2n$ 55. $-m + 2n$ 57. $-x + y - z$ 59. $2x^4 + 3x^3 + 7x^2 - x - 8$ 61. $-3x^3 + x^2 + 3x - 2$ 63. $-2m^3 - 5$ 65. $P = 2x + 2(x - 5) = 4x - 10$ 67. $-t + 27$ 69. $2x - w$ 71. Value in cents $= 5x + 10(x - 5) + 25[(x - 5) + 2] = 40x - 125$

EXERCISE 8

1. y^5 3. $10y^5$ 5. $-24x^{20}$ 7. $6u^{16}$ 9. c^3d^4 11. $15x^2y^3z^5$ 13. $y^2 + 7y$ 15. $10y^2 - 35y$ 17. $3a^5 + 6a^4$ 19. $2y^3 + 4y^2 - 6y$ 21. $7m^6 - 14m^5 - 7m^4 + 28m^3$ 23. $10u^4v^3 - 15u^2v^4$ 25. $2c^3d^4 - 4c^2d^4 +$

$8c^4d^5$ **27.** $6y^3 + 19y^2 + y - 6$ **29.** $m^3 - 2m^2n - 9mn^2 - 2n^3$ **31.** $6m^4 + 2m^3 - 5m^2 + 4m - 1$ **33.** $a^3 + b^3$ **35.** $2x^4 + x^3y - 7x^2y^2 + 5xy^3 - y^4$ **37.** $x^2 + 5x + 6$ **39.** $a^2 + 4a - 32$ **41.** $t^2 - 16$ **43.** $m^2 - n^2$ **45.** $4t^2 - 11t + 6$ **47.** $3x^2 - 7xy - 6y^2$ **49.** $4m^2 - 49$ **51.** $30x^2 - 2xy - 12y^2$ **53.** $6s^2 - 11st + 3t^2$ **55.** $9x^2 + 12x + 4$ **57.** $9x^2 + 24xy + 16y^2$ **59.** $x^3 + 6x^2y + 12xy^2 + 8y^3$ **61.** $-x^2 + 17x - 11$ **63.** $2x^3 - 13x^2 + 25x - 18$ **65.** $9x^3 - 9x^2 - 18x$ **67.** $12x^5 - 19x^3 + 12x^2 + 4x - 3$ **69.** $y(y - 8)$; $y^2 - 8y$

EXERCISE 9

1. $7(m + 1)$ **3.** $7x^2y(2x - 3)$ **5.** $3y(1 - 2y - 4y^2)$ **7.** $5xy(2x^2 - xy - 4y^2)$ **9.** $(x + 2)(x + 5)$ **11.** $(x - 2)(x - 5)$ **13.** Not factorable **15.** $(a + 3b)(a + 5b)$ **17.** $(x - 2y)(x - 8y)$ **19.** Not factorable **21.** $(2x - 3)(x - 2)$ **23.** $(2x - 3y)(x - 2y)$ **25.** $(n + 4)(n - 2)$ **27.** $(m + 2n)(m - 6n)$ **29.** $(3x + 2y)(4x + 3y)$ **31.** Not factorable **33.** $(x + 2y)(x - 6y)$ **35.** $(3x - y)(2x + 3y)$ **37.** $6(m + 2)(m + 6)$ **39.** $x(x - 3)(x - 8)$ **41.** $3m(m^2 - 2m + 5)$ **43.** $(4x - 3)(2x + 3)$ **45.** Not factorable **47.** $2uv(2u + v)(u + 3v)$ **49.** $(5x - y)(3x + 4y)$ **51.** $3y(3x - 5y)(x + 2y)$ **53.** $(8x + 3y)(3x - 5y)$ **55.** $4x^2(5x - y)(3x + 4y)$ **57.** 6, 10, 12

EXERCISE 10

1. $(y - 5)(y + 5)$ **3.** $(3x - 2)(3x + 2)$ **5.** Not factorable **7.** $(5x - 1)(5x + 1)$ **9.** $3(x - 1)(x + 1)$ **11.** $(x + 1)(x^2 - x + 1)$ **13.** $(x + 3)(x^2 - 3x + 9)$ **15.** $(x + y)(y + 2)$ **17.** $(x - 5)(x + y)$ **19.** $(ab - c)(ab + c)$ **21.** Not factorable **23.** $xy(2x - y)(2x + y)$ **25.** $2x(x^2 + 4)$ **27.** $(2y - 1)(4y^2 + 2y + 1)$ **29.** $(ab + 2)(a^2b^2 - 2ab + 4)$ **31.** $4(u + 2v)(u^2 - 2uv + 4v^2)$ **33.** $(x - y)(a - 2b)$ **35.** $(3c - 4d)(5a + b)$ **37.** $(x - 2)(x^2 + 1)$ **39.** $(y - x)(y - x - 1)$ **41.** $(r^2 + s^2)(r - s)(r + s)$ **43.** $(x^2 - 4)(x^2 + 1) = (x - 2)(x + 2)(x^2 + 1)$ **45.** $[(a - b) - 2(c - d)][(a - b) + 2(c - d)] = (a - b - 2c + 2d)(a - b + 2c - 2d)$ **47.** $[5(2x - 3y) - 3ab][5(2x - 3y) + 3ab]$ **49.** $(x^3 - 1)(x^3 + 1) = (x - 1)(x^2 + x + 1)(x + 1)(x^2 - x + 1)$ **51.** $(2x - 1)(x - 2)(x + 2)$ **53.** $[5 - (a + b)][5 + (a + b)]$ **55.** $[4x^2 - (x - 3y)][4x^2 + (x - 3y)]$

EXERCISE 11

1. $3x + 1$ **3.** $2y^2 + y - 3$ **5.** $3x + 1, R = 3$ **7.** $4x - 1$ **9.** $3x - 4, R = -1$ **11.** $x + 2$ **13.** $4x + 1, R = -4$ **15.** $4x + 6, R = 25$ **17.** $x - 4, R = 3$ **19.** $x^2 + x + 1$ **21.** $x^3 + 3x^2 + 9x + 27$ **23.** $4a + 5, R = -7$ **25.** $x^2 + 3x - 5$ **27.** $x^2 + 3x + 8, R = 27$ **29.** $3x^3 + x^2 - 2, R = -4$ **31.** $4x^2 - 2x - 1, R = -2$ **33.** $2x^3 + 6x^2 + 32x + 84, R = 186x - 170$ **35.** $Q(x) = x - 3, R(x) = x + 2$

EXERCISE 12

1. $5x^2 + x - 5$ **2.** $-2x^2 + x - 7$ **3.** $6x^5y - 9x^4y^2 + 3x^2y^3$ **4.** $6x^2 + 11x - 10$ **5.** $2x^3 - 7x^2 + 13x - 5$ **6.** $3x + 4, R = 2$ **7.** $2xy(2x - 3y)$ **8.** $(x - 2)(x - 7)$ **9.** $(3x - 2)^2$ **10.** Not factorable **11.** $(u + 8)(u - 8)$ **12.** $(3x - 4)(x - 2)$ **13.** $x(x - 2)(x - 3)$ **14.** $(x + y)^2$ **15.** $27x^4 + 63x^3 - 66x^2 - 28x + 24$ **16.** $2x^2 + 5x + 5$ **17.** $3x^2 + 2x - 2, R = -2$ **18.** (A) 5; (B) 3 **19.** $(m - 4n)(m + n)$ **20.** $2(m - 2n)(m + 2n)$ **21.** $3xy(4x^2 + 9y^2)$ **22.** $(2x - 3y)(x + y)$ **23.** $3n(2n - 5)(n + 1)$ **24.** Not factorable **25.** $(p + q)(x + y)$

26. $(x-y)(x-4)$ **27.** $(y-b)(y-b-1)$ **28.** $3(x-2y)(x^2+2xy+4y^2)$ **29.** $12x^5-19x^3+12x^2+4x-3$ **30.** $4x^2-2x+3$, $R=-2$ **31.** $-2x+20$ **32.** $2x^3-4x^2+12x$ **33.** $3xy(6x-5y)(2x+3y)$ **34.** $4u^2(3u^2-3uv-5v^2)$ **35.** $y(y-2x)$ **36.** $(a^2-b^2)^2=(a-b)^2(a+b)^2$ **37.** $(m-n)(m^2+mn+n^2)(m+n)(m^2-mn+n^2)$ **38.** $(2x-3m+1)(2x+3m-1)$ **39.** 9, −9, 6, −6 **40.** The set of all real numbers **41.** $Q(x)=2x-3$, $R(x)=x+4$

EXERCISE 13

1. −5 **3.** −2 **5.** 5 **7.** −3 **9.** 3 **11.** 0 **13.** 3 **15.** −2 **17.** −7 **19.** 5 **21.** No solution **23.** 18 **25.** 9 **27.** 10 **29.** 1 **31.** 4 **33.** No solution **35.** −3 **37.** 150 **39.** $2x=x-8$, $x=-8$ **41.** T **43.** T **45.** $ca=ca$ (identity property *of equality),* $a=b$ (given), $ca=cb$ (substitution principle)

EXERCISE 14

1. 31, 32, 33 **3.** 12, 14, 16 **5.** 50 miles/hour **7.** 19 inches and 38 inches **9.** 45°, 45°, 90° **11.** 5, 7, 9 **13.** 14 inches by 50 inches **15.** 8 dimes and 12 nickels **17.** 5 seconds **19.** Mechanic: 10 hours, assistant: 8 hours **21.** 1,311 brown-eyed, 437 blue-eyed **23.** 10:00 A.M.; 24 miles **25.** 8 hours **27.** 130 minutes (or 2 hours 10 minutes)

EXERCISE 15

1. 3,000 books **3.** 75% **5.** 350 cubic centimeters **7.** 1 liter **9.** 300 miles **11.** 18 four-cent stamps and 12 five-cent stamps **13.** 90 miles **15.** 20 cents **17.** 78 by 27 feet **19.** 4 feet **21.** 315; 105; 105; 35 **23.** 8 seconds **25.** 550 pounds **27.** 120 pounds **29.** (*A*) 15 ohms; (*B*) 0.3 ampere **31.** 141.2 centimeters **33.** 15 minutes or $\frac{1}{4}$ hour

EXERCISE 16

1. $x<3$ **3.** $x>-7$ **5.** $x<2$ **7.** $x>-2$ **9.** $x<-21$ **11.** $x<10$ **13.** $x>4$

15. $x<5$ (5, x) **17.** $x\geq 3$ (3, x)

19. $n\leq -3$ (−3, n) **21.** $N<-8$ (−8, N)

23. $x<-3$ (−3, x) **25.** $x>-4$ (4, x)

27. $m>3$ (3, m) **29.** $x\geq 3$ (3, x)

31. $2\leq x\leq 13$ (2, 13, x) **33.** $-4\leq x\leq 1$ (−4, 1, x)

35. $-2<t\leq 3$ (−2, 3, t) **37.** $-2\leq m<3$ (−2, 3, m)

39. $2x-3\geq -6$, $x\geq -\frac{3}{2}$ **41.** $15-3x<6$, $x>3$ **43.** $10w>65$, $w>6.5$ **45.** $2x>300+1.5x$, $x>600$ **47.** $220\leq 110I\leq 2{,}750$, $2\leq I\leq 25$ **49.** Negative **51.** *1.* Given; *2.* Definition of >; *3.* Definition of <; *4.* Property of equality; *5.* Definition of <; *6.* Definition of > **53.** $m-n-p=0$, and division by 0 is not permitted

EXERCISE 17

1. $x = \pm 5$

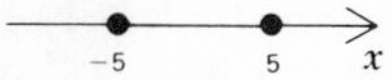

3. $t = -1$ or 7

5. $x = -6$ or 4

7. $-5 \le t \le 5$

9. $-1 < t < 7$

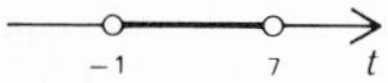

11. $-6 \le x \le 4$

13. $x = -1, 4$

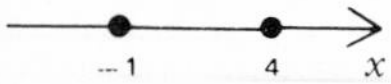

15. $-1 \le x \le 4$

17. $m = -\frac{11}{3}, \frac{2}{3}$

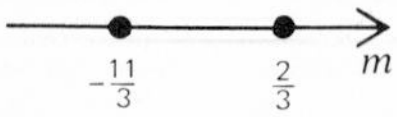

19. $-\frac{8}{9} < M < \frac{22}{9}$

21. $x \le -7$ or $x \ge 7$

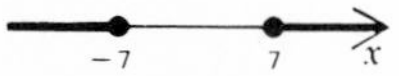

23. $t < -1$ or $t > 7$

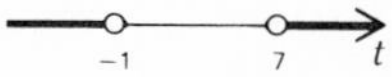

25. $x \le -6$ or $x \ge 4$

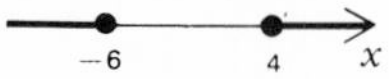

27. $u \le -\frac{7}{3}$ or $u \ge -\frac{1}{3}$

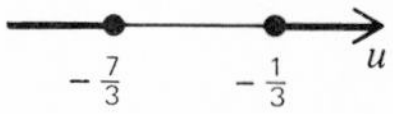

29. $x \ge -7$ **31.** $x < 11$ **33.** Case 1: Assume $x \ge 0$; then by definition of absolute value $|x| < p$ can be replaced with $x < p$ and $x \ge 0$, or $0 \le x < p$. Case 2: Assume $x < 0$; then by definition of absolute value, $|x| < p$ can be replaced with $-x < p$ and $x < 0$, or $-p < x < 0$. Putting cases 1 and 2 together we see that if $|x| < p$, then $-p < x < p$.

EXERCISE 18

1. $x = -2$ **2.** 52, 53, 54

3. $x \le -3$

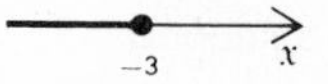

4. $x = \pm 6$

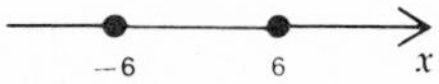

5. $-6 < x < 6$

6. $-14 < y < -4$

7. $x \le 3$

8. $x = 2$ **9.** 8 ft by 13 ft **10.** 10 nickels and 5 quarters

11. $x \ge 1$

12. $-4 \le x < 3$

13. $x = \frac{1}{2}, 3$

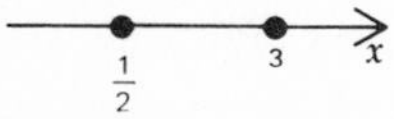

14. $\frac{1}{2} \le x \le 3$

15. (A) T; (B) T **16.** $a(-b/a) + b = -b + b = 0$ **17.** 1 **18.** $x = 1$
19. 0.45 gal

20. $x < -1$ or $x > 5$

21. $x < 2$ or $x > \frac{10}{3}$

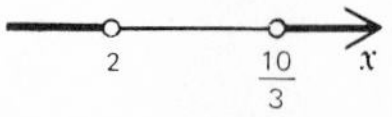

EXERCISE 19

1. $A(-10, 10)$, $B(10, -10)$, $C(16, 14)$, $D(16, 0)$, $E(-14, -16)$, $F(0, 4)$, $G(6, -16)$, $H(-14, 0)$

3.

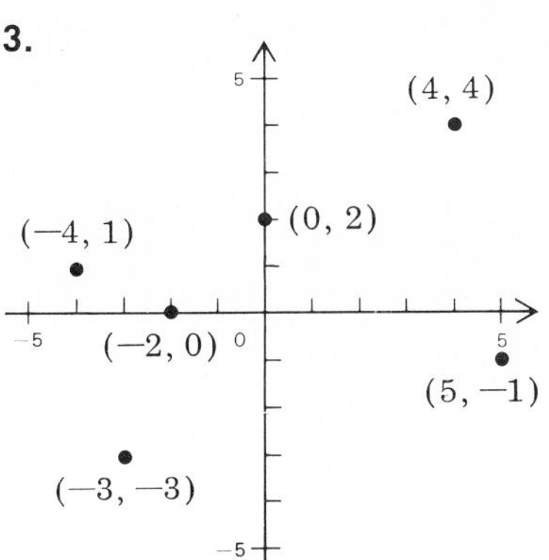

5.

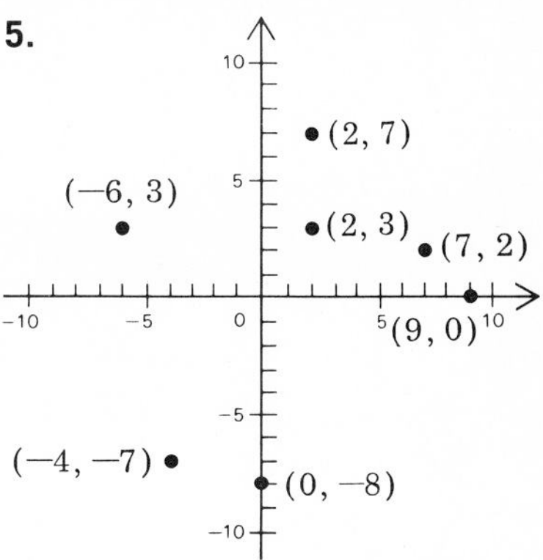

7. $A(-3\frac{1}{2}, 2)$, $B(-2, -4\frac{1}{2})$, $C(0, -2\frac{1}{2})$, $D(2\frac{3}{4}, 0)$, $E(3\frac{3}{4}, 2\frac{3}{4})$, $F(4\frac{1}{4}, -3\frac{1}{2})$

9.

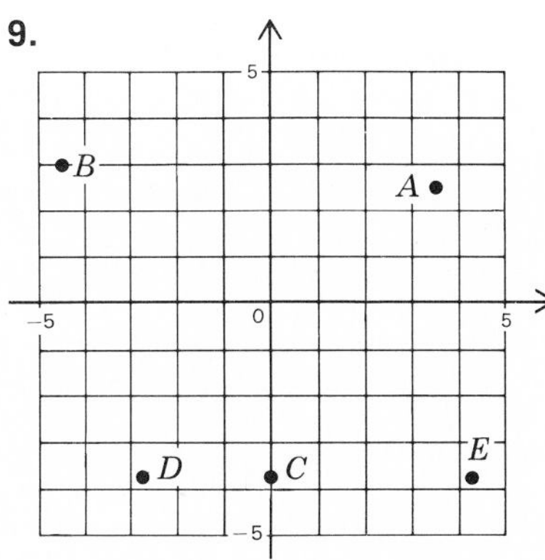

11. (*A*) II; (*B*) IV; (*C*) I; (*D*) III
13. 0 **15.** I (both positive), III (both negative) **17.** I, IV
19. (0, 0), (5, 0), (5, 3), (0, 3)
21. (6, 4), (−6, 4), (−6, −4), (6, −4)

EXERCISE 20

1.

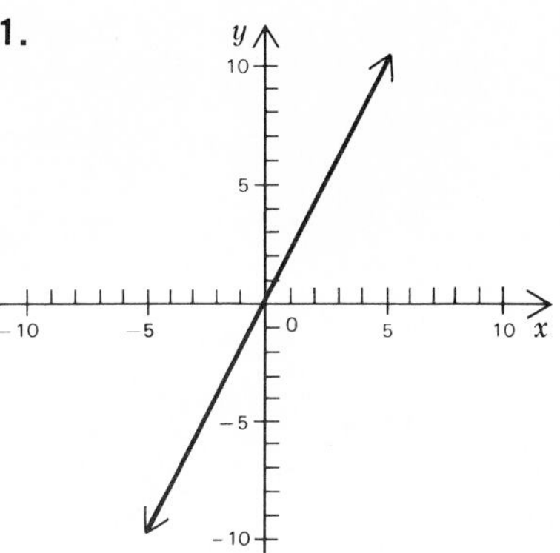

3.

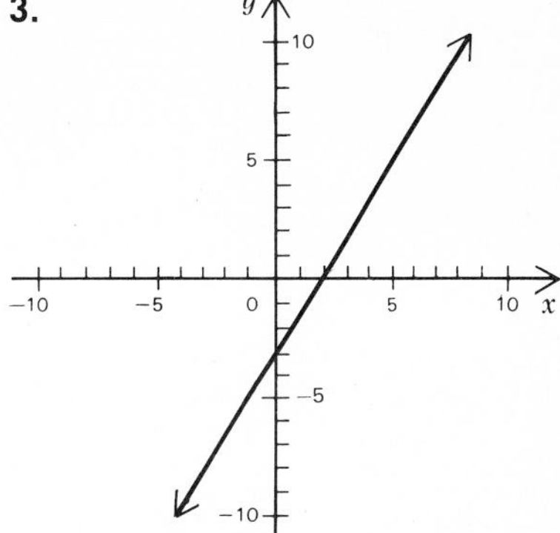

5.

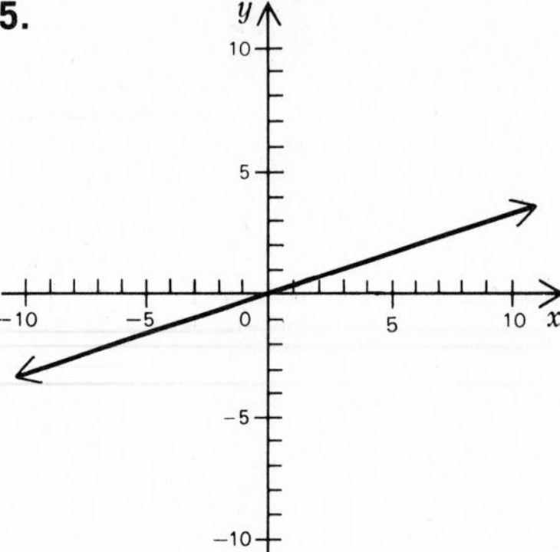
y
10
5
-10
-5
0
5
10
x
-5
-10

7.

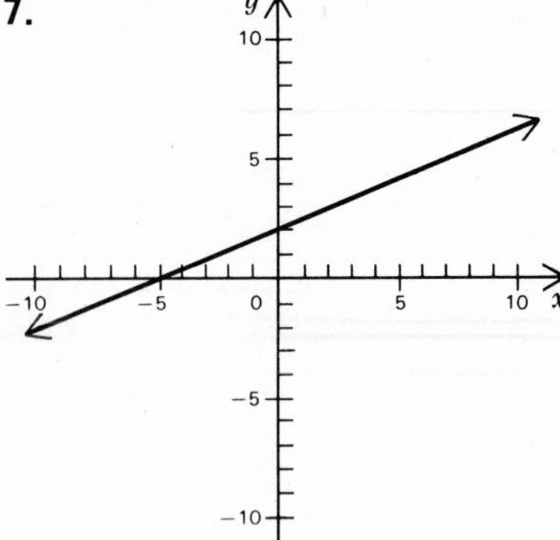
y
10
5
-10
-5
0
5
10
x
-5
-10

9.

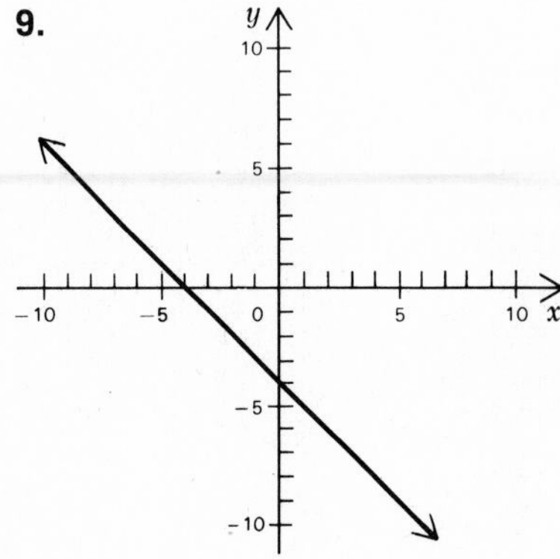
y
10
5
-10
-5
0
5
10
x
-5
-10

11.

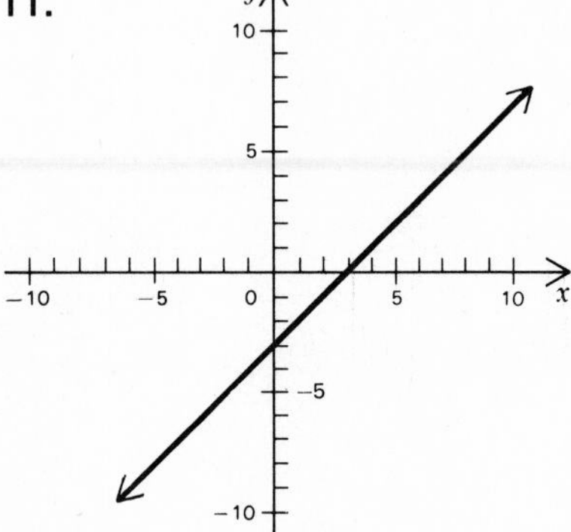
y
10
5
-10
-5
0
5
10
x
-5
-10

13.

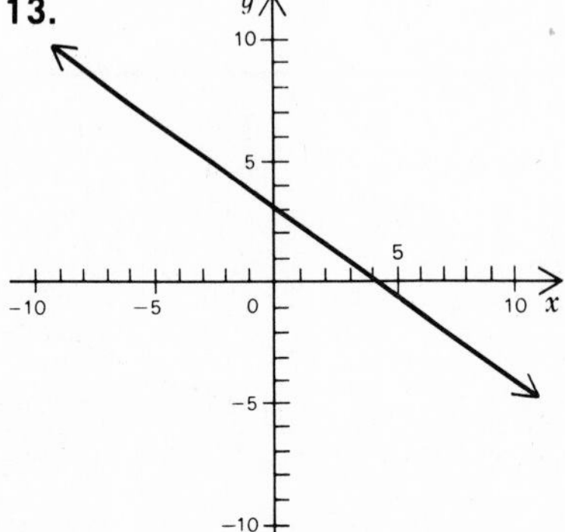
y
10
5
5
-10
-5
0
10
x
-5
-10

15.

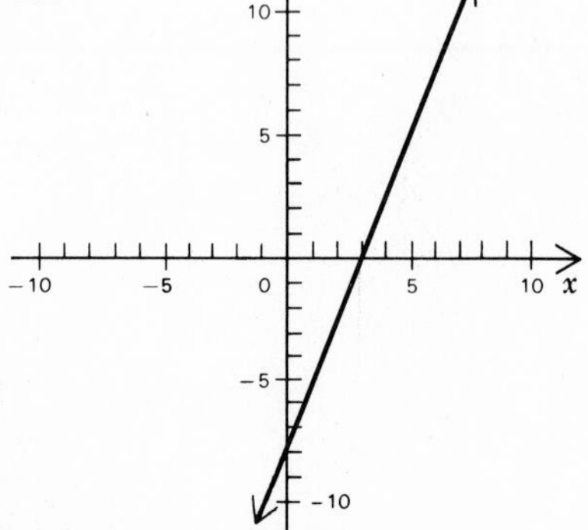
y
10
5
-10
-5
0
5
10
x
-5
-10

17.

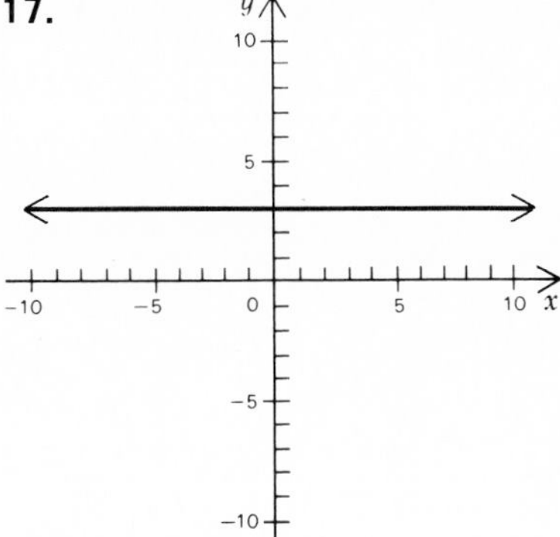

19.

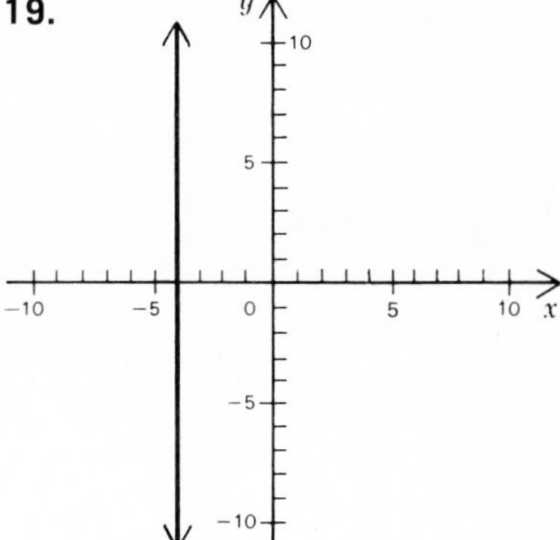

21.

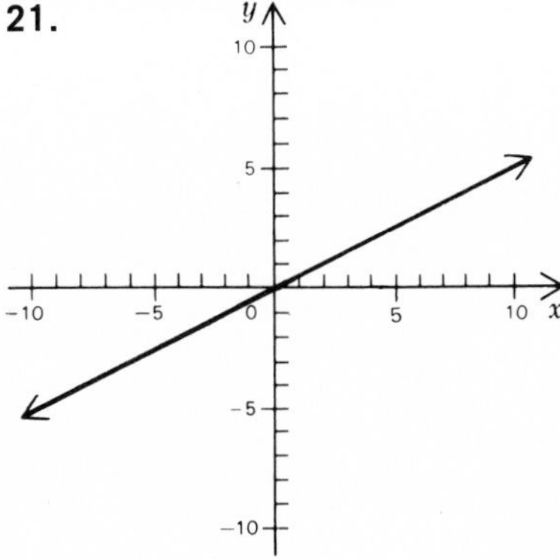

23.

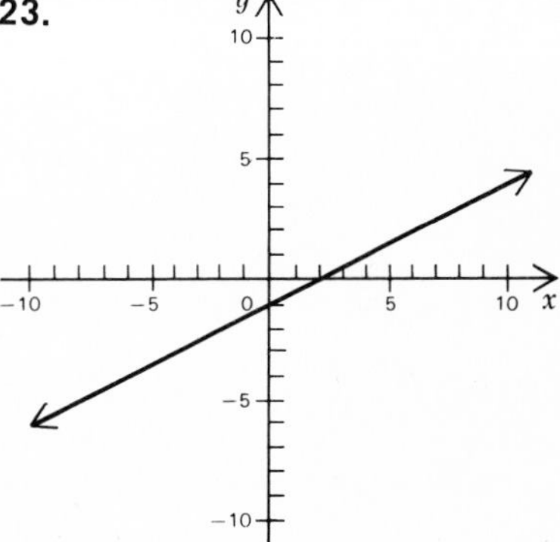

25.

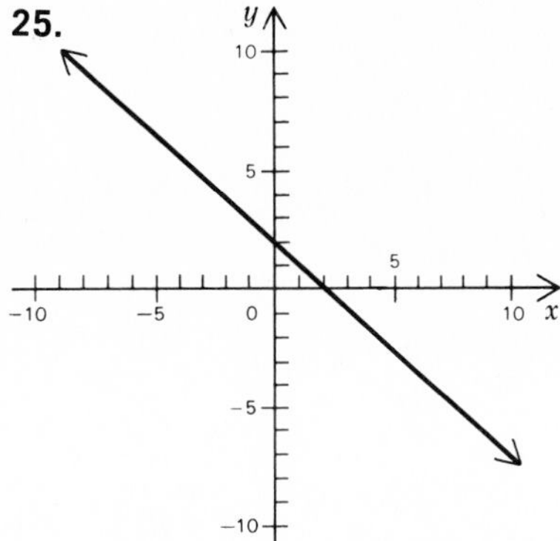

27.

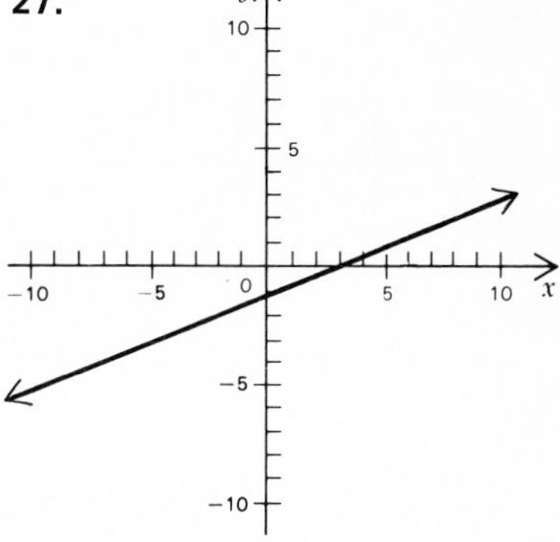

29.

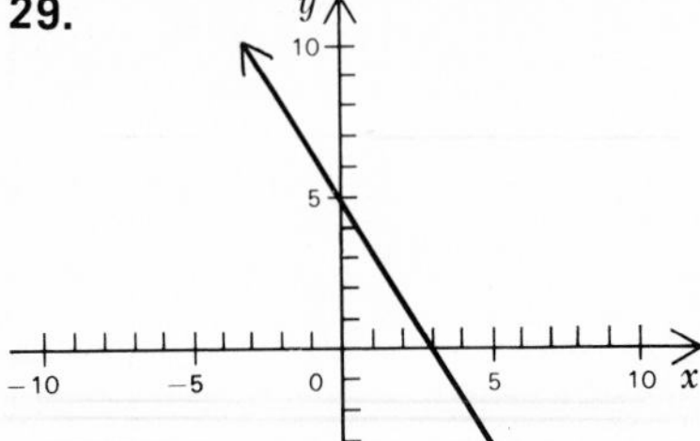

31.

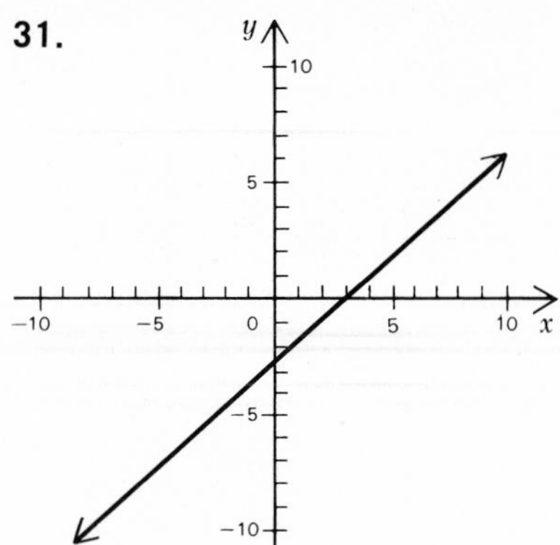

33.

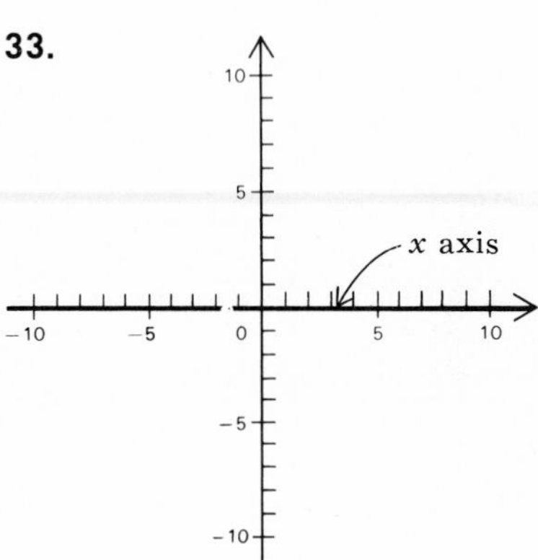

35.

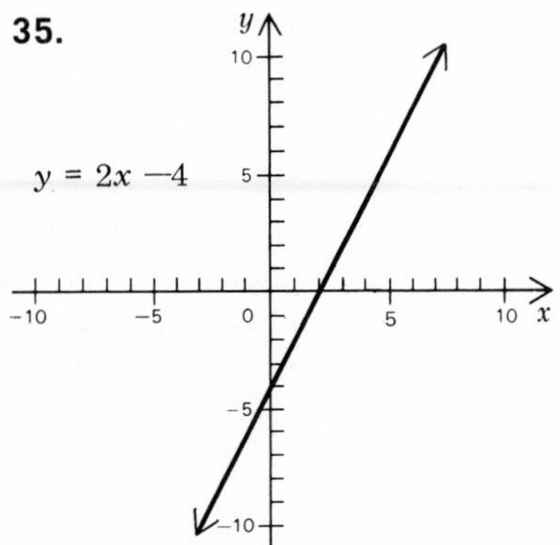

37.

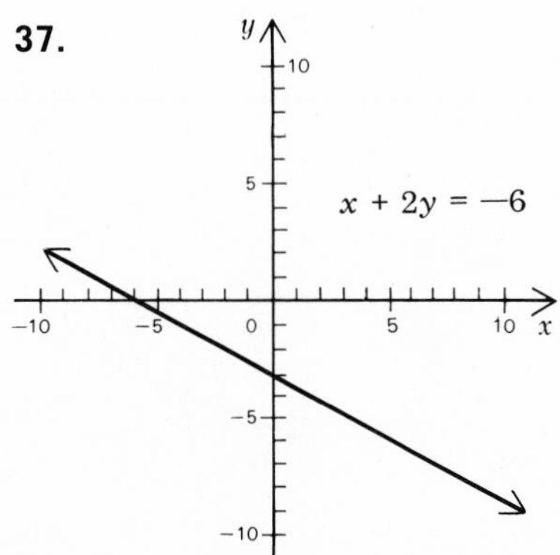

39.

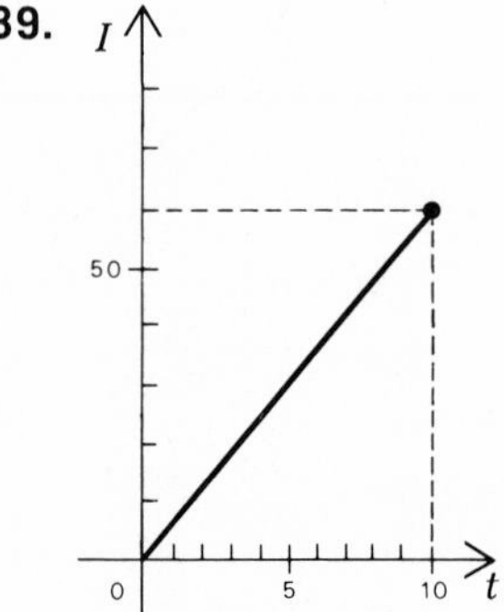

41.

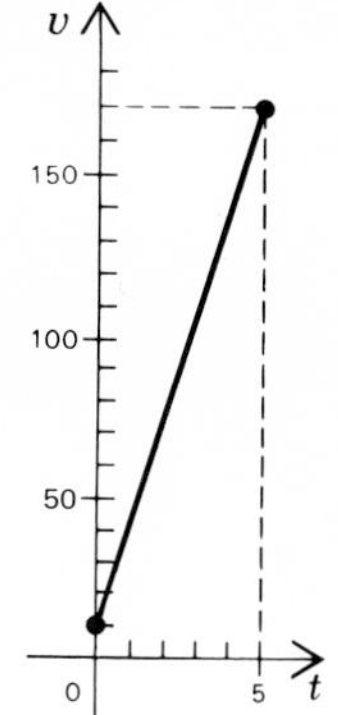

43.

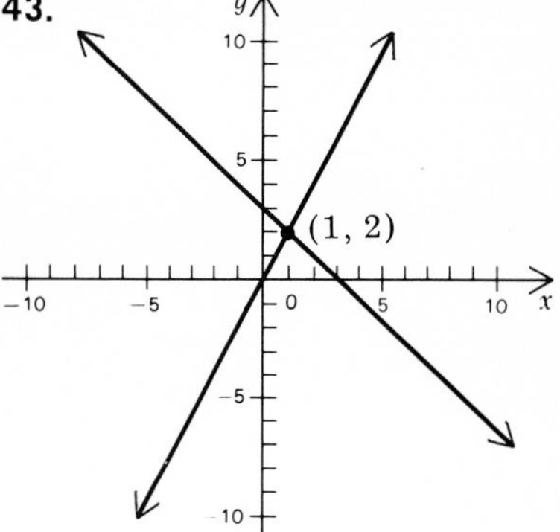

45.

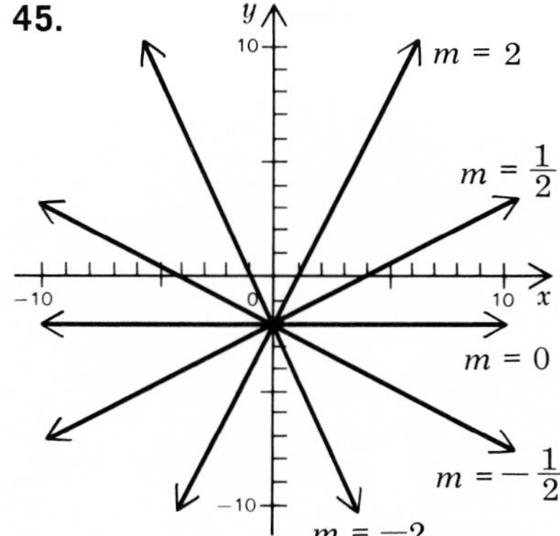

47.

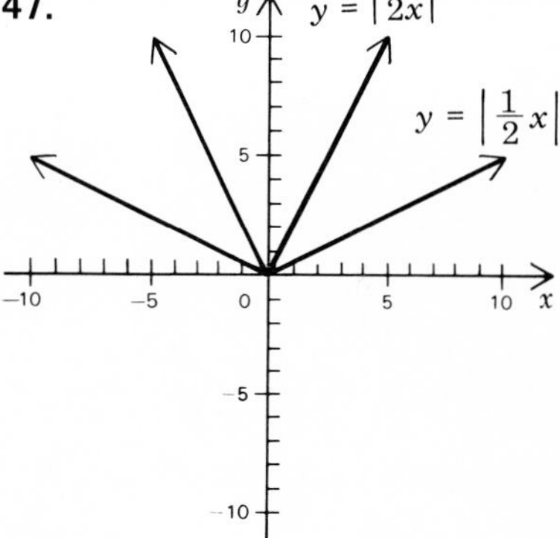

49.

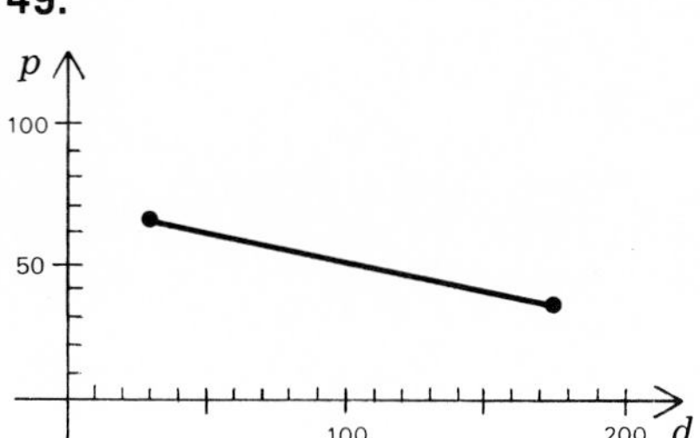

51.

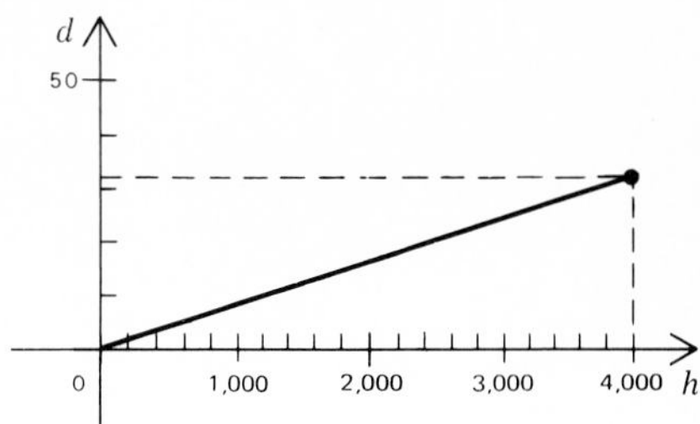

EXERCISE 21

1.

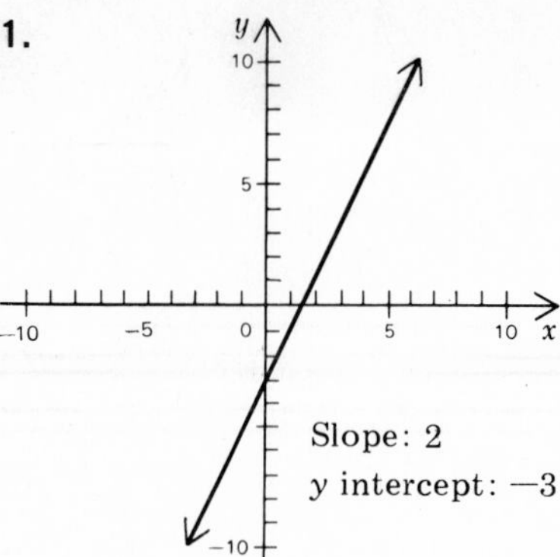

3.

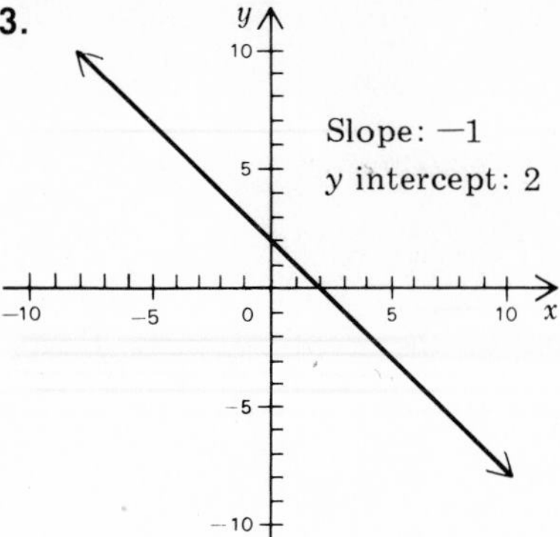

5. $y = 5x - 2$ **7.** $y = -2x + 4$ **9.** $y - 4 = 2(x - 5)$ **11.** $y - 1 = -2(x - 2)$ **13.** 2 **15.** $\frac{1}{2}$ **17.** $y - 6 = 2(x - 5)$ or $y - 2 = 2(x - 3)$ **19.** $y - 5 = \frac{1}{2}(x - 10)$ or $y - 1 = \frac{1}{2}(x - 2)$

21.

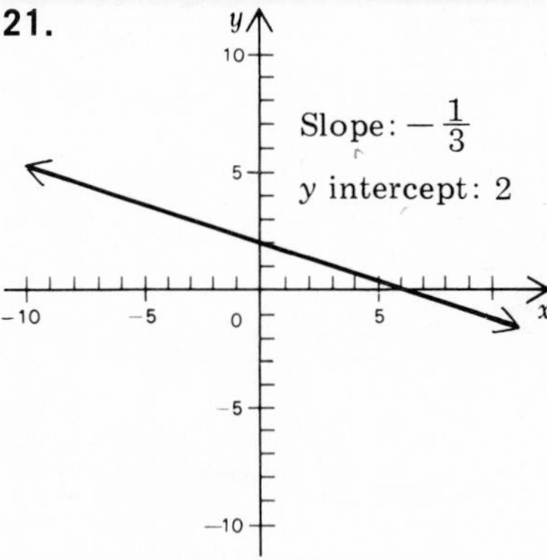

23.

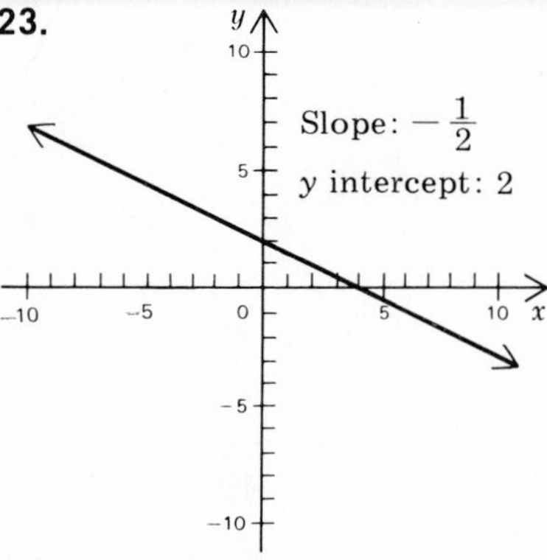

25.

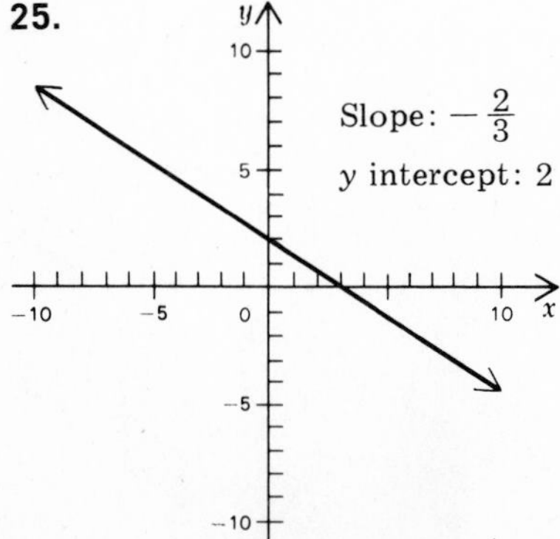

27. $y = -x/2 - 2$
29. $y = \frac{2}{3}x + \frac{3}{2}$
31. $y - 2 = -2(x + 3)$, $y = -2x - 4$
33. $y - 3 = \frac{1}{2}(x + 4)$, $y = x/2 + 5$
35. $\frac{1}{3}$ **37.** $-\frac{1}{4}$
39. $y - 4 = \frac{1}{3}(x + 6)$ or $y - 7 = \frac{1}{3}(x - 3)$, $y = x/3 + 6$
41. $y = -\frac{1}{4}(x + 4)$ or $y + 2 = -\frac{1}{4}(x - 4)$, $y = -x/4 - 1$
43. $x = -3$, $y = 5$
45. $x = -1$, $y = 22$ **47.** $x = 3$
49. $x - 2y = -4$

51. (A) $R = \frac{3}{2}C + 3$
(C) \$158

(B)
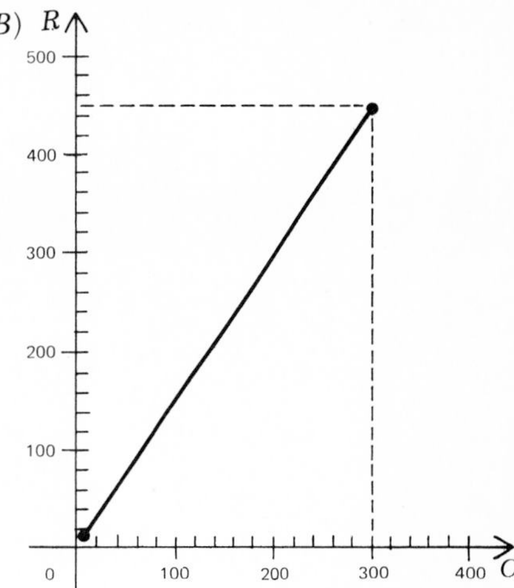

EXERCISE 22

1.
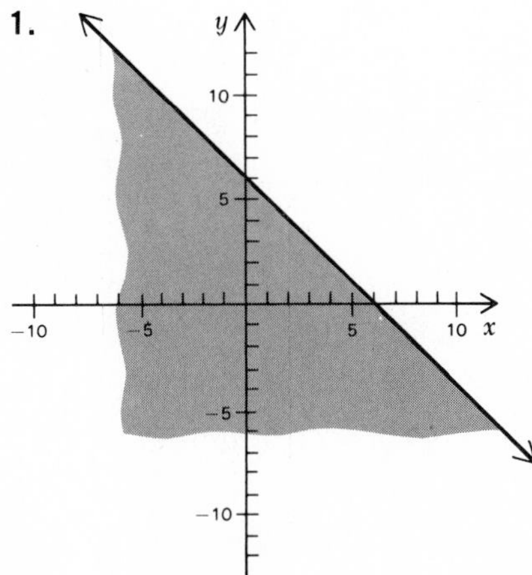

3.
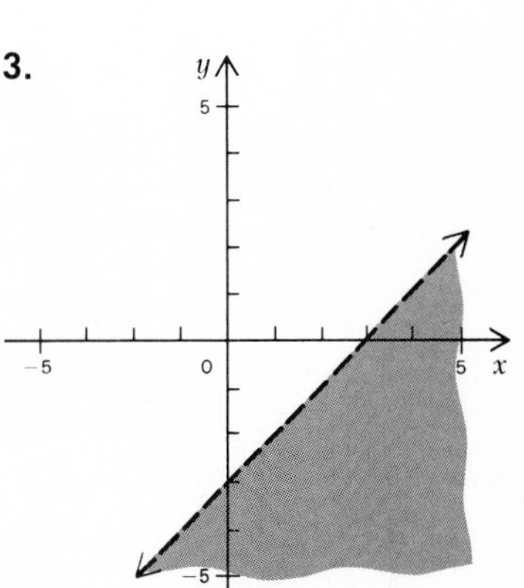

5.
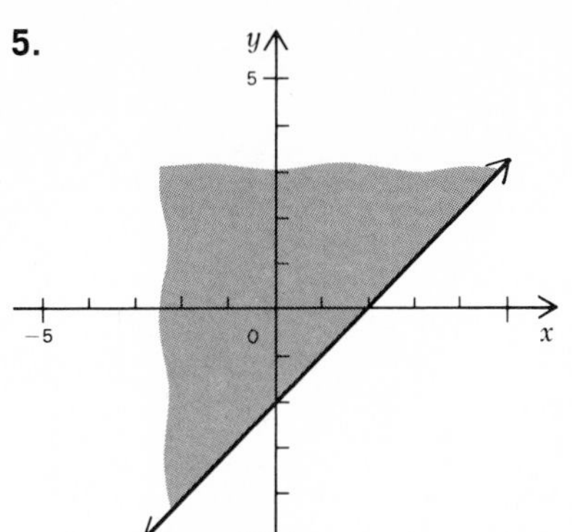

7.
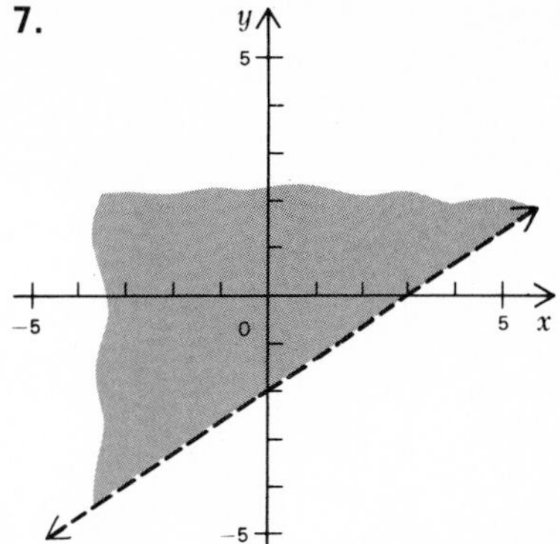

9.

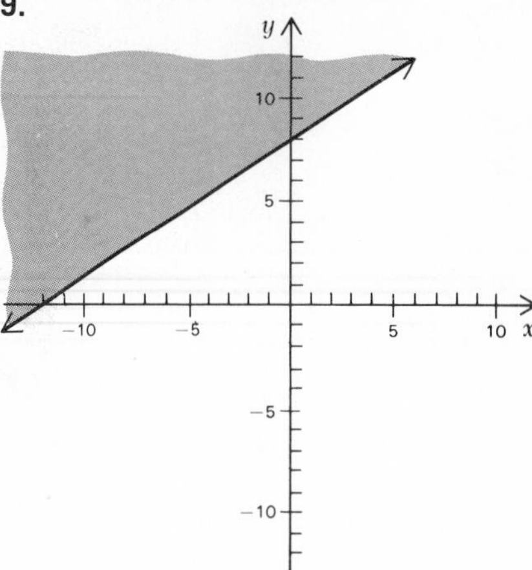

11.

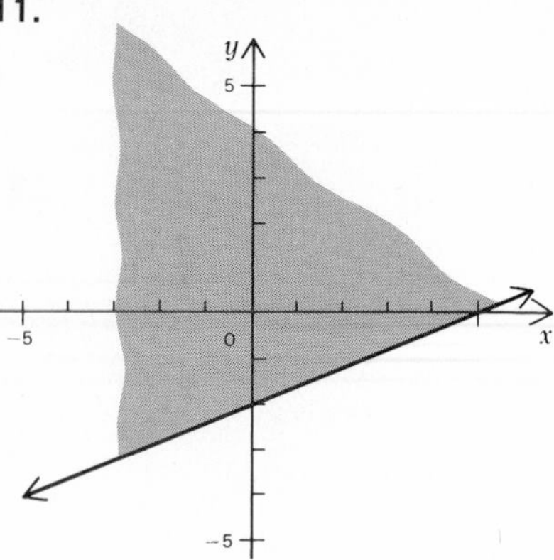

13.

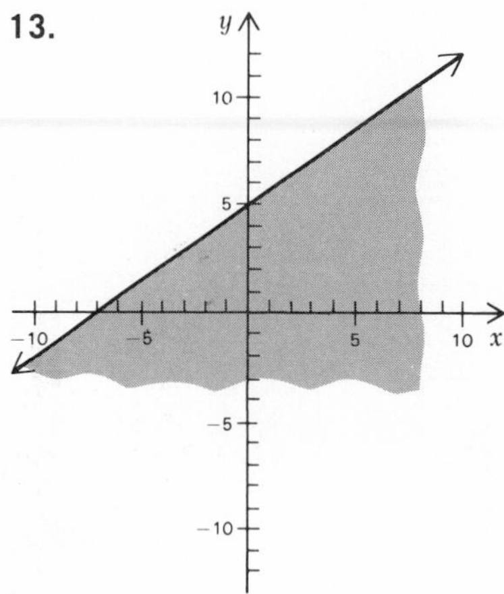

15.

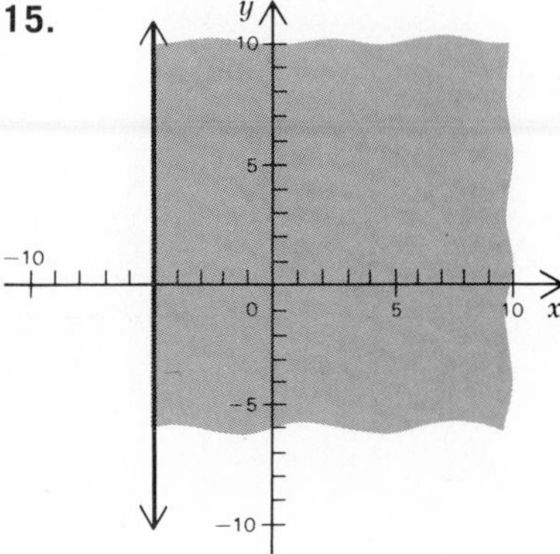

17.

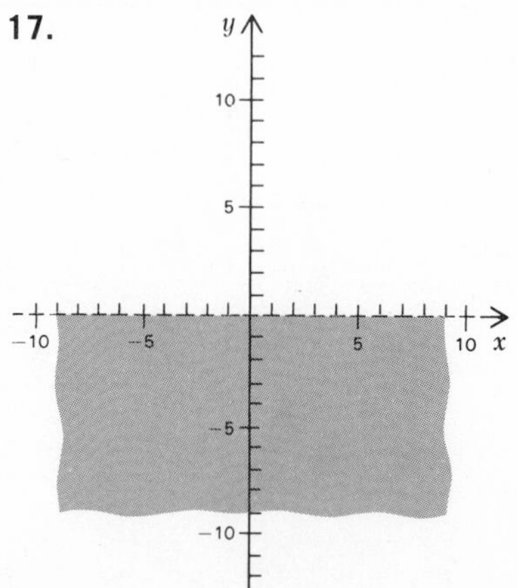

19.

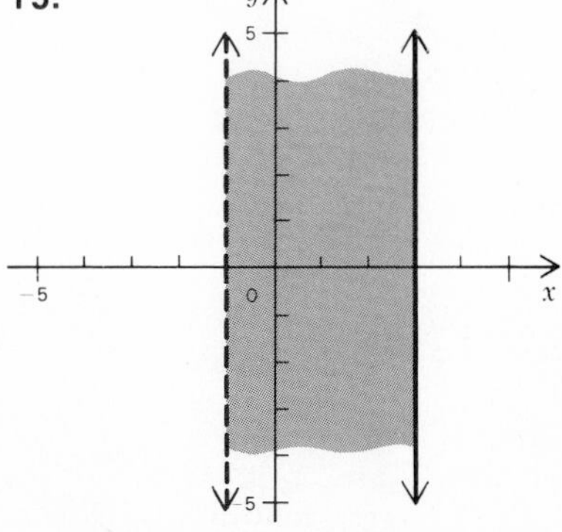

21.

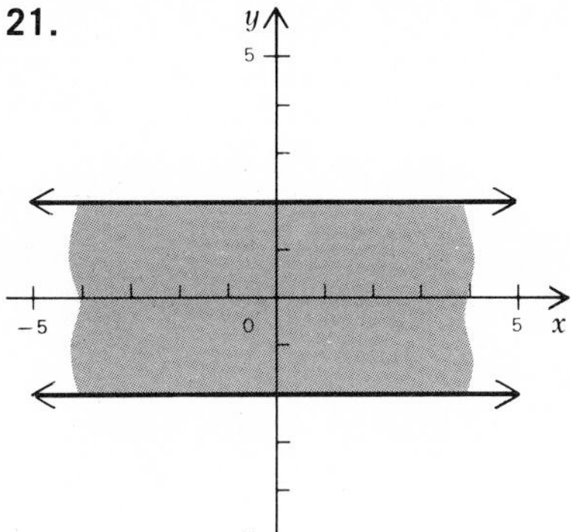

23.

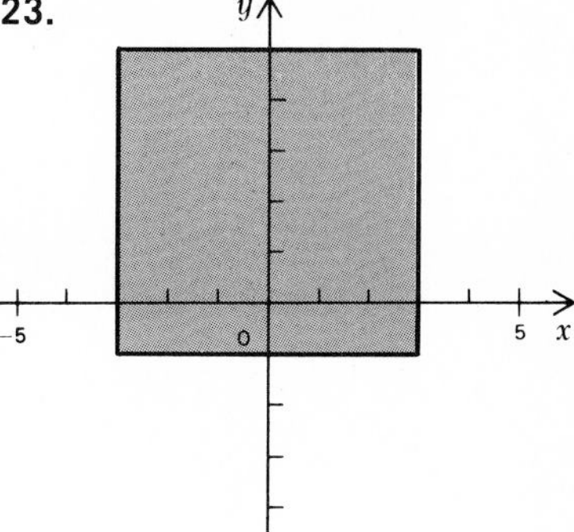

25.

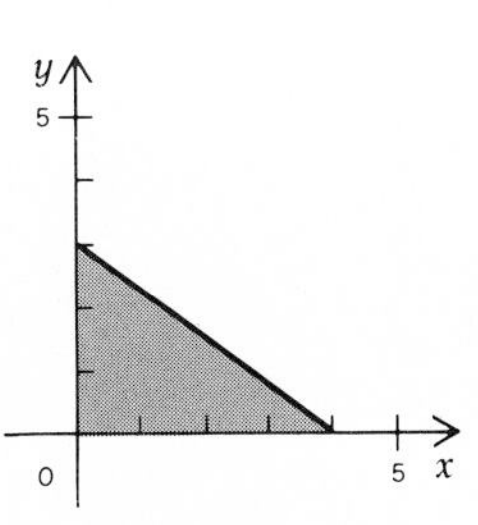

27.

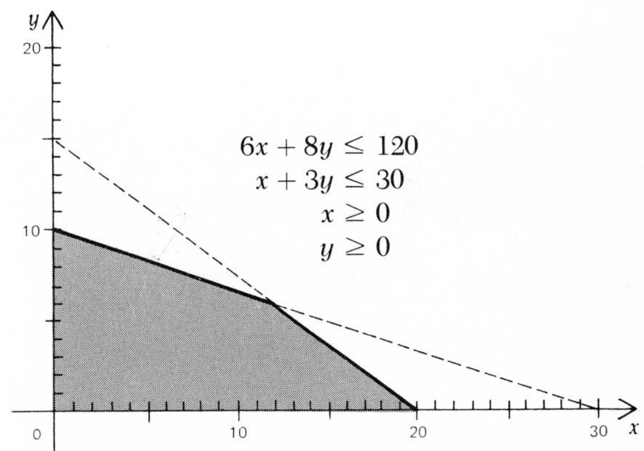

EXERCISE 23

1.

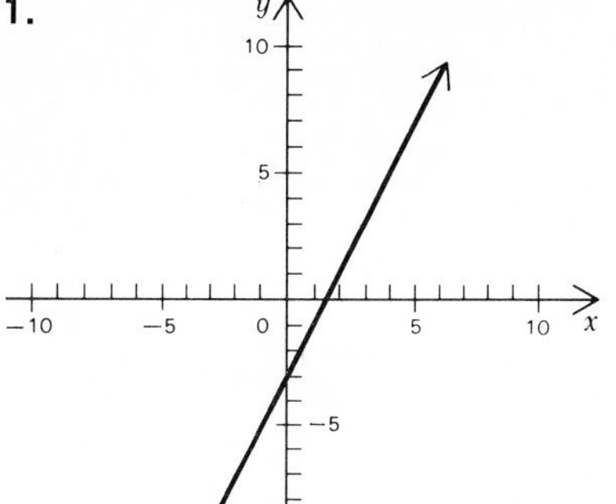

2.

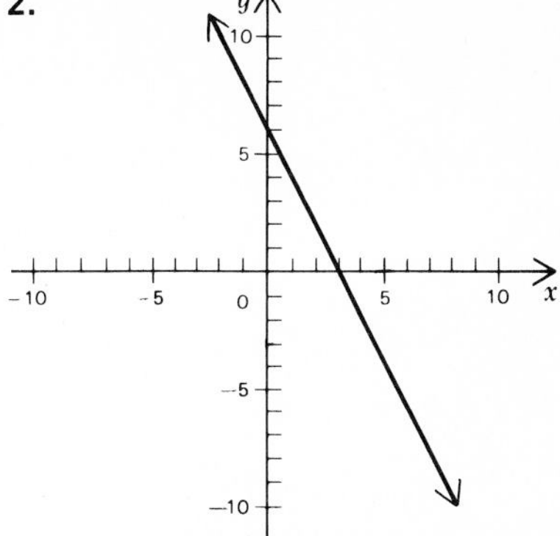

3.

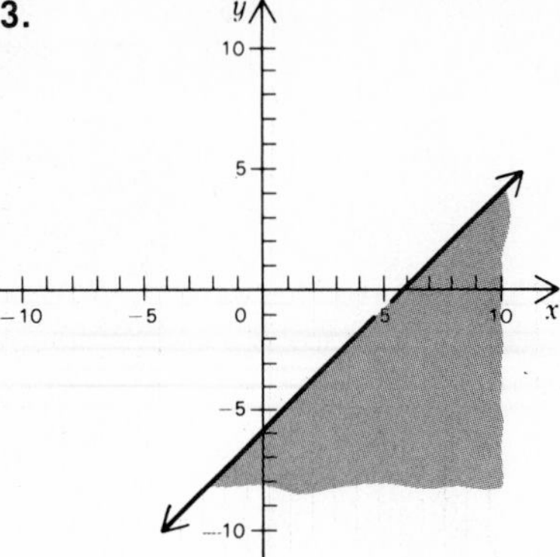

4.

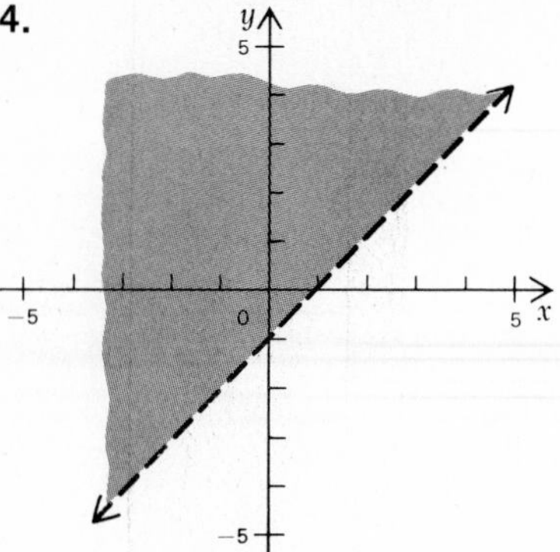

5. Slope $= -2$, y intercept $= -3$ **6.** $y - 4 = -2(x - 2)$ **7.** 2
8. $y - 7 = 2(x - 3)$ or $y - 3 = 2(x - 1)$

9.

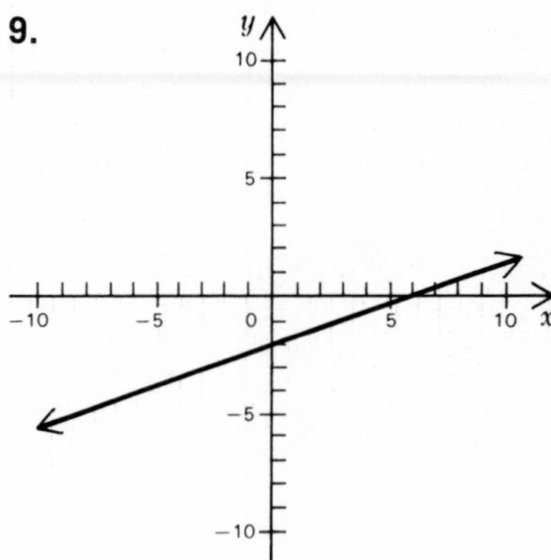

10.

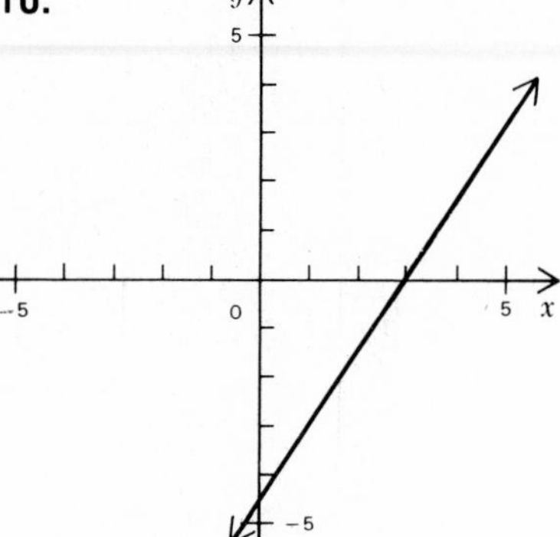

11.

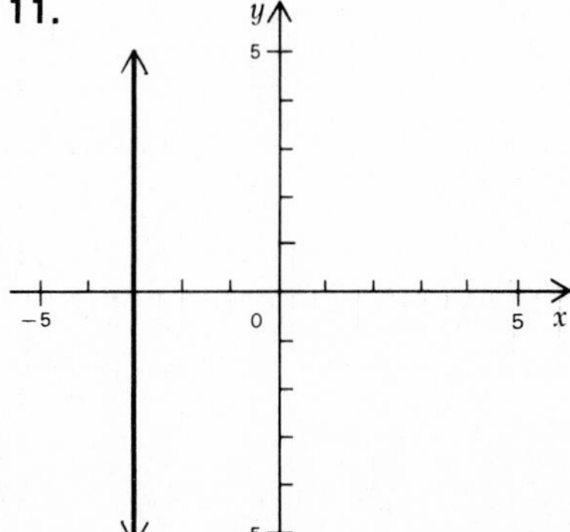

12.

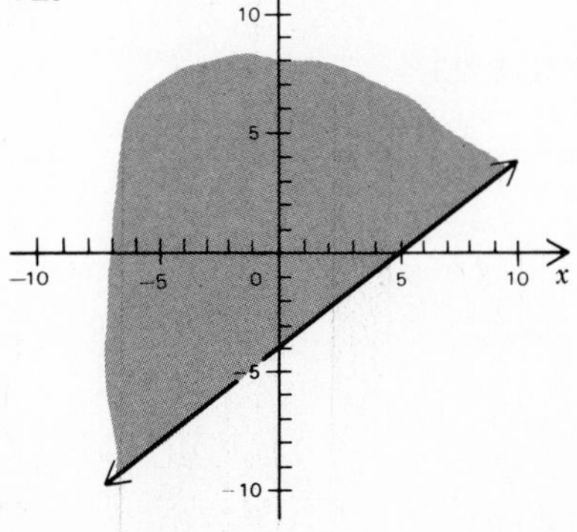

13.

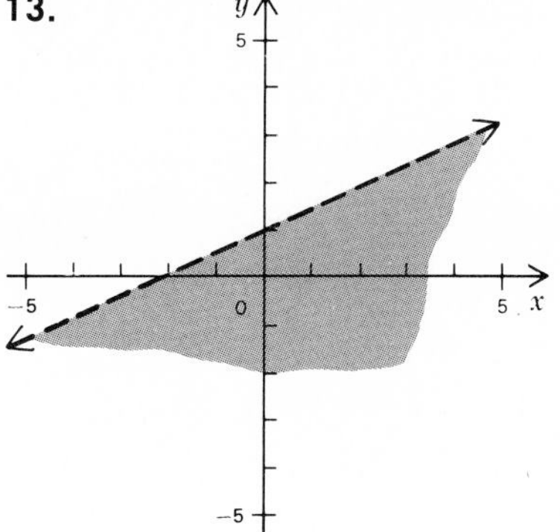

14.

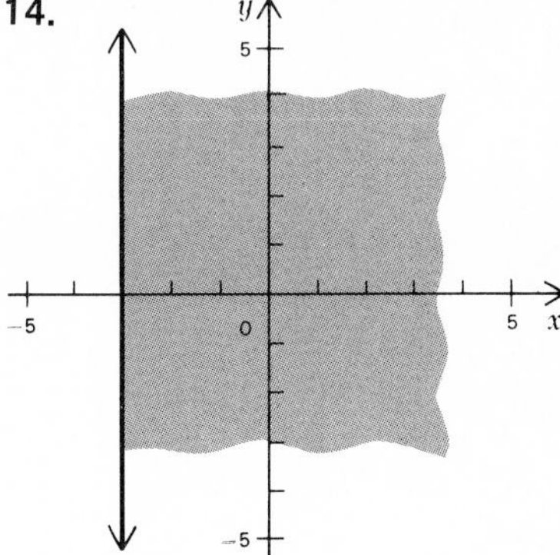

15.

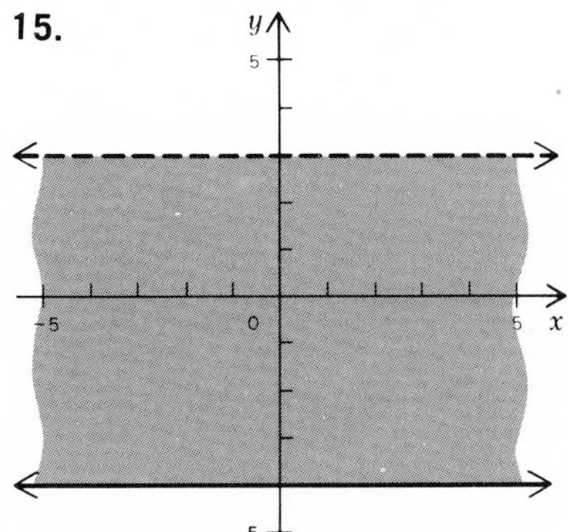

16. Slope $= -\frac{1}{2}$, y intercept $= -3$

17. $y = -\frac{1}{3}x + 1$ **18.** $-\frac{2}{3}$

19. $y = -\frac{2}{3}x$ **20.** $x = -2$

21. III and IV **22.** II and III

23.

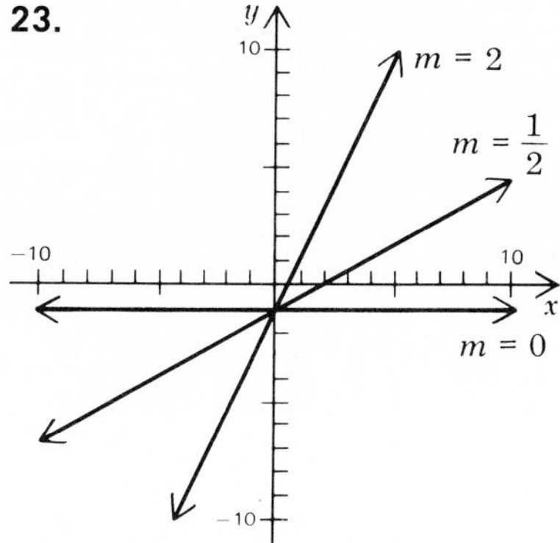

24.

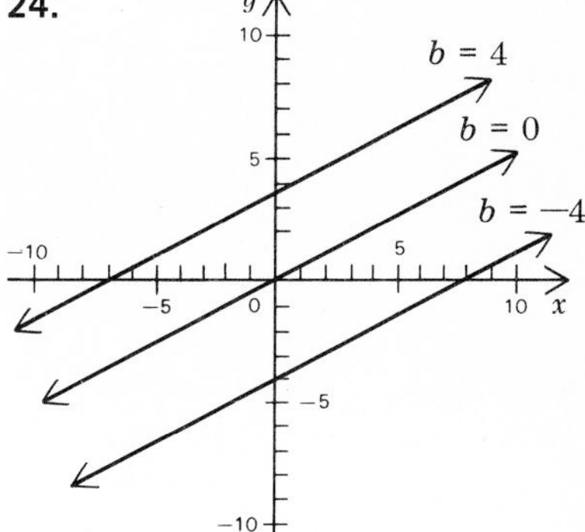

25. Slope $= \frac{3}{2}$, y intercept $= -\frac{5}{2}$ **26.** Not defined **27.** $x = 3$

28. $y = 3$ **29.** $7x + 5y = 11$

30.

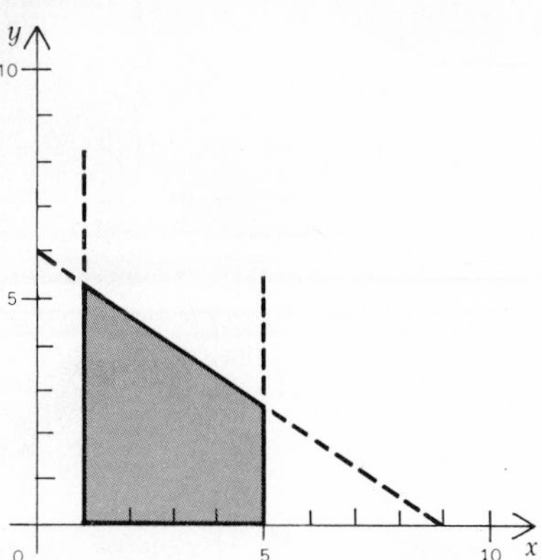

EXERCISE 24

1. $x = 4, y = 2$ **3.** $x = 8, y = 6$ **5.** $x = 3, y = 4$ **7.** $x = -2, y = -3$ **9.** $x = -2, y = -2$ **11.** $x = 4, y = 1$ **13.** $x = 2, y = -3$ **15.** $x = 5, y = 2$ **17.** No solution **19.** $m = 1, n = -\frac{2}{3}$ **21.** Infinite number of solutions **23.** No solution **25.** An infinite number of solutions **27.** $(\frac{1}{3}, -2)$ **29.** $(-2, 2)$ **31.** $(1, 0.2)$ **33.** $(-2, 4)$ **35.** Both have slope $-\frac{1}{2}$ **37.** $m(Aa + Bb + C) + n(Da + Eb + F) = m \cdot 0 + n \cdot 0 = 0$

EXERCISE 25

1. Limes: 11 cents each; lemons: 4 cents each **3.** Both companies pay \$136 on sales of \$1,700. The straight commission company pays better to the right of this point, and the other company pays better to the left. **5.** 60 cubic centimeters of 80% solution and 40 cubic centimeters of 50% solution **7.** 700 student, 300 adult **9.** (*A*) $1\frac{1}{4}$ hours, $\frac{3}{4}$ hour; (*B*) $112\frac{1}{2}$ miles; (*C*) 1 hour, 120 miles **11.** 40 seconds; 24 seconds; 120 miles **13.** Breakeven: (*A*) 3,000 at \$21,000; (*B*) Left: loss, right: profit **15.** (*A*) and (*B*) At \$1.10 per pound, $d = s = 1{,}080$ pounds; (*C*) Left: prices are forced up; right: prices are forced down **17.** 52°, 38° **19.** 927 brown, 309 blue **21.** 60 ounces of mix *A* and 80 ounces of mix *B* **23.** 16 inches, 20 inches **25.** Gemini, 1 hour; 18,700 miles **27.** 14 nickels and 8 dimes **29.** 84 $\frac{1}{4}$-pound packages; 60 $\frac{1}{2}$-pound packages

EXERCISE 26

1. $x = -3, y = -1$ **3.** $x = -1, y = -1, z = -2$ **5.** $u = -2, v = 1$ **7.** $x = 3, y = -2, z = 0$ **9.** $r = 0, s = -2, t = 5$ **11.** $x = 2, y = 0, z = -1$ **13.** $a = -1, b = 2, c = 0$ **15.** $x = 0, y = 2, z = -3$ **17.** $x^2 + y^2 - 4x - 4y - 17 = 0$ **19.** No solution; the system is inconsistent. **21.** $x = 1, y = -\frac{1}{2}, z = -\frac{1}{3}$ **23.** $w = 1, x = -1, y = 0, z = 2$ **25.** *A*: 60 grams; *B*: 50 grams; *C*: 40 grams

EXERCISE 27

1. -14 **3.** 2 **5.** -2.6 **7.** $\begin{vmatrix} a_{22} & a_{23} \\ a_{32} & a_{33} \end{vmatrix}$ **9.** $\begin{vmatrix} a_{11} & a_{12} \\ a_{31} & a_{32} \end{vmatrix}$

11. $(-1)^{1+1}\begin{vmatrix} a_{22} & a_{23} \\ a_{32} & a_{33} \end{vmatrix}$ **13.** $(-1)^{2+3}\begin{vmatrix} a_{11} & a_{12} \\ a_{31} & a_{32} \end{vmatrix}$ **15.** $\begin{vmatrix} 1 & -2 \\ -4 & 8 \end{vmatrix}$

17. $\begin{vmatrix} -2 & 0 \\ 5 & -2 \end{vmatrix}$ **19.** $(-1)^{1+1}\begin{vmatrix} 1 & -2 \\ -4 & 8 \end{vmatrix} = 0$ **21.** $(-1)^{3+2}\begin{vmatrix} -2 & 0 \\ 5 & -2 \end{vmatrix} = -4$

23. 10 **25.** -21 **27.** -120 **29.** -40 **31.** -12 **33.** 0 **35.** -8 **37.** 48 **39.** Expand the determinant about the first row to obtain $x - 3y + 7 = 0$, then show that the two points satisfy this linear equation. **41.** $\frac{23}{2}$

EXERCISE 28

1. $x = 5, y = -2$ **3.** $x = 1, y = -1$ **5.** $x = -1, y = 1$ **7.** $x = 2, y = -2, z = -1$ **9.** $x = 2, y = -1, z = 2$ **11.** $x = 2, y = -3, z = -1$ 13. $x = 1, y = -1, z = 2$ **15.** Since $D = 0$, the system either has no solution or infinitely many. Since $x = 0, y = 0, z = 0$ is a solution; the second must hold.

EXERCISE 29

1. $x = 6, y = 1$ **2.** $x = 2, y = 1$ **3.** -17 **4.** 0 **5.** $x = 2, y = 1$ **6.** $x = 3, y = 3$ **7.** 16 dimes, 14 nickels **8.** $x = 4, y = 3$

9. $m = -1, n = -3$ **10.** $\dfrac{\begin{vmatrix} a_1 & k_1 \\ a_2 & k_2 \end{vmatrix}}{\begin{vmatrix} a_1 & b_1 \\ a_2 & b_2 \end{vmatrix}}$ **11.** $m = -1, n = -3$ **12.** No solution

13. -33 **14.** 35 **15.** $x = 2, y = -1, z = 2$ **16.** $x = -1, y = 2, z = 1$ **17.** \$4,000 at 6% and \$2,000 at 10% **18.** Lines coincide—infinite number of solutions **19.** No solution **20.** $x = 2, y = -1, z = 2$ **21.** Yes, since $D = 0$ **22.** 30 grams of 70%, 70 grams of 40% **23.** 48 $\frac{1}{2}$-pound packages; 72 $\frac{1}{3}$-pound packages

EXERCISE 30

1. 3 **3.** $35m^3$ **5.** $4y$ **7.** $\frac{8}{9}$ **9.** $6/b$ **11.** y/x **13.** $\frac{3}{2}$ **15.** $3c/a$ **17.** $x/9y^2$ **19.** $16xy/3$ **21.** $9xy/8c$ **23.** $-45u^2/16v^2$ **25.** c^3d^2/a^6b^6 **27.** $x/2$ **29.** $x/(x-3)$ **31.** $3y/(x+3)$ **33.** $1/2y$ **35.** $t(t-4)$ **37.** $1/m$ **39.** $2x - x^2$ **41.** $a^2/2$ **43.** 2 **45.** -1 **47.** $\dfrac{(x-y)^2}{y^2(x+y)}$ **49.** $x = 3$ **51.** $\dfrac{R}{S} \cdot \left(\dfrac{P}{Q} \cdot \dfrac{S}{R}\right) = \dfrac{\cancel{R}P\cancel{S}}{\cancel{S}Q\cancel{R}} = \dfrac{P}{Q}$

EXERCISE 31

1. $6xy$ **3.** $4(x+2)$ **5.** $(x-3)(x+2)$ **7.** $\dfrac{2x+5}{3y}$

9. $(7x-2)/5x^2$ **11.** $(12x+y)/4y$ **13.** $(2+x)/x$ **15.** $\dfrac{u^2+uv-v^2}{v^3}$

17. 2 **19.** $\dfrac{1}{2x+3}$ **21.** $\dfrac{5x-1}{(x+1)(x-2)}$ **23.** $\dfrac{6-7y}{3y(y+3)}$

25. $4/3n^3$ **27.** $\dfrac{2(m+n)}{m^2n}$ **29.** $\dfrac{a+4b}{b}$ **31.** $\dfrac{3-2x}{k}$

33. $\dfrac{15t^3+14t-6}{36t^3}$ **35.** $\dfrac{-4}{(x-1)(x+3)}$ **37.** $\dfrac{-2n}{(m-n)(m+n)^2}$

39. $\dfrac{3x-5}{x-3}$ **41.** $\dfrac{x^2-6x+7}{x-2}$ **43.** $\dfrac{5a^2-2a-5}{(a+1)(a-1)}$ **45.** $\dfrac{2}{x+y}$

47. $\dfrac{2s^2+s-2}{2s(s-2)(s+2)}$ **49.** $\dfrac{-2n}{(m-n)(m+n)^2}$ **51.** $\dfrac{7}{y-3}$

53. $\dfrac{7y-9x}{xy(a-b)}$ **55.** $\dfrac{x+3}{(x-2)(x+7)}$ **57.** $\dfrac{xy^2-xy+y^2}{x^3-y^3}$

59. $-\dfrac{1}{3-x}=\dfrac{1}{x-3}$ **61.** $x-y$

EXERCISE 32

1. 4 : 1 **3.** 3 : 1 **5.** 1 : 4 **7.** 20 **9.** 7 **11.** 20 **13.** 490
15. 1,020 feet **17.** 4 inches **19.** 45 minutes **21.** 1.4 inches
23. 7.5 pounds **25.** 24,000 miles

EXERCISE 33

1. 13 **3.** 8 **5.** −6 **7.** $x > 4$ **9.** $-\frac{1}{12}$ **11.** $x > -\frac{3}{4}$
13. 20 **15.** 10 **17.** 3 **19.** $-\frac{7}{4}$ **21.** $x > 10$ **23.** $B \ge -4$
25. 10 **27.** 3 **29.** $p \ge 12$ **31.** $-20 \le c \le 20$ **33.** 4
35. No solution **37.** 5 **39.** $n = \frac{53}{11}$ **41.** 1 **43.** $\frac{31}{24}$ **45.** $x \ge 4\frac{1}{2}$
47. $23 \le F \le 50$ **49.** $t = -4$ **51.** No solution **53.** $\frac{2}{3}$

EXERCISE 34

1. 10 inches **3.** $2x > 300 + 1.5x,\ x > 600$ **5.** $30 \le \frac{5}{9}(F-32) \le 35$, $86° \le F \le 95°$ **7.** 10 A.M.; 24 miles **9.** $\dfrac{55+73+x}{3} \ge 70,\ x \ge 82$

11. 85.8 feet **13.** 250 **15.** (A) 15 inches; (B) 20 inches; (C) 22.5 inches; (D) 24 inches; (E) 25 inches; (F) 18 inches; (G) 18.75 inches
17. 1 square inch **19.** $70 \le \text{MA} \cdot 100/12 \le 120,\ 8.4 \le \text{MA} \le 14.4$
21. 75 feet

EXERCISE 35

1. $r = d/t$ **3.** $r = C/2\pi$ **5.** $\pi = C/D$ **7.** $x = -b/a,\ a \ne 0$
9. $x = (y+5)/2$ **11.** $y = \frac{3}{4}x - 3$ **13.** $R = E/I$ **15.** $B = CL/100$
17. $G = Fd^2/m_1m_2$ **19.** $C = \frac{5}{9}(F-32)$ **21.** $f = ab/(a+b)$
23. $n = (a_n - a_1 + d)/d$ **25.** $T_2 = T_1P_2V_2/P_1V_1$ **27.** $x = (5y-3)/(3y-2)$

EXERCISE 36

1. $1/(x-3)$ **3.** $(x+y)/x$ **5.** $1/(y-x)$ **7.** $(x-3)/(x-1)$ **9.** 1
11. $-\frac{1}{2}$ **13.** $1/(1-x)$ **15.** $-x$ **17.** $(3x+5)/(2x+3)$
19. $r = 2r_Rr_G/(r_R + r_G)$

EXERCISE 37

1. $5y/6$ **2.** $(3x+2)/3x$ **3.** $(2x+11)/6x$ **4.** $2/5x^2$ **5.** $2xy/ab$
6. $-1/(x+2)(x+3)$ **7.** $(d-2)^2/(d+2)$ **8.** $(y-2)/(y+1)$ **9.** $-\frac{10}{9}$
10. 60 **11.** 9 **12.** $x \le -12$ **13.** $m = 5$ **14.** No solution

15. 660 **16.** $x = -12$ **17.** $R = W/I^2$ **18.** $b = 2A/h$ **19.** $(12a^3b - 40b^2 - 5a)/30a^3b^2$ **20.** $2y^4/9a^4$ **21.** $2/(m+1)$ **22.** $(7s-4)/6s(s-4)$ **23.** $(5x-12)/3(x-4)(x+4)$ **24.** $x/(x+1)$ **25.** $(x-y)/x$ **26.** $x/y(x+y)$ **27.** -12 **28.** 41 **29.** $x \geq -19$ **30.** $x = -2$ **31.** $x = -5$ **32.** No solution **33.** $L = 2S/n - a$ or $L = (2S - an)/n$ **34.** 50 gallons **35.** \$80 **36.** 175 miles/hour **37.** 450 grams **38.** 1 **39.** $-1/(s+2)$ **40.** -1 **41.** y^2/x **42.** $(x+1)(x-2)/2x$ **43.** $(x-y)/(x+y)$ **44.** $-\frac{13}{5}$ **45.** $x \geq \frac{25}{7}$ **46.** -15 **47.** $W = E^2/R$ **48.** $f_1 = ff_2/(f_2 - f)$ **49.** 264 cycles per second, 330 cycles per second **50.** Seventy $\frac{1}{2}$-lb packages; thirty $\frac{1}{3}$-lb packages **51.** 250 grams

EXERCISE 38

1. 9 **3.** 5 **5.** 12 **7.** 2 **9.** u^7v^7 **11.** 4 **13.** a^8/b^8 **15.** 3 **17.** 6 **19.** 2 **21.** 7 **23.** 12 **25.** $10x^{11}$ **27.** $3x^2$ **29.** $3/4m^2$ **31.** $x^{10}y^{10}$ **33.** m^5/n^5 **35.** $12y^{10}$ **37.** 35×10^{17} **39.** 10^{14} **41.** x^6 **43.** m^6n^{15} **45.** c^6/d^{15} **47.** $3u^4/v^2$ **49.** $2^4s^8t^{16}$ or $16s^8t^{16}$ **51.** $6x^5y^{15}$ **53.** m^4n^{12}/p^8q^4 **55.** u^3/v^9 **57.** $9x^4$ **59.** -1 **61.** $-1/x^5$

EXERCISE 39

1. 1 **3.** 1 **5.** $1/3^3$ **7.** $1/m^7$ **9.** 4^3 **11.** y^5 **13.** 10^2 **15.** y **17.** 1 **19.** 10^{10} **21.** x^{11} **23.** $1/z^5$ **25.** $1/10^7$ **27.** 10^{12} **29.** y^8 **31.** $u^{10}v^6$ **33.** x^4/y^6 **35.** x^2/y^3 **37.** 1 **39.** 10^2 **41.** y **43.** 10 **45.** 3×10^{16} **47.** y^9 **49.** $3^2m^2n^2$ **51.** $2^3m^3/n^9$ **53.** n^{15}/m^{12} **55.** $3^3/2^2$ **57.** 1 **59.** $4y^3/3x^5$ **61.** $a^9/8b^4$ **63.** $1/x^7$ **65.** n^8/m^{12} **67.** $m^3n^3/8$ **69.** $1/(a^2 - b^2)$ **71.** $1/xy$ **73.** $-cd$ **75.** $xy/(x+y)$ **77.** $(y-x)^2/(x^2y^2)$

EXERCISE 40

1. 7×10 **3.** 8×10^2 **5.** 8×10^4 **7.** 8×10^{-3} **9.** 8×10^{-8} **11.** 5.2×10 **13.** 6.3×10^{-1} **15.** 3.4×10^2 **17.** 8.5×10^{-2} **19.** 6.3×10^3 **21.** 6.8×10^{-6} **23.** 800 **25.** 0.04 **27.** 300,000 **29.** 0.0009 **31.** 56,000 **33.** 0.0097 **35.** 430,000 **37.** 0.00000038 **39.** 5.46×10^9 **41.** 7.29×10^{-8} **43.** 10^{13} **45.** 10^{-5} **47.** 83,500,000,000 **49.** 0.00000000000614 **51.** 865,000 **53.** 0.0000000000000000000000017 **55.** 9×10^4 **57.** 6×10^{-4} **59.** 3×10^5 **61.** 5×10^4 **63.** 3×10 or 30 **65.** 3×10^{-4} or 0.0003 **67.** 6.6×10^{21} tons **69.** 10^7; 6×10^8

EXERCISE 41

1. 5 **3.** Not a real number **5.** 2 **7.** -2 **9.** -2 **11.** 64 **13.** 4 **15.** x **17.** $1/x^{1/5}$ **19.** x^2 **21.** ab^3 **23.** x^3/y^4 **25.** x^2y^3 **27.** $\frac{2}{5}$ **29.** $\frac{8}{125}$ **31.** $\frac{1}{4}$ **33.** $\frac{1}{6}$ **35.** $\frac{1}{125}$ **37.** 25 **39.** $\frac{1}{9}$ **41.** $1/x^{1/2}$ **43.** $n^{1/12}$ **45.** x^4 **47.** $2v^2/u$ **49.** $1/x^2y^3$ **51.** x^4/y^3 **53.** $\frac{5}{4}x^4y^2$ **55.** $64y^{1/3}$ **57.** $12m - 6m^{35/4}$ **59.** $2x + 3x^{1/2}y^{1/2} + y$ **61.** $x + 2x^{1/2}y^{1/2} + y$ **63.** Not defined **65.** $\dfrac{2}{a} + \dfrac{5}{a^{1/2}b^{1/2}} - \dfrac{3}{b}$ **67.** $a^{1/2}b^{1/3}$ **69.** x **71.** $1/x^m$ **73.** (*A*) Any negative number; (*B*) n even and x any negative number

EXERCISE 42

1. 1 **2.** $\frac{1}{9}$ **3.** 8 **4.** $\frac{1}{2}$ **5.** Not a real number **6.** 4
7. (A) 4.28×10^9; (B) 3.18×10^{-5} **8.** (A) 729,000; (B) 0.000603
9. $6x^4y^7$ **10.** $3u^4/v^2$ **11.** $6x^5y^{15}$ **12.** c^6/d^{15} **13.** $4x^4/9y^6$
14. x^{12} **15.** y^2 **16.** y^3/x^2 **17.** x^3 **18.** $1/x^2$ **19.** $1/x^{1/3}$
20. u **21.** 2×10^{-3} or 0.002 **22.** $m^2/2n^5$ **23.** x^6/y^4 **24.** $4x^4/y^6$
25. c/a^2b^4 **26.** $\frac{1}{4}$ **27.** $n^{10}/9m^{10}$ **28.** $1/(x-y)^2$ **29.** $3a^2/b$
30. $3x^2/2y^2$ **31.** $1/m$ **32.** $6x^{1/6}$ **33.** $x^{1/12}/2$ **34.** $\frac{5}{9}$

35. $x + 2x^{1/2}y^{1/2} + y$ **36.** $a^2 = b$ **37.** $\dfrac{xy}{x+y}$ **38.** $\dfrac{a^2b^2}{a^3+b^3}$

39. x^{m-1} **40.** x^{m-1} **41.** 1.036 cm^3

EXERCISE 43

1. $\sqrt{11}$ **3.** $\sqrt[3]{5}$ **5.** $\sqrt[5]{u^3}$ **7.** $4\sqrt[7]{y^3}$ **9.** $\sqrt[7]{(4y)^3}$ **11.** $\sqrt[5]{(4ab^3)^2}$
13. $\sqrt{a+b}$ **15.** $6^{1/2}$ **17.** $m^{1/4}$ **19.** $y^{3/5}$ **21.** $(xy)^{3/4}$
23. $(x^2-y^2)^{1/2}$ **25.** $-5\sqrt[5]{y^2}$ **27.** $\sqrt[7]{(1+m^2n^2)^3}$ **29.** $1/\sqrt[3]{w^2}$

31. $\dfrac{1}{\sqrt[5]{(3m^2n^3)^3}}$ **33.** $\sqrt{a}+\sqrt{b}$ **35.** $\sqrt[3]{(a^3+b^3)^2}$ **37.** $(a+b)^{2/3}$

39. $-3x(a^3b)^{1/4}$ **41.** $(-2x^3y^7)^{1/9}$ **43.** $\dfrac{3}{y^{1/3}}$ or $3y^{-1/3}$ **45.** $\dfrac{-2x}{(x^2+y^2)^{1/2}}$ or

$-2x(x^2+y^2)^{-1/2}$ **47.** $m^{2/3} - n^{1/2}$ **49.** $(x+y)^2 \neq x^2+y^2$ **51.** (A) Any negative number; (B) n even and x any negative number

EXERCISE 44

1. y **3.** $2u$ **5.** $7x^2y$ **7.** $3\sqrt{2}$ **9.** $m\sqrt{m}$ **11.** $2x\sqrt{2x}$
13. $\frac{1}{3}$ **15.** $1/y$ **17.** $\sqrt{5}/5$ **19.** $\sqrt{5}/5$ **21.** $\sqrt{y}/y$ **23.** $\sqrt{y}/y$
25. $3xy^2\sqrt{xy}$ **27.** $3x^4y^2\sqrt{2y}$ **29.** $\sqrt{2x}/2x$ **31.** $2x\sqrt{3x}$
33. $3\sqrt{2ab}/2b$ **35.** $\sqrt{42xy}/7y$ **37.** $3m^2\sqrt{2mn}/2n$ **39.** $2x^2y$
41. $2xy^2\sqrt[3]{2xy}$ **43.** $\sqrt{x}$ **45.** 4 **47.** $6m^3n^3$ **49.** $2\sqrt[3]{9}$
51. $2a\sqrt{3ab}/3b$ **53.** Is in the simplest radical form **55.** $2x/3y^2$
57. $-3m^2n^2\sqrt[5]{3m^2n}$ **59.** $\sqrt[3]{x^2(x-y)}$ **61.** $x^2y\sqrt[3]{6xy}$
63. $2x^2y\sqrt[3]{4x^2y}$ **65.** $-\sqrt[3]{6x^2y^2}$ **67.** $\sqrt[3]{(x-y)^2}$ **69.** $\sqrt[4]{12x^3y^3}/2x$
71. $-x\sqrt{x^2+2}$ **73.** $4x^9y^2/\sqrt[3]{2y}$ **75.** $mn\sqrt[12]{3^7m^5n^2}$ **77.** $x^n(x+y)^{n+2}$

EXERCISE 45

1. $9\sqrt{3}$ **3.** $-5\sqrt{a}$ **5.** $-5\sqrt{n}$ **7.** $4\sqrt{5}-2\sqrt{3}$ **9.** $\sqrt{m}-3\sqrt{n}$
11. $4\sqrt{2}$ **13.** $-6\sqrt{2}$ **15.** $8\sqrt{2mn}$ **17.** $6\sqrt{2}-2\sqrt{5}$ **19.** $-\sqrt[5]{a}$
21. $5\sqrt[3]{x}-\sqrt{x}$ **23.** $9\sqrt{2}/4$ **25.** $-3\sqrt{6uv}/2$ **27.** $3\sqrt{3}$ **29.** $\frac{10}{3}\sqrt[3]{9}$

EXERCISE 46

1. $7-2\sqrt{7}$ **3.** $3\sqrt{2}-2$ **5.** $y-8\sqrt{y}$ **7.** $4\sqrt{n}-n$
9. $3+3\sqrt{2}$ **11.** $3-\sqrt{3}$ **13.** $9+4\sqrt{5}$ **15.** $m-7\sqrt{m}+12$

17. $\sqrt{5}-2$ **19.** $\dfrac{\sqrt{5}-1}{2}$ **21.** $\dfrac{\sqrt{5}+\sqrt{2}}{3}$ **23.** $\dfrac{y-3\sqrt{y}}{y-9}$

25. $38-11\sqrt{3}$ **26.** $x-y$ **29.** $10m-11\sqrt{m}-6$
31. $5+\sqrt[3]{18}+\sqrt[3]{12}$ **33.** $x+2\sqrt[3]{xy}-\sqrt[3]{x^2y^2}-2y$ **37.** $-7-4\sqrt{3}$
39. $5+2\sqrt{6}$

41. $\dfrac{x+5\sqrt{x}+6}{x-9}$ **43.** $\dfrac{6x+9\sqrt{x}}{4x-9}$ **45.** $\dfrac{62-19\sqrt{10}}{117}$

47. $\dfrac{\sqrt[3]{x^2}-\sqrt[3]{xy}+\sqrt[3]{y^2}}{x+y}$ **49.** $\dfrac{(\sqrt{x}+\sqrt{y}+\sqrt{z}\,[(x+y-z)-2\sqrt{xy}]}{(x+y-z)^2-4xy}$

EXERCISE 47

1. $8+3i$ **3.** $-5+3i$ **5.** $5+3i$ **7.** $6+13i$ **9.** $3-2i$
11. -15 or $-15+0i$ **13.** $-6-10i$ **15.** $15-3i$ **17.** $-4-33i$
19. 65 or $65+0i$ **21.** $\frac{2}{5}-\frac{1}{5}i$ **23.** $\frac{3}{13}+\frac{11}{13}i$ **25.** $5+3i$
27. $7-5i$ **29.** $-3+2i$ **31.** $8+25i$ **33.** $\frac{5}{3}-\frac{2}{3}i$
35. $\frac{2}{23}+\frac{3}{13}i$ **37.** $-\frac{2}{5}i$ **39.** $\frac{3}{2}-\frac{1}{2}i$ **41.** $4-5i$
43. 0 or $0+0i$ **45.** $-1, -i, 1, i, -1, -i, 1$ **47.** $x=1, y=3$
49. $(a+c)+(b+d)i$ **51.** a^2+b^2 or $(a^2+b^2)+0i$
53. $(ac-bd)+(ad+bc)i$ **55.** 1 **57.** $\pm 6i$ **59.** $9\pm 3i$
61. For $x \geq 10$

EXERCISE 48

1. (A) $\sqrt{3m}$; (B) $3\sqrt{m}$; (C) $(2x)^{1/2}$; (D) $(a+b)^{1/2}$ **2.** $2xy^2$
3. $5/y$ **4.** $6x^2y^3\sqrt{y}$ **5.** $\sqrt{2y}/2y$ **6.** $2b\sqrt{3a}$ **7.** $6x^2y^3\sqrt{xy}$
8. $\sqrt{2xy}/2x$ **9.** $-3\sqrt{x}$ **10.** $\sqrt{7}-2\sqrt{3}$ **11.** $5+2\sqrt{5}$

12. $1+\sqrt{3}$ **13.** $\dfrac{5+3\sqrt{5}}{4}$ **14.** $3-6i$ **15.** $15+3i$ **16.** $2+i$

17. (A) $\sqrt[3]{(2mn)^2}$; (B) $3\sqrt[5]{x^2}$; (C) $x^{5/7}$; (D) $-3(xy)^{2/3}$ **18.** $2x^2y$
19. $3x^2y\sqrt{x^2y}$ **20.** $n^2\sqrt{6m}/3$ **21.** $\sqrt[4]{y^3}$ **22.** $-6x^2y^2\sqrt[5]{3x^2y}$

23. $x\sqrt[3]{2x^2}$ **24.** $\dfrac{\sqrt[5]{12x^3y^2}}{2x}$ **25.** $2x-3\sqrt{xy}-5y$ **26.** $\dfrac{x-4\sqrt{x}+4}{x-4}$

27. $\dfrac{6x+3\sqrt{xy}}{4x-y}$ **28.** $5\sqrt{6}/6$ **29.** $-1-i$ **30.** $\frac{4}{13}-\frac{7}{13}i$ **31.** $5+4i$

32. $-i$ **33.** $y\sqrt[3]{2x^2y}$ **34.** 0 **35.** $4\sqrt[12]{xy}$ **36.** $3m\sqrt{m^2+n^2}$

37. $2xy\sqrt[6]{2xy}$ **38.** $x^{(n+1)}$ **39.** $\dfrac{x\sqrt[3]{x}+\sqrt[3]{x^2y^2}+y\sqrt[3]{y}}{x^2-y^2}$ **40.** All real numbers x **41.** 1, multiplicative inverse property

EXERCISE 49

1. ± 4 **3.** $\pm 4i$ **5.** $\pm 3\sqrt{5}$ **7.** $\pm\frac{3}{2}$ **9.** $\pm\frac{3}{4}$ **11.** $0, -5$
13. $0, -4$ **15.** $6, -2$ **17.** $1, -5$ **19.** $-\frac{2}{3}, 4$ **21.** $\pm\sqrt{2}$

23. $\pm\frac{3}{4}i$ **25.** $\pm\sqrt{\frac{7}{9}}$ or $\pm\dfrac{\sqrt{7}}{3}$ **27.** $8, -2$ **29.** $-1\pm 3i$ **31.** $-\frac{1}{3}, 1$
33. $-1, 3$ **35.** $-\frac{2}{3}, 1$ **37.** Not factorable in the integers **39.** $-\frac{1}{2}, 3$
41. $-2, 2$ **43.** $3, -4$ **45.** $-\frac{1}{2}, 2$ **47.** $\frac{1}{2}, 2$ **49.** 11 by 3 inches

51. $\dfrac{-5\pm\sqrt{10}}{2}$ **53.** $a=\sqrt{c^2-b^2}$ **55.** 60 cents per gallons

EXERCISE 50

1. $a=2, b=-5, c=3$ **3.** $a=3, b=1, c=-1$ **5.** $a=3, b=0, c=-5$

7. $-4 \pm \sqrt{13}$ **9.** $5 \pm 2\sqrt{7}$ **11.** $\dfrac{-3 \pm \sqrt{13}}{2}$ **13.** $1 \pm i\sqrt{2}$

15. $\dfrac{3 \pm \sqrt{3}}{2}$ **17.** $\dfrac{-1 \pm \sqrt{13}}{6}$ **19.** $5 \pm \sqrt{7}$ **21.** $-1 \pm \sqrt{3}$

23. $0, -\frac{3}{2}$ **25.** $2 \pm 3i$ **27.** $1 \pm i\sqrt{2}$ **29.** $\frac{2}{5}, 3$ **31.** $t = \sqrt{2d/g}$

33. $r = -1 + \sqrt{A/P}$ **35.** $5 \pm 2\sqrt{7}$ **37.** $\dfrac{2 \pm \sqrt{2}}{2}$

39. $\dfrac{-m \pm \sqrt{m^2 - 4n}}{2}$ **41.** $\left(\dfrac{-b + \sqrt{b^2 - 4ac}}{2a}\right)\left(\dfrac{-b - \sqrt{b^2 - 4ac}}{2a}\right) =$

$\dfrac{b^2 - (b^2 - 4ac)}{4a^2} = \dfrac{c}{a}$

EXERCISE 51

1. 12, 14 **3.** 0, 2 **5.** 127 miles **7.** 5.12 inches by 3.12 inches **9.** 1 foot **11.** 5.66 feet/second **13.** 50 miles/hour **15.** 1.41 minutes **17.** 2 hours; 3 hours **19.** 2 miles/hour **21.** \$2

EXERCISE 52

1. 4 **3.** 18 **5.** $\pm 1, \pm 3$ **7.** 4 **9.** $\pm 3, \pm\sqrt{2}i$ **11.** $-1, 2$ **13.** $-8, 125$ **15.** 1, 16 **17.** $\frac{2}{3}, -\frac{3}{2}$ **19.** $1, -3, -1 \pm i$ **21.** 19 **23.** 9 **25.** 5, 13 **27.** $3 \pm 2i, 4, 2$ **29.** $\pm 2, \pm\frac{1}{2}$

EXERCISE 53

1. $(-3, -4), (3, -4)$ **3.** $(\frac{1}{2}, 1)$ **5.** $(4, -2), (-4, 2)$ **7.** $(3 - i, 4 - 3i), (3 + i, 4 + 3i)$ **9.** $(2, 1), (2, -1), (-2, 1), (-2, -1)$ **11.** $(-3, -2), (-3, 2), (3, -2), (3, 2)$ **13.** $(2 + \sqrt{10}, -2 + \sqrt{10}), (2 - \sqrt{10}, -2 - \sqrt{10})$ **15.** $(-1, 2), (4, 7)$ **17.** $(i, 2i), (i, -2i), (-i, 2i), (-i, -2i)$ **19.** $(2, 4), (-2, 4), (i\sqrt{5}, -5), (-i\sqrt{5}, -5)$ **21.** $(4, 0), (-3, \sqrt{7}), (-3, -\sqrt{7})$ **23.** 2 ft by 16 ft **25.** $(2\sqrt{2}, \sqrt{2}), (-2\sqrt{2}, -\sqrt{2}), (1, 4), (-1, -4)$ **27.** $(3, 3), (-3, -3), (0, 6), (0, -6)$, **29.** $(4, 4), (-4, -4), (\frac{4}{5}\sqrt{5}, -\frac{4}{5}\sqrt{5}), (-\frac{4}{5}\sqrt{5}, \frac{4}{5}\sqrt{5})$

EXERCISE 54

1. $-4 < x < 3$ 

3. $x \leq -4$ or $x \geq 3$ 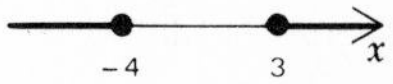

5. $-4 < x < 3$ 

7. $x < 3$ or $x > 7$ 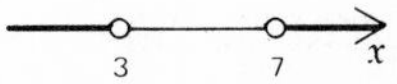

9. $x \leq -6$ or $x \geq 0$ 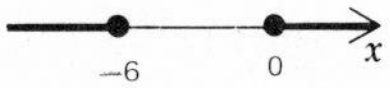

11. $x \leq -3$ or $x \geq 3$ 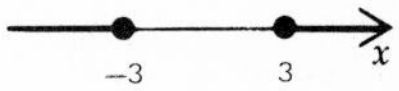

13. $-2 < x \leq 5$ 

15. $x < -2$ or $x > 5$ 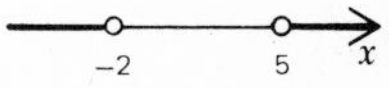

17. $x < -2$ or $0 < x \leq 4$ 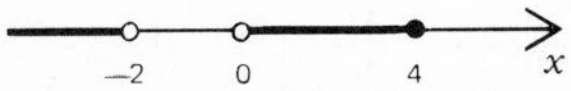

19. $x < 0$ or $x > \frac{1}{4}$ 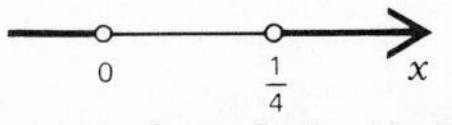

21. All real numbers; graph: whole real line **23.** No solution

25. $x \leq -\sqrt{3}$ or $x \geq \sqrt{3}$ 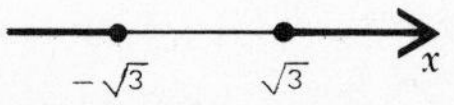

27. $2 \leq x < 3$ 

29. $-2 < x < 3$

31. $x \geq 2$ or $x \leq 1$ **33.** 6 seconds

35.
$ax + b = 0$, $ax = -b$, $x = -b/a$
$ax + b < 0$, $ax < -b$, $x < -b/a$
$ax + b > 0$, $ax > -b$, $x > -b/a$

EXERCISE 55

1. 0, 3 **2.** ± 5 **3.** 2, 3 **4.** -3, 5 **5.** $\pm\sqrt{7}$ **6.** $a = 3$, $b = 4$, $c = -2$ **7.** $x = \dfrac{-b \pm \sqrt{b^2 - 4ac}}{2a}$ **8.** $\dfrac{-3 \pm \sqrt{5}}{2}$ **9.** 3, 9

10. $-5 < x < 4$ **11.** $x \leq -5$ or $x \geq 4$

12. $(1, -1)$, $(\frac{7}{5}, -\frac{1}{5})$ **13.** $(4, 3)$, $(-4, 3)$, $(4, -3)$, $(-4, -3)$ **14.** 0, 2
15. $\pm 2\sqrt{3}$ **16.** $\pm 3i$ **17.** -2, 6 **18.** $-\frac{1}{3}$, 3 **19.** $\frac{1}{2}$, -3

20. $\dfrac{1 \pm \sqrt{7}}{3}$ **21.** $\dfrac{1 \pm \sqrt{7}}{2}$ **22.** -4, 5 **23.** $\frac{3}{4}$, $\frac{5}{2}$ **24.** $\dfrac{-1 \pm \sqrt{5}}{2}$

25. $1 \pm i\sqrt{2}$ **26.** 2, 3 **27.** 9, 25 **28.** ± 2, $\pm 3i$ **29.** 64, $-\frac{27}{8}$

30. $x \leq -3$ or $x \geq 7$ **31.** $x < 0$ or $x > \frac{1}{2}$

32. No solution **33.** All real numbers; graph: real line. **34.** $(1, 3)$, $(1, -3)$, $(-1, 3)$, $(-1, -3)$ **35.** $(1 + i, 2i)$, $(1 - i, -2i)$ **36.** 6 by 5 in.

37. $3 \pm 2\sqrt{3}$ **38.** $\dfrac{3 \pm i\sqrt{6}}{2}$ **39.** $\dfrac{-3 \pm \sqrt{57}}{6}$ **40.** ± 1, ± 2, $\pm 2i$, $\pm i$

41. 3, $\frac{9}{4}$

42. $x \leq 1$ or $3 < x < 4$

43. 9 and 12 **44.** (*A*) 2,000 and 8,000; (*B*) 5,000

EXERCISE 56

1.

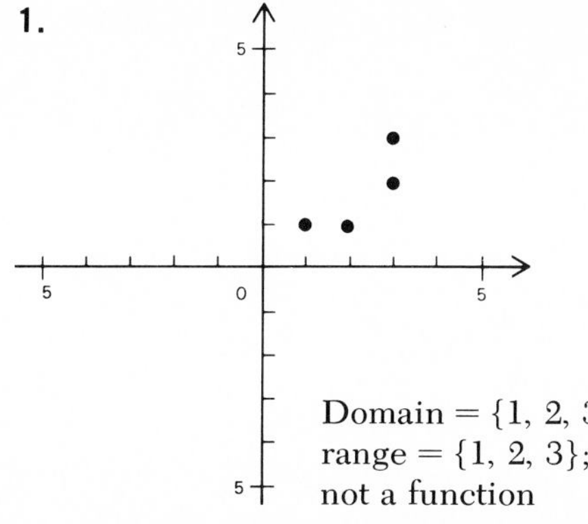

3.

Domain = {−1, 0, 1, 2, 3, 4};
range = {−2, −1, 0, 1, 2};
a function

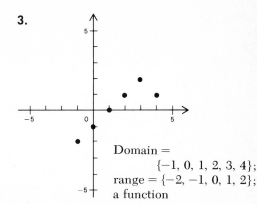

5.

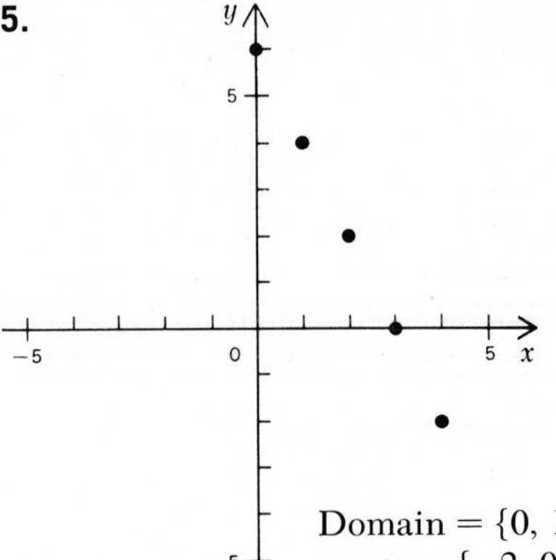

Domain = {0, 1, 2, 3, 4};
range = {−2, 0, 2, 4, 6};
a function

7. Domain = {−6, −4, −2, 0, 2, 4, 6}
= range;
not a function

9. Domain: $-8 \le x \le 8$;
range: $-4 \le y \le 4$;
not a function

11. Domain: all real numbers;
range: −4;
a function

13.

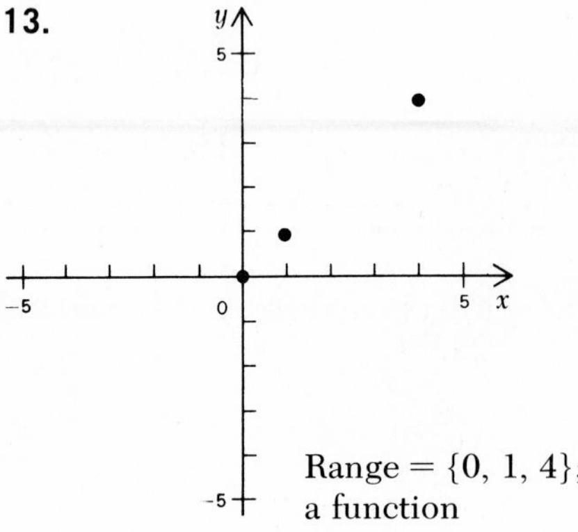

Range = {0, 1, 4};
a function

15.

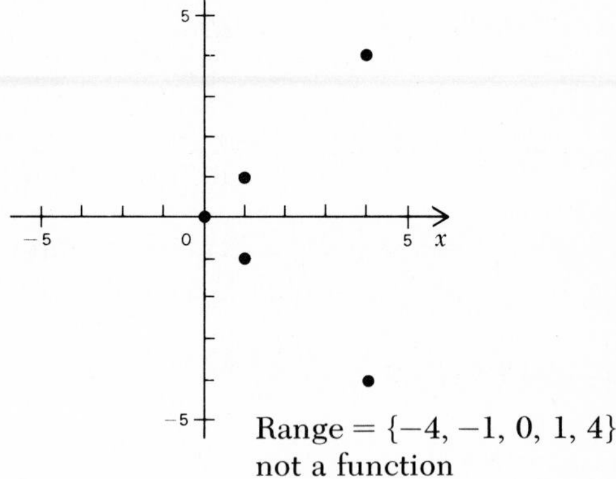

Range = {−4, −1, 0, 1, 4};
not a function

17.

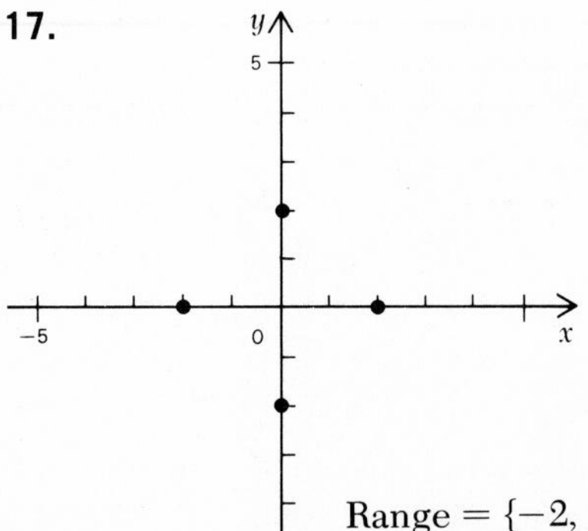

Range = {−2, 0, 2};
not a function

19.

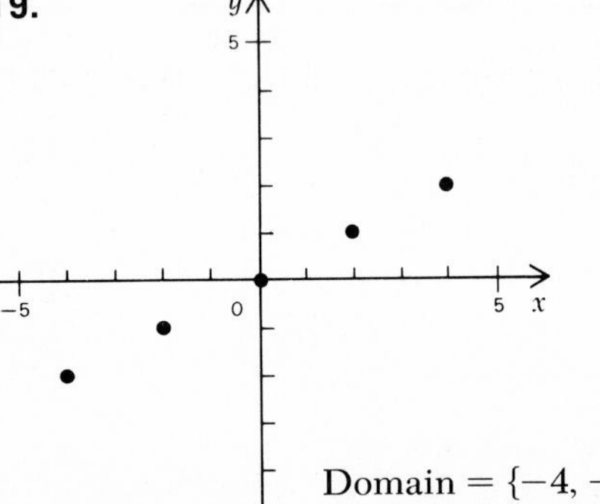

Domain = {−4, −2, 0, 2, 4};
range = {−2, −1, 0, 1, 2};
a function

21.

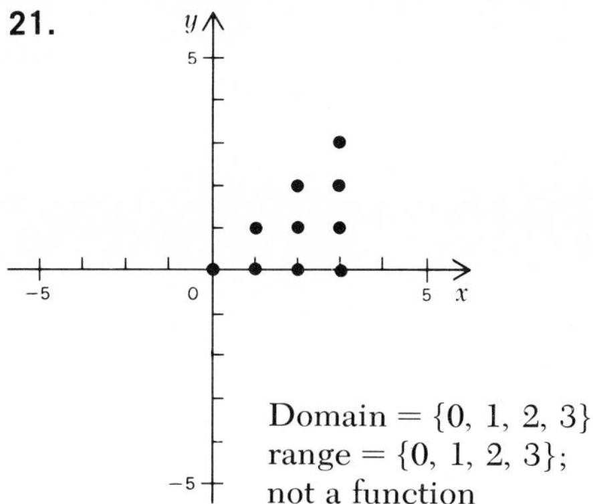

Domain = {0, 1, 2, 3};
range = {0, 1, 2, 3};
not a function

EXERCISE 57

1. 4 **3.** −8 **5.** −2 **7.** −2 **9.** −12 **11.** −6 **13.** −27 **15.** 2 **17.** 6 **19.** 25 **21.** 22 **23.** −91 **25.** $2-2h$ **27.** −2 **29.** 10 **31.** 48 **33.** 0 **35.** −7 **37.** $\frac{1}{5}$, $-\frac{3}{5}$, not defined **39.** (*A*) 30, 300; (*B*) 30 **41.** $C(x)=10+0.1x$, $x \geq 0$
43. (*A*) Yes; (*B*) yes; (*C*) no

EXERCISE 58

1. $R^{-1}=\{1,-2),(3,0),(2,2)\}$
3. $G^{-1}=\{(4,-2),(1,-1),(0,0),(1,1),(4,2)\}$

5.

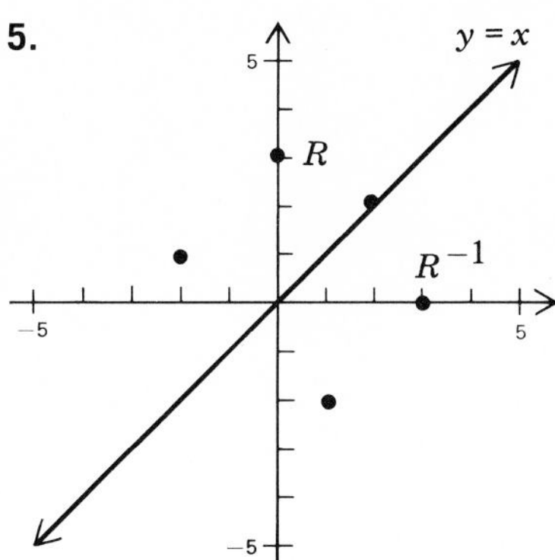

7.

9. Both are functions **11.** G is a function; G^{-1} is not.
13. f^{-1}: $x=3y-2$ or $y=(x+2)/3$ **15.** F^{-1}: $x=y/3-2$ or $y=3(x+2)$
17. h^{-1}: $x=y^2/2$ or $y=\pm\sqrt{2x}$

19.

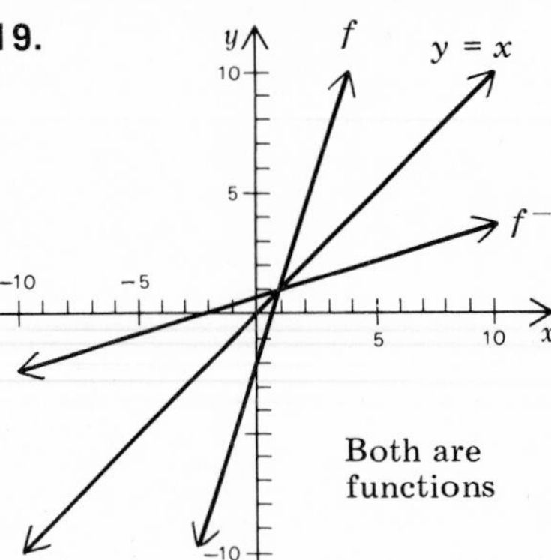

21.

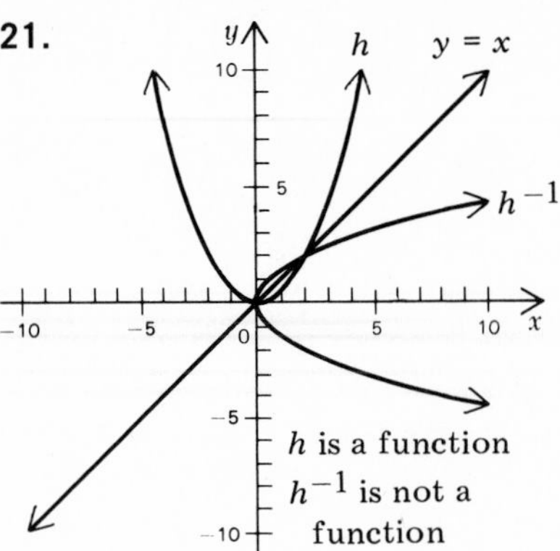

23. (A) $f^{-1}(x) = (x + 2)/3$; (B) $\frac{4}{3}$; (C) 3 **25.** (A) $F^{-1}(x) = 3(x + 2)$; (B) 3; (C) 4 **27.** (A) $f^{-1}(x) = 3(x - 2)$; (B) a **29.** x

EXERCISE 59

1.

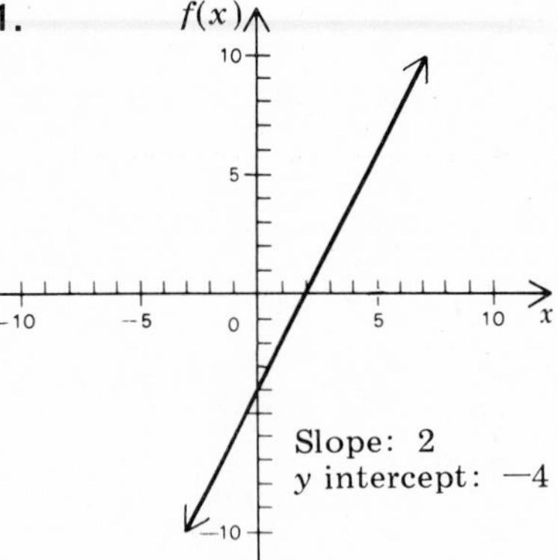

3.

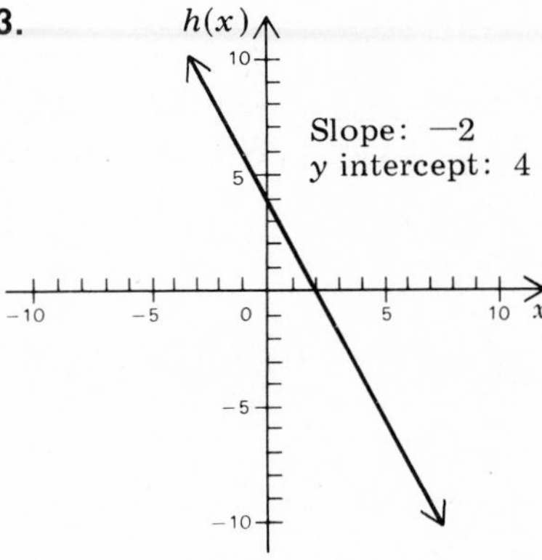

5.

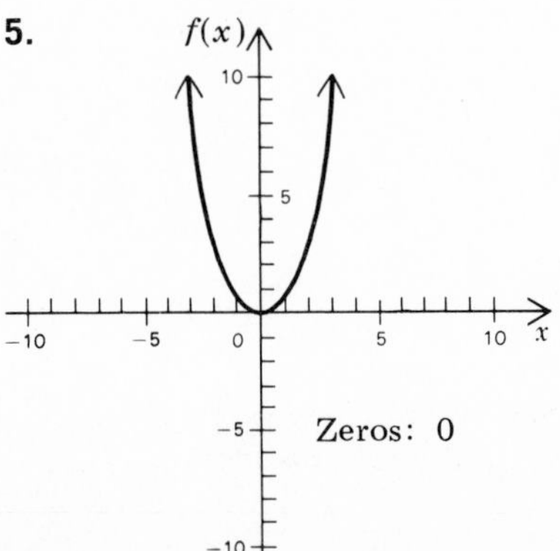

7.

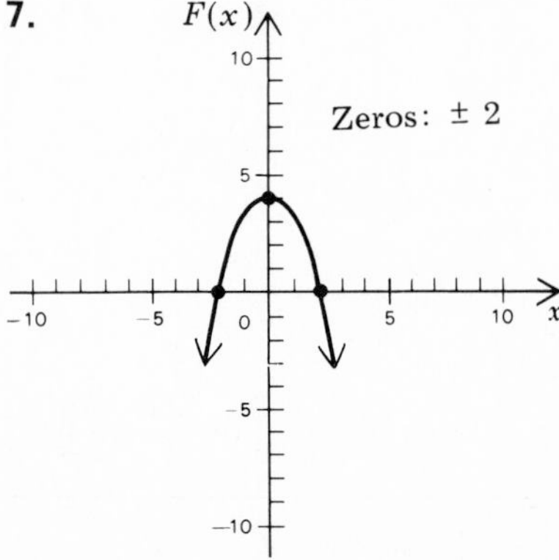

9.

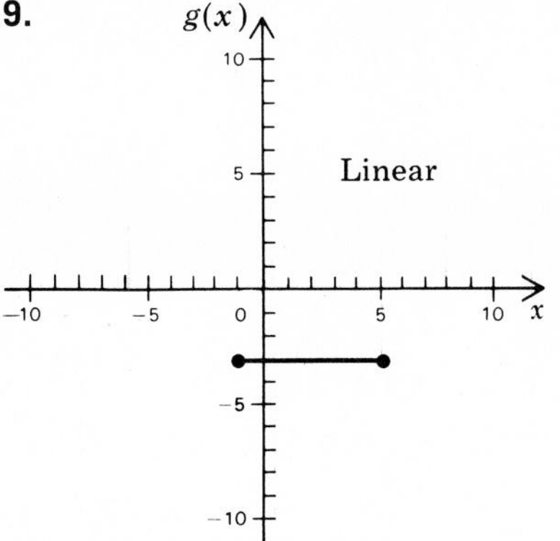

11.

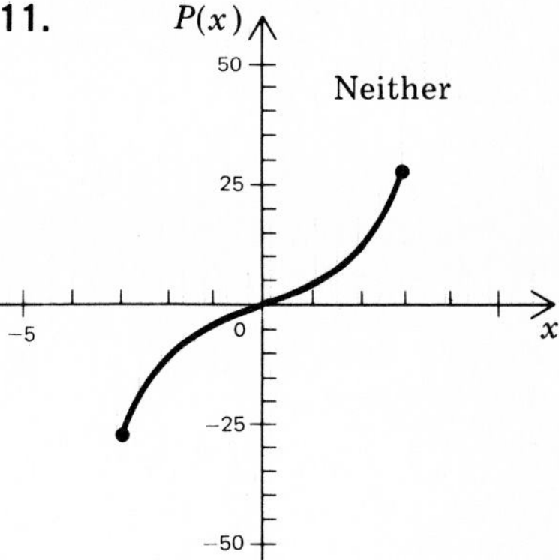

13.

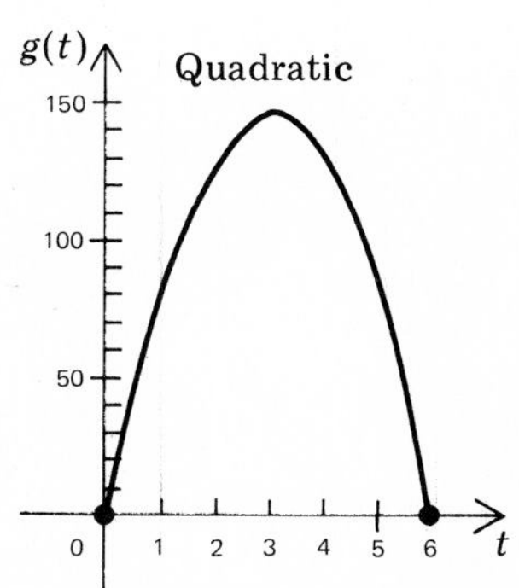

15.

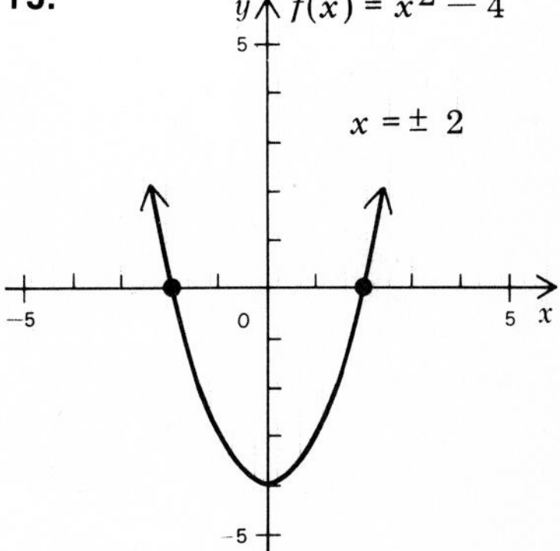

17.

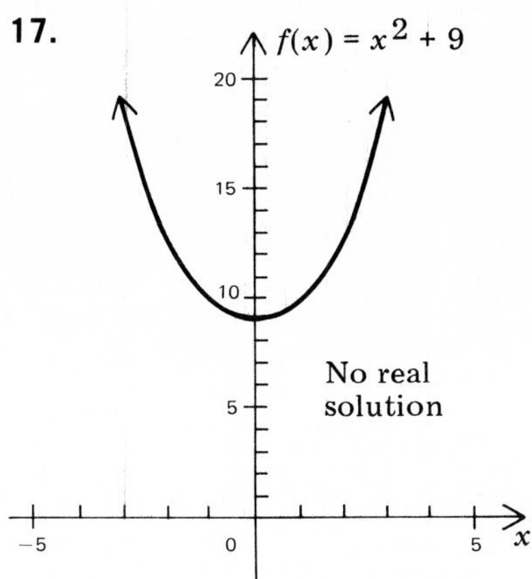

19.

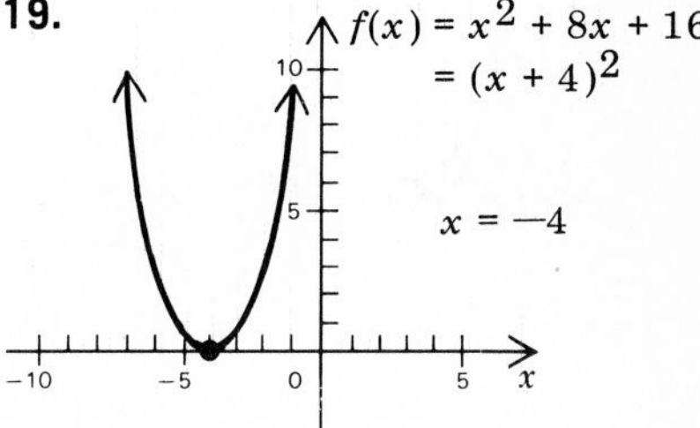

21.

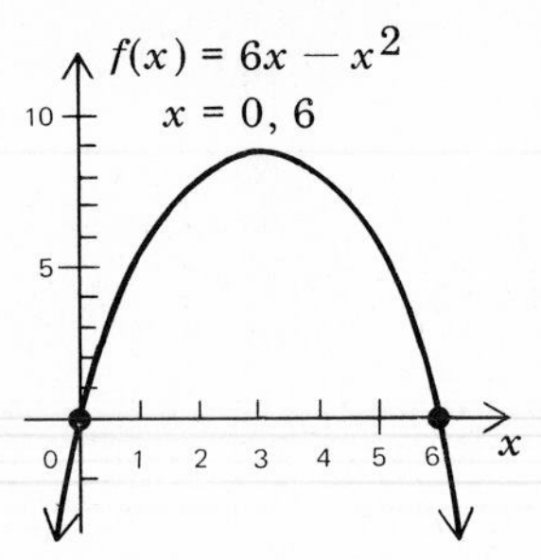

23.

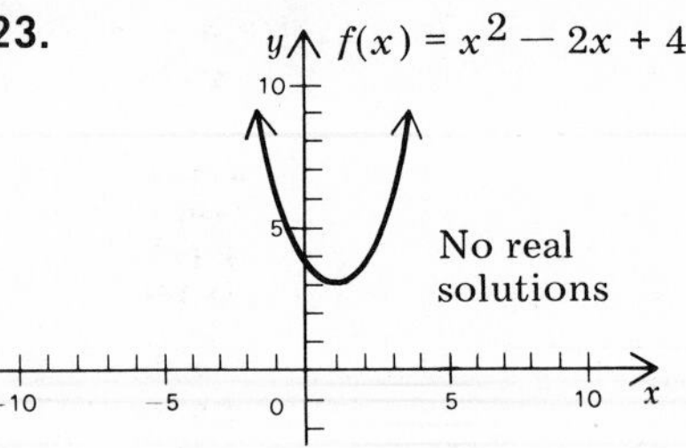

25. (A)

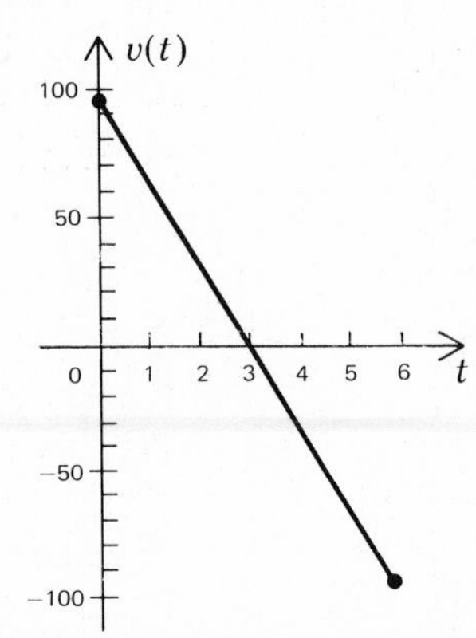

(B) $t = 3$ seconds, the time when the object reaches its maximum height

27. (A) $A(x) = x(50 - x)$
(B) $0 < x < 50$
(C)

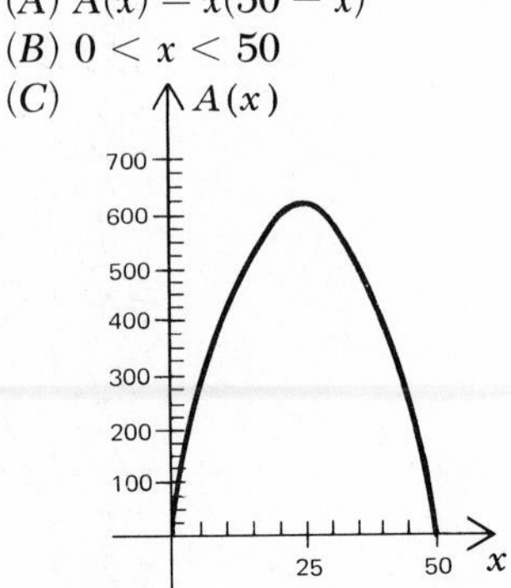

(D) 25 by 25 feet

EXERCISE 60

1. $F = kv^2$ **3.** $f = k\sqrt{T}$ **5.** $y = k/\sqrt{x}$ **7.** $t = k/T$ **9.** $R = kSTV$

11. $v = khr^2$ **13.** 4 **15.** $9\sqrt{3}$ **17.** $U = k\dfrac{ab}{c^3}$ **19.** $L = k(wh^2/l)$

21. −12 **23.** 83 pounds **25.** 20 amperes **27.** The new horsepower must be 8 times the old **29.** No effect

EXERCISE 61

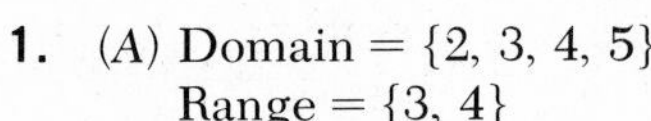

1. (A) Domain = {2, 3, 4, 5}
Range = {3, 4}
(C) Yes

(B)

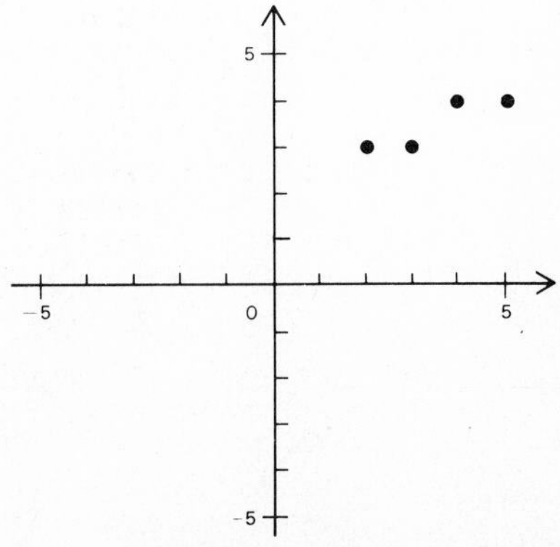

2. $0, 6, 9, 6 - m$ **3.** $-6, 0, -3, c - 2c^2$

4.

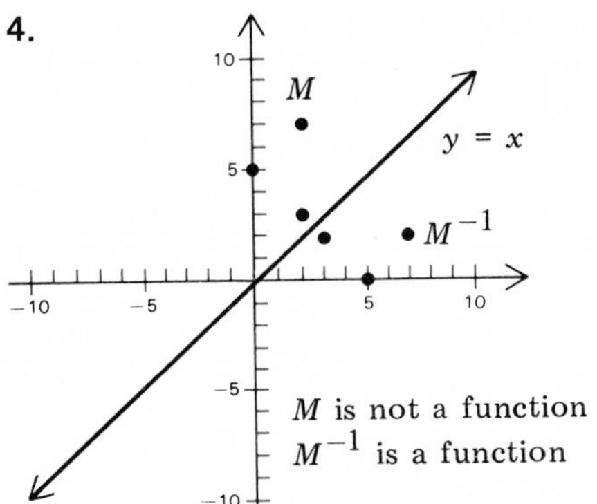

5. Domain = {3, 5, 7}
Range = {0, 2}

6.

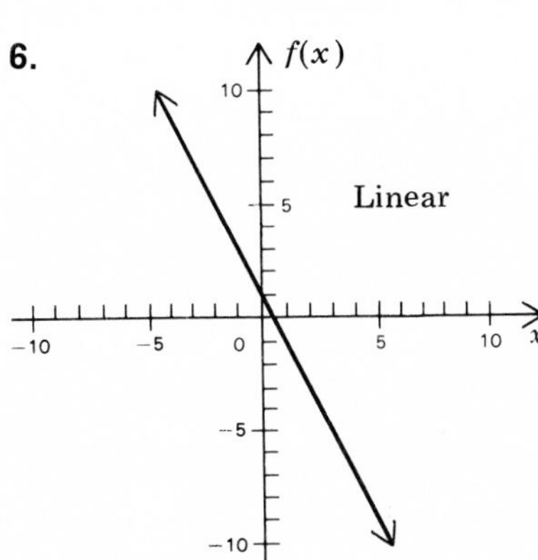

7. $f(x) = x^2 - 9$

$x = \pm 3$

8. $m = kn^2$ **9.** $P = k/Q^2$ **10.** $A = kab$ **11.** $y = kx^3/\sqrt{z}$ **12.** No. A function is a relation, but with the added property that each element in the domain is associated with one and only one element in the range. **13.** (A) 7; (B) 1; (C) $x^2 + x - 7$; (D) 3 **14.** $-2h - h^2; -2 - h$

15.

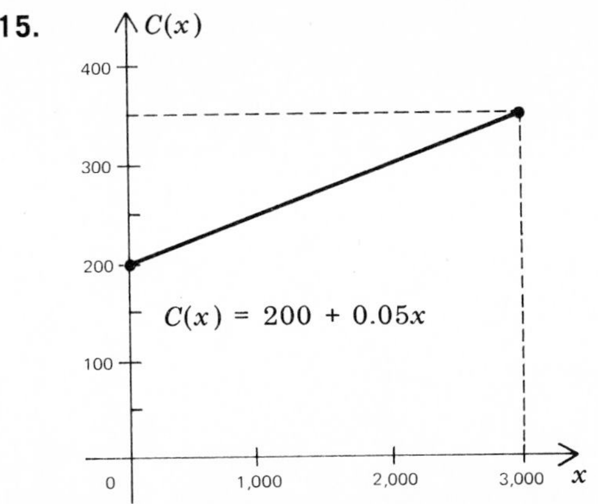

16. (A) $M^{-1}(x) = 2x - 3$; (B) Yes; (C) 3

17.

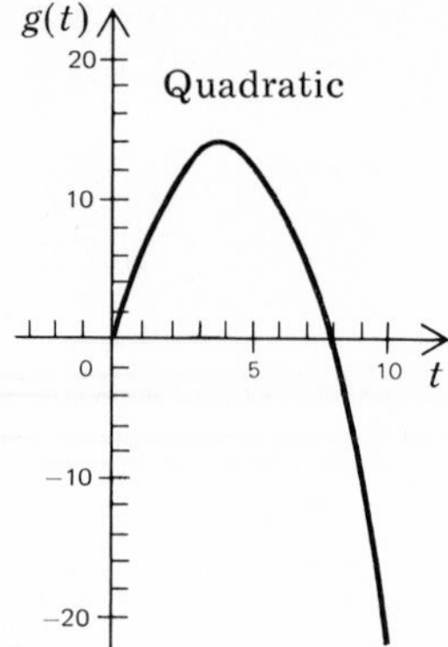

18.

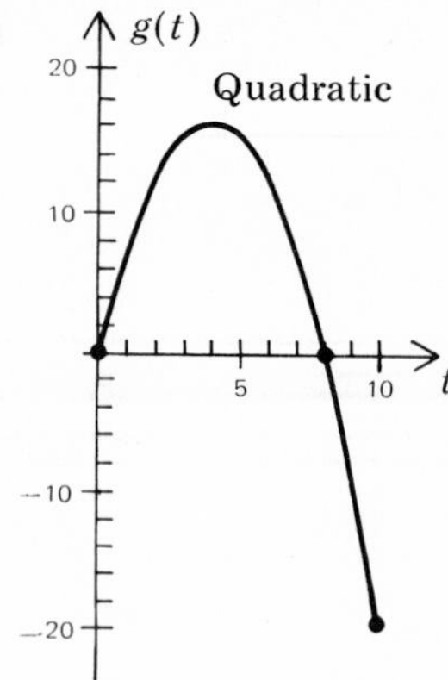

19. $t = k\dfrac{wd}{p}$

20. (A) $y = k\dfrac{x}{z}$; (B) $y = \frac{4}{3}$

21.

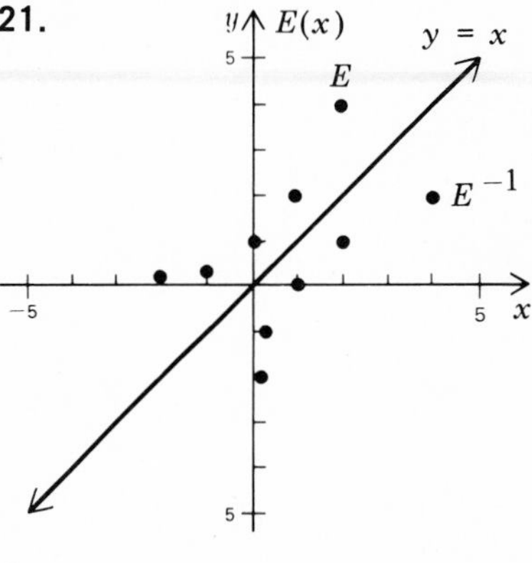

22. $f(x) = 3x$ **23.** $f(x) = 3x$

24. All but $g(x) = x^2$

25. 8 footcandles; $\sqrt{2}$ feet

26. The total force is doubled

EXERCISE 62

1.

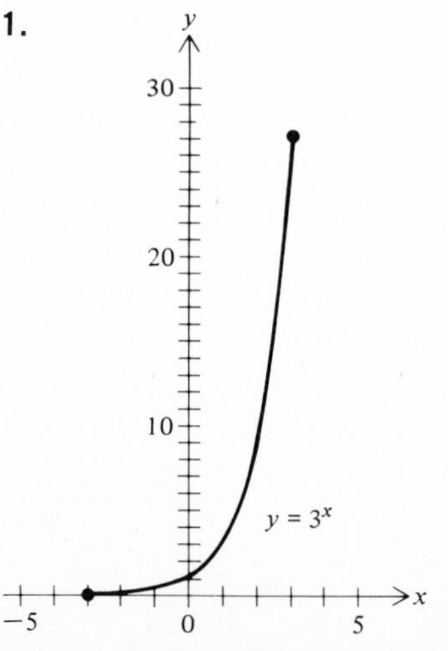

3.

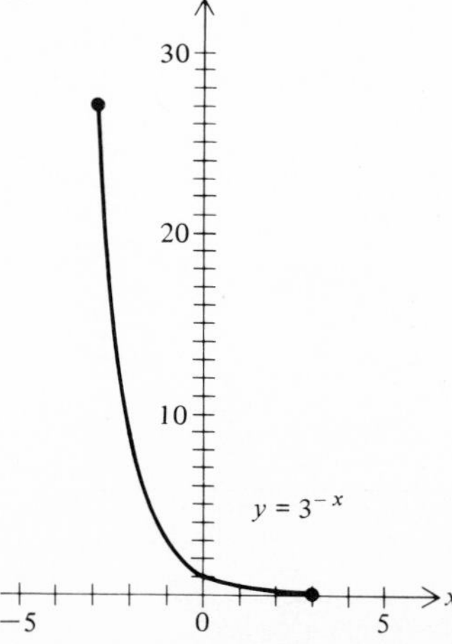

5.

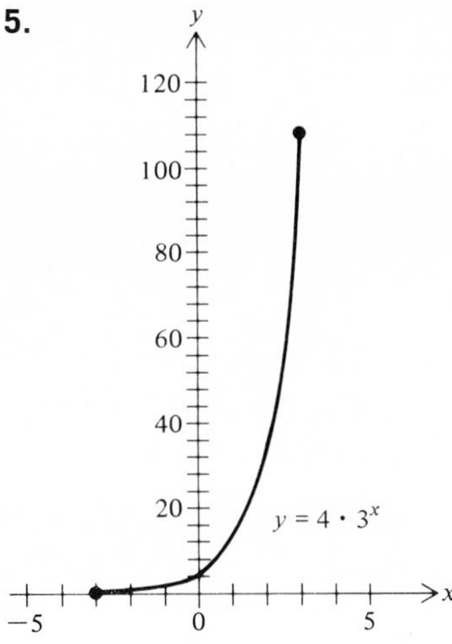

y
120
100
80
60
40
20
y = 4 · 3^x
−5
0
5
x

7.

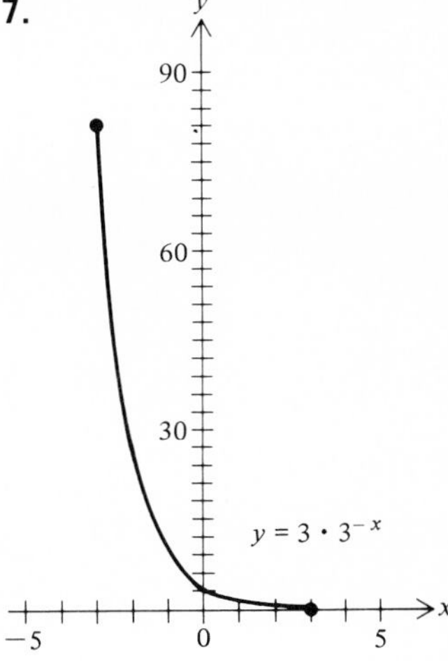

y
90
60
30
y = 3 · 3^−x
−5
0
5
x

9.

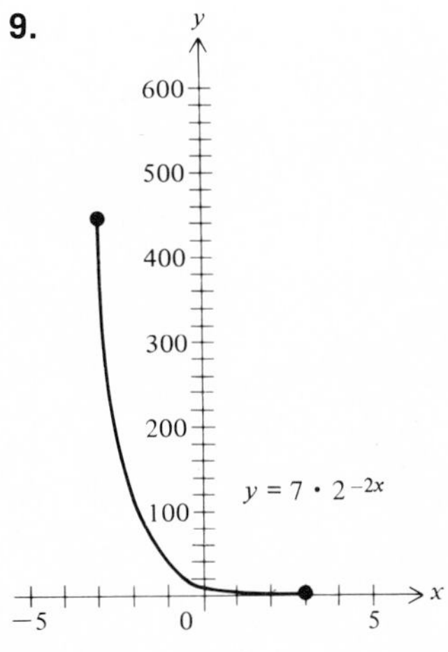

y
600
500
400
300
200
100
y = 7 · 2^−2x
−5
0
5
x

11.

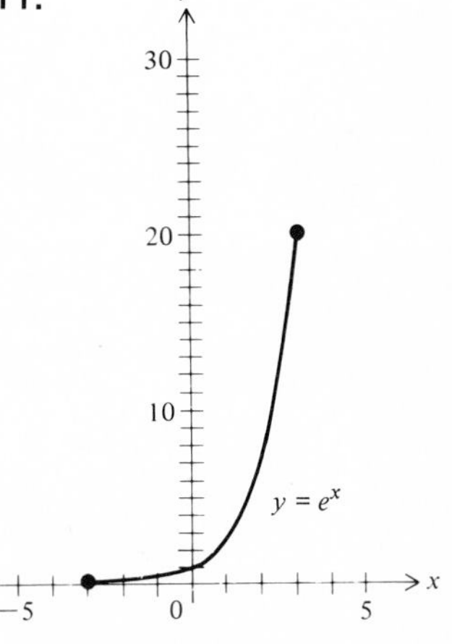

y
30
20
10
y = e^x
−5
0
5
x

13.

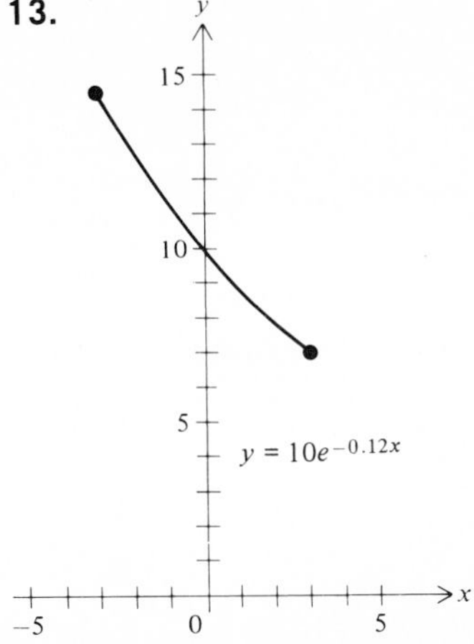

y
15
10
5
y = 10e^−0.12x
−5
0
5
x

15.

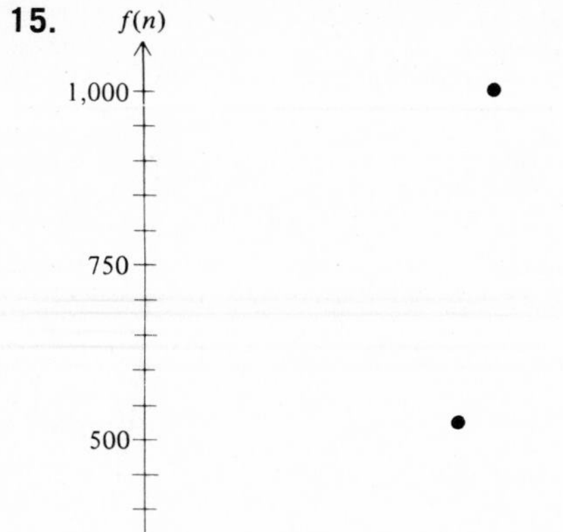

f(n)
1,000
750
500
250
0
5
10
n

17.

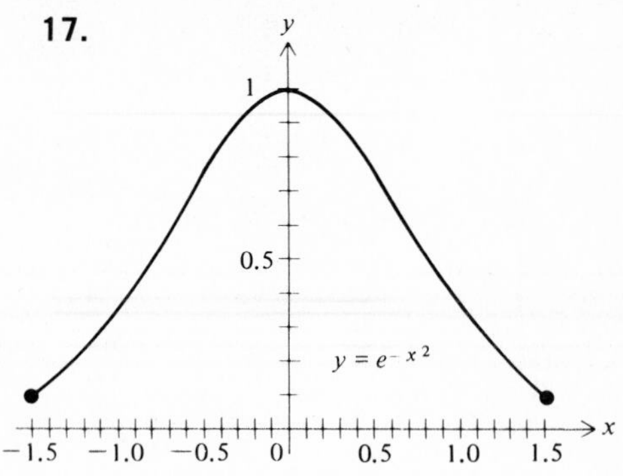

y
1
0.5
y = e^{-x^2}
−1.5
−1.0
−0.5
0
0.5
1.0
1.5
x

19.

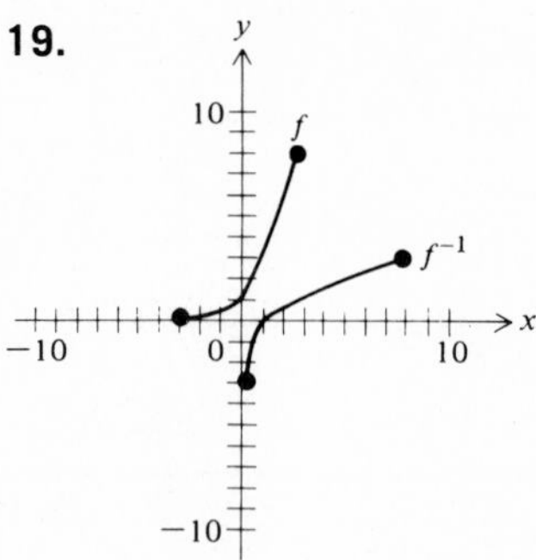

y
10
f
f^{-1}
−10
0
10
x
−10

21.

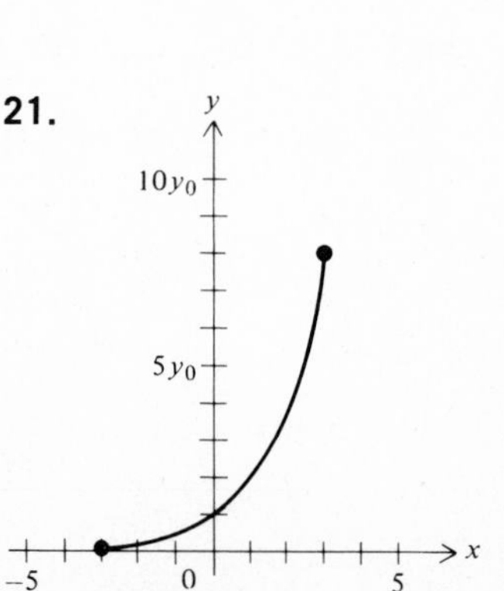

y
10y_0
5y_0
−5
0
5
x

23.

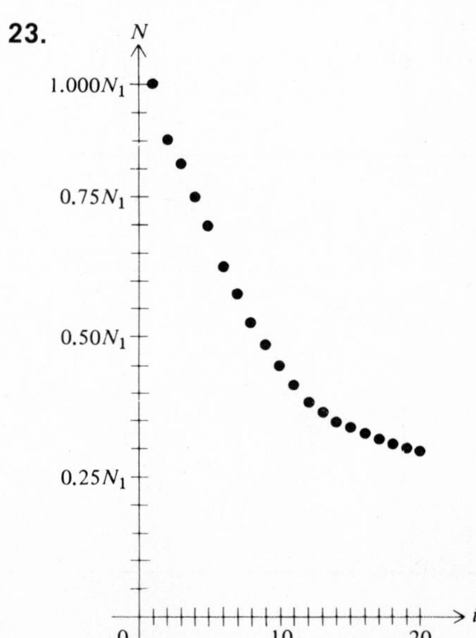

N
1.000N_1
0.75N_1
0.50N_1
0.25N_1
0
10
20
i

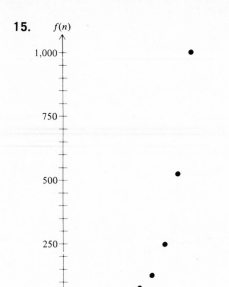

EXERCISE 63

1. $9 = 3^2$ **3.** $81 = 3^4$ **5.** $1{,}000 = 10^3$ **7.** $1 = e^0$ **9.** $\log_8 64 = 2$
11. $\log_{10} 10{,}000 = 4$ **13.** $\log_v u = x$ **15.** $\log_{27} 9 = \frac{2}{3}$ **17.** 5
19. -4 **21.** 2 **23.** 3 **25.** $x = 4$ **27.** $y = 2$ **29.** $b = 4$
31. $0.001 = 10^{-3}$ **33.** $3 = 81^{1/4}$ **35.** $16 = (\frac{1}{2})^{-4}$ **37.** $N = a^e$
39. $\log_{10} 0.01 = -2$ **41.** $\log_e 1 = 0$ **43.** $\log_2 \frac{1}{8} = -3$
45. $\log_{81} \frac{1}{3} = -\frac{1}{4}$ **47.** $\log_{49} 7 = \frac{1}{2}$ **49.** u **51.** $\frac{1}{2}$ **53.** $\frac{3}{2}$ **55.** 0
57. $\frac{3}{2}$ **59.** 2 **61.** $y = -2$ **63.** $b = 100$ **65.** Any positive real number except 1 **67.** Domain of f in the set of all real numbers; the range of f is 1. The domain of f^{-1} is 1; the range of f^{-1} is the set of all real numbers. No, f^{-1} is not.
69. (*A*)

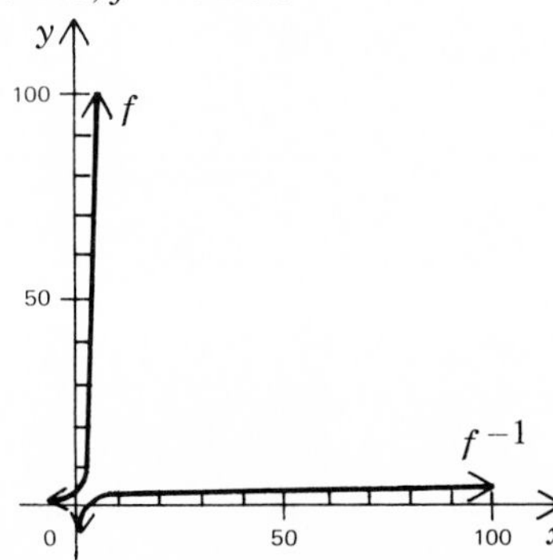

(*B*) Domain of f is the set of real numbers; range of f is the set of positive real numbers. The domain of f is the range of f^{-1} and the range of f is the domain of f^{-1}. (*C*) f^{-1} is called the logarithmic function with base 10
71. -3

EXERCISE 64

1. $\log_b u + \log_b v$ **3.** $\log_b A - \log_b B$ **5.** $5 \log_b u$ **7.** $\frac{3}{5} \log_b N$

9. $\frac{1}{2} \log_b Q$ **11.** $\log_b u + \log_b v + \log_b w$ **13.** $\log_b AB$ **15.** $\log_b \frac{X}{Y}$

17. $\log_b \frac{wx}{y}$ **19.** 3.40 **21.** -0.92 **23.** 3.30

25. $2 \log_b u + 7 \log_b v$ **27.** $-\log_b a$ **29.** $\frac{1}{3} \log_b N - 2 \log_b p - 3 \log_b q$

31. $\frac{1}{4}(2 \log_b x + 3 \log_b y - \frac{1}{2} \log z)$ **33.** $\log_b \frac{x^2}{y}$ **35.** $\log_b \frac{x^3y^2}{z^4}$
37. $\log_b \sqrt[5]{x^2y^3}$ **39.** 2.02 **41.** 0.23 **43.** -0.05 **45.** 8
47. $y = cb^{-kt}$ **49.** Let $u = \log^b M$ and $v = \log_b N$, then $M = b^u$ and $N = b^v$. Thus, $\log_b M/N = \log_b b^u/b^v = \log_b b^{u-v} = u - v = \log_b M - \log_b N$
51. $MN = b^{\log_b M} b^{\log_b N} = b^{\log_b M + \log_b N}$; hence, by definition of logarithm $\log_b MN = \log_b M + \log_b N$

EXERCISE 65

1. 0.3711 **3.** 0.7016 **5.** 1.8692 **7.** 6.4914 **9.** $0.8035 - 3$
11. $0.6902 - 5$ **13.** 5.28 **15.** 9.36 **17.** $8.31 \times 10^4 = 83{,}100$
19. $4.07 \times 10^{-1} = 0.407$ **21.** $2.78 \times 10 = 27.8$

23. $9.14 \times 10^{-4} = 0.000914$ **25.** $4.73 \times 10^{-3} = 0.00473$
27. $7.27 \times 10^{-1} = 0.727$ **29.** $6.8 \times 10^{-4} = 0.00068$ **31.** 0.3649
33. 5.8472 **35.** 1.8131 **37.** $0.6027 - 3$ **39.** $0.9546 - 1$
41. 5.8403 **43.** 5.204 **45.** $3.514 \times 10^5 = 351{,}400$
47. $2.693 \times 10^3 = 2.693$ **49.** $4.017 \times 10^{-3} = 0.004017$
51. $6.906 \times 10^{-1} = 0.6906$ **53.** $1.474 \times 10^{-1} = 0.1474$ **55.** -0.7949
57. 4.8887 **59.** 6.0200 **61.** $0.5809 - 2$ **63.** $0.1449 - 1$
65. $y = 41.1$ **67.** 4.3062; $2.024 \times 10^4 = 20{,}240$ **69.** $0.8552 - 8$; 7.165×10^{-8}

EXERCISE 66

1. 77,340 **3.** 13.82 **5.** 168,600 **7.** 0.006745 **9.** 99.54
11. 3,581 **13.** 2.4×10^{-11} **15.** 4.495 **17.** 0.6240 **19.** 0.0191
21. $-(7.98 \times 10^{-6})$ **23.** 151.2 **25.** 9.790 **27.** $-(4.462 \times 10^{-6})$
29. 1.37×10^{12} **31.** 0.02546 **33.** 7.6777 **35.** 2.475 seconds
37. 1.947×10^{25} dollars **39.** -1.415 **41.** (A) $t = kn^{3/2}$; (B) 39 or 40

EXERCISE 67

1. 2.9 **3.** 0.91 **5.** 6.0 **7.** 2.4 **9.** 80 **11.** 19 **13.** -2.4
15. 5.9 **17.** 0.063 **19.** 10 **21.** 4 **23.** 390.2 milligrams
25. 28 years **27.** \$21,430,000 **29.** 1.5117 **31.** 5.459

33. $\log_e 10 = \dfrac{\log_{10} 10}{\log_{10} e} = \dfrac{1}{\log_{10} e}$; $(\log_e 10)(\log_{10} e) = 1$ **35.** 2.593; 2.691

EXERCISE 68

1.

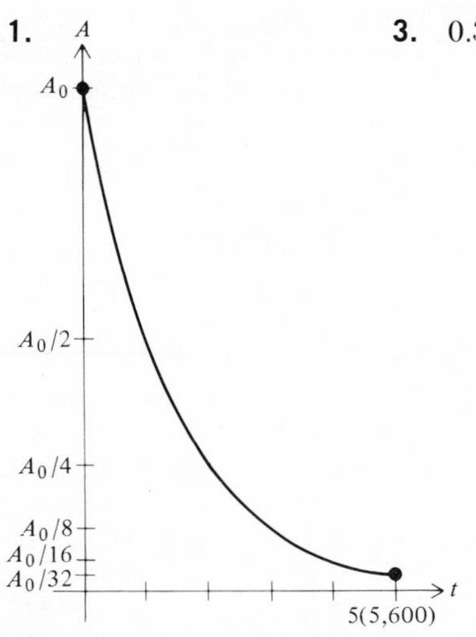

3. 0.3981

5.

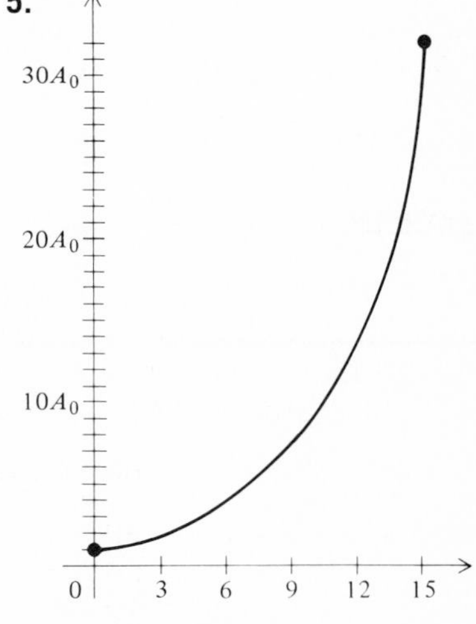

7. 35 years, 55.5 years **9.** 9.55×10^{-7} **11.** \$1,338 **13.** 7.2%
15. 95 feet, 489 feet **17.** 3.568 inches
19.

$$\begin{aligned} I &= I_0\, 10^{N/10} \\ \log I &= \log I_0 + \log 10^{N/10} \\ \log I &= \log I_0 + N/10 \\ N/10 &= \log I - \log I_0 \\ N/10 &= \log I/I_0 \\ N &= 10 \log I/I_0 \end{aligned}$$

21. 

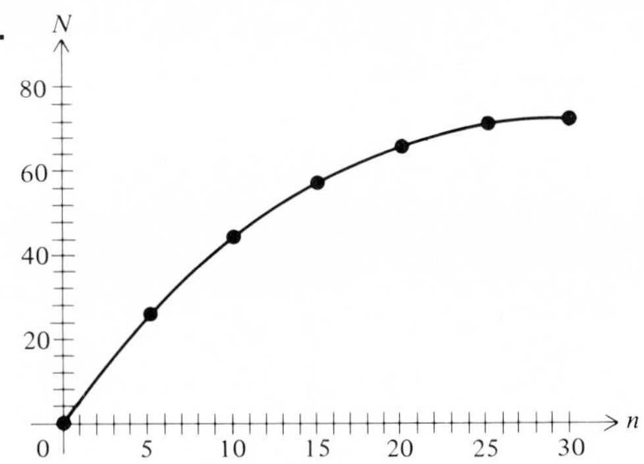

23. Factor of 8 (approximately)
25. 18,000,000 miles

EXERCISE 69

1. $x = 10^y$ **2.** $\log_{10} m = n$ **3.** 4 **4.** 5 **5.** 3 **6.** 114.1
7. 7.575 **8.** 884 **9.** 1.2 **10.** 25
11.

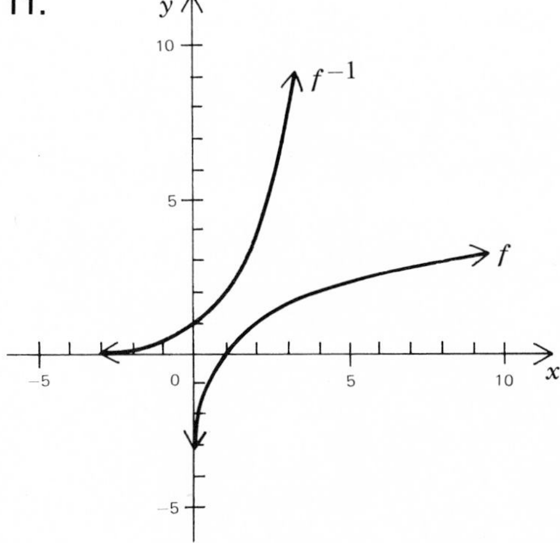

12. $y = e^x$ **13.** $\log_e y = x$
14. -2 **15.** 1/3 **16.** 64
17. e **18.** 33 **19.** 0.02145
20. 0.2416 **21.** 0.839
22. 11.41 **23.** 42.36
24. 0.00247 **25.** 2.3 **26.** 93
27. $x = 1$ **28.** 30 **29.** ± 2
30. 1.948 **31.** 23 years
32. If $\log_1 x = y$, then we would have to have $1^y = x$; that is, $1 = x$ for arbitrary positive x, which is impossible. **33.** Let $u = \log_b M$ and $v = \log_b N$, then $M = b^u$ and $N = b^v$. Thus, $\log_b M/N = \log_b b^u/b^v = \log_b b^{u-v} = u - v = \log_b M - \log_b N$ **34.** $y = ce^{-5t}$ **35.** 4.605

36.

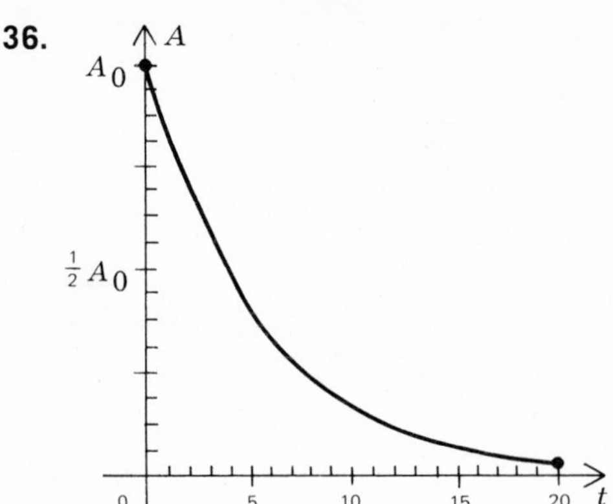

37. 38,000 years

EXERCISE 70

1. $-1, 0, 1, 2$ **3.** $0, \frac{1}{3}, \frac{1}{2}, \frac{3}{5}$ **5.** $4 -8, 16, -32$ **7.** 6 **9.** $\frac{99}{101}$
11. $S_5 = 1 + 2 + 3 + 4 + 5$ **13.** $S_3 = \frac{1}{10} + \frac{1}{100} + \frac{1}{1000}$
15. $S_4 = -1 + 1 - 1 + 1$ **17.** $1, -4, 9, -16, 25$ **19.** 0.3, 0.33, 0.333, 0.3333, 0.33333 **21.** $7, 3, -1, -5, -9$ **23.** $4, 1, \frac{1}{4}, \frac{1}{16}, \frac{1}{64}$ **25.** $a_n = n + 3$
27. $a_n = 3n$ **29.** $a_n = \dfrac{n}{n+1}$ **31.** $a_n = (-1)^{n+1}$ **33.** $a_n = (-2)^n$
35. $a_n = x^n/n$ **37.** $\frac{4}{1} - \frac{8}{2} + \frac{16}{3} - \frac{32}{4}$ **39.** $S_3 = x^2 + \dfrac{x^3}{2} + \dfrac{x^4}{3}$
41. $x - \dfrac{x^2}{2} + \dfrac{x^3}{3} - \dfrac{x^4}{4} + \dfrac{x^5}{5}$ **43.** $S_4 = \sum_{k=1}^{4} k^2$ **45.** $S_5 = \sum_{k=1}^{5} 1/2^k$
47. $S_n = \sum_{k=1}^{n} 1/k^2$ **49.** $S_n = \sum_{k=1}^{n} (-1)^{k+1}k^2$ **53.** (A) 3, 1.83, 1.46, 1.415; (B) Table, $\sqrt{2} = 1.414$; (C) 1, 1.5, 1.417, 1.414

EXERCISE 71

1. (B) $d = -0.5$; 5.5, 5; (C) $d = -5$; $-26, -31$ **3.** $a_2 = -1$, $a_3 = 3$, $a_4 = 7$
5. $a_{15} = 67$; $S_{11} = 242$ **7.** $S_{21} = 861$ **9.** $a_{15} = -21$ **11.** $d = 6$; $a_{101} = 603$ **13.** $S_{40} = 200$ **15.** $a_{11} = 2$, $S_{11} = \frac{77}{6}$ **17.** $a_1 = 1$
19. $S_{51} = 4{,}131$ **21.** $-1{,}071$ **23.** 4,446 **27.** $a_n = -3 + (n-1)3$
29. (A) 336 feet; (B) 1,936 feet; (C) $16t^2$

EXERCISE 72

1. (A) $r = -2$; $-16, 32$; (D) $r = \frac{1}{3}$; $\frac{1}{54}, \frac{1}{162}$ **3.** $a_2 = 3$, $a_3 = -\frac{3}{2}$, $a_4 = \frac{3}{4}$
5. $a_{10} = \frac{1}{243}$ **7.** $S_7 = 3{,}279$ **9.** $r = 0.398$ **11.** $S_{10} = -1{,}705$
13. $a_2 = 6$, $a_3 = 4$ **15.** $S_7 = 547$ **17.** $\frac{1023}{1024}$ **19.** $x = 2\sqrt{3}$
21. $S_\infty = \frac{9}{2}$ **23.** No sum **25.** $S_\infty = \frac{8}{5}$ **27.** $\frac{7}{9}$ **29.** $\frac{6}{11}$
31. $3\frac{8}{37}$ or $\frac{119}{37}$ **33.** $A = P(1 + r)^n$; approx. 12 years **35.** 900
37. $4,000,000

EXERCISE 73

1. 720 **3.** 20 **5.** 720 **7.** 15 **9.** 1 **11.** 28 **13.** 9!/8!
15. 8!/5! **17.** 126 **19.** 6 **21.** 1 **23.** 2,380
25. $u^5 + 5u^4v + 10u^3v^2 + 10u^2v^3 + 5uv^4 + v^5$ **27.** $y^4 - 4y^3 + 6y^2 - 4y + 1$
29. $32x^5 - 80x^4y + 80x^3y^2 - 40x^2y^3 + 10xy^4 - y^5$ **31.** $5{,}005u^9v^6$

33. $264m^2n^{10}$ **35.** $924w^6$ **37.** 1.1045

39. $\binom{n}{r} = \dfrac{n!}{r!(n-r)!} = \dfrac{n!}{(n-r)![n-(n-r)]!} = \binom{n}{n-r}$

EXERCISE 74

1. Arithmetic: (B) and (C); Geometric: (A) and (E) **2.** (A) 5, 7, 9, 11; (B) $a_{10} = 23$; (C) $S_{10} = 140$ **3.** (A) 16, 8, 4, 2; (B) $a_{10} = \frac{1}{32}$; (C) $S_{10} = 31\frac{31}{32}$ **4.** (A) −8, −5, −2, 1; (B) $a_{10} = 19$; (C) $S_{10} = 55$ **5.** (A) −1, 2, −4, 8; (B) $a_{10} = 512$; (C) $S_{10} = 341$ **6.** $S_\infty = 32$ **7.** 720
8. $20 \cdot 21 \cdot 22 = 9{,}240$ **9.** 21
10. $S_{10} = -6 - 4 - 2 + 0 + 2 + 4 + 6 + 8 + 10 + 12 = 30$
11. $S_7 = 8 + 4 + 2 + 1 + \frac{1}{2} + \frac{1}{4} + \frac{1}{8} = 15\frac{7}{8}$ **12.** $S_\infty = \frac{81}{5}$

13. $S_n = \sum_{k=1}^{n} \dfrac{(-1)^{k+1}}{3^k}$, $S_\infty = \frac{1}{4}$ **14.** $d = 3$, $a_5 = 25$ **15.** 190

16. 1,820 **17.** 1 **18.** $x^5 - 5x^4y + 10x^3y^2 - 10x^2y^3 + 5xy^4 - y^5$

19. $-1{,}760x^3y^9$ **20.** $\frac{8}{11}$ **21.** $49g/2$ feet; $625g/2$ feet

22. $x^6 + 6ix^5 - 15x^4 - 20ix^3 + 15x^2 + 6ix - 1$

Index